Detlef Klimm
Thermal Analysis and Thermodynamics

Also of Interest

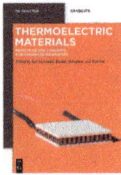

Thermoelectric Materials
Principles and Concepts for Enhanced Properties
Ken Kurosaki, Yoshiki Takagiwa, Xun Shi (Eds.), 2021
ISBN 978-3-11-059648-9, e-ISBN 978-3-11-059652-6

Non-equilibrium Thermodynamics and Physical Kinetics
Halid Bikkin, Igor I. Lyapilin, 2021
ISBN 978-3-11-072706-7, e-ISBN 978-3-11-072719-7

Thermal Control Thin Films
Spacecraft Technology
Jia-wen Qiu, Yu-Dong Feng, Chun-Hua Wu, 2022
ISBN 978-3-11-061286-8, e-ISBN 978-3-11-061489-3

Einführung in die Kristallographie
Joachim Bohm, Detlef Klimm, Manfred Mühlberg, Björn Winkler,
Founded by: Will Kleber, 2021
ISBN 978-3-11-046023-0, e-ISBN 978-3-11-046024-7

Detlef Klimm

Thermal Analysis and Thermodynamics

In Materials Science

DE GRUYTER

Author
Dr. Detlef Klimm
Leibniz-Institut für Kristallzüchtung (IKZ)
Max-Born-Str. 2
12489 Berlin
Germany
detlef.klimm@ikz-berlin.de

ISBN 978-3-11-074377-7
e-ISBN (PDF) 978-3-11-074378-4
e-ISBN (EPUB) 978-3-11-074384-5

Library of Congress Control Number: 2022931650

Bibliographic information published by the Deutsche Nationalbibliothek
The Deutsche Nationalbibliothek lists this publication in the Deutsche Nationalbibliografie;
detailed bibliographic data are available on the Internet at http://dnb.dnb.de.

© 2022 Walter de Gruyter GmbH, Berlin/Boston
Cover image: Sample crucibles for differential thermal analysis (DTA).
Photograph courtesy of NETZSCH-Gerätebau GmbH, Selb, Bavaria, Germany.
Typesetting: VTeX UAB, Lithuania
Printing and binding: CPI books GmbH, Leck

www.degruyter.com

Preface

The term "thermal analysis" in its widest sense includes the measurement of one of a broad range of physical properties (e. g., thermal conductivity, dielectric properties, electrical transport properties, mechanical strength, or mechanical damping, thermal expansion) either as a function of temperature (dynamic measurement) or as a function of time (isothermal measurement). In a closer sense, "thermal analysis" means a group of methods derived from simple experiments where the temperature of a sample was recorded as a function of time if a constant rate of heat flows into or out of the sample ("heating curve", "cooling curve"). In contemporary thermal analyzers the sample together with a reference is subjected to a temperature program, and either their temperature difference is measured, or the difference of heat energy flowing in or out of them. Commercial devices are offered by different companies; for special applications, also home-made setups are used occasionally. Often, the measurement of the thermal signals is combined with the continuous measurement of the sample mass or with the analysis of gaseous particles emanating from the sample.

Meanwhile, a variety of textbooks about thermal analysis are available on the market, and to some of them we refer in this book. It was not the intention of the author to repeat all things that were already written; rather a practical guide should be given how some typical problems occurring in materials science can be solved by methods of thermal analysis. This comprises not only the experimental part. Whenever reasonable, the holistic interpretation of a thermoanalytic measurement should include an explanation of all experimental observations on a thermodynamic basis, which is nowadays often possible with commercial or freeware software. One possible representation of such "thermodynamic assessment" of a system is a phase diagram (or constitution diagram). At least some basic knowledge about phase diagrams is inevitable; unfortunately, such a knowledge is often sparsely taught at universities, at least outside the mineralogical departments.

The book is divided into three sections. "Thermodynamic basis" describes the physical quantities that are most relevant for the interpretation of thermal analysis measurements and gives an introduction to typical phase diagram features like mixed crystals, eutectics, and peritectics. Some crystal structures that are shown there were drawn with the freeware program VESTA by Momma and Izumi (2011). Already here the relation between thermodynamic properties and measurements is shown. The second section "Techniques" describes the methods "Differential Thermal Analysis" (DTA), "Differential Scanning Calorimetry" (DSC), "Thermogravimetry" (TG), and "Evolved Gas Analysis" (EGA), with examples and together with their combinations. The last section "Applications" brings examples (mainly from the author's own work) for typical practical problems that can be solved by thermal analysis. This is not always the "big science", because some tasks like the characterization of raw material batches seem perhaps somewhat boring; nevertheless, it is often necessary and well done by thermal analysis.

https://doi.org/10.1515/9783110743784-201

Here I want to mention two old (german) textbooks, which gave me a guide into the field of thermal analysis: The book "Differentialthermoanalyse" by Dietrich Schultze (1969),[1] who was working in the field of crystal growth of inorganics for decades, can hold as a paragon for this book. It is, however, one half century old, and since that, technical options and numerical methods of analysis were much improved. I am grateful to Dr. Dietrich Schultze that he read major parts of the book manuscript and gave me valuable hints. The booklet "Phasendiagramme" by Peter Paufler (1981)[2] is quite mathematical, even if written in a time when computers were room-filling mainframes, but the quantitative presentations found there are the basis of current software packages like FactSage 8.0 (2020), which are available nowadays.

I have to acknowledge the work of numerous undergraduate students from Humboldt-Universität zu Berlin and other universities, who did practical courses in my laboratory: some of their measurements of phase diagrams are included in this book as examples. Also, I have to thank the students who made internships with me, as well as many Bachelor, Master, and PhD students who worked in the thermoanalytic laboratory. The proofreading and comments by Dr. Nora Wolff and Prof. Dr. Joachim Bohm are highly acknowledged. Last but not least, I want to thank my dear colleagues from Leibniz-Institut für Kristallzüchtung Berlin (Germany) for the great working opportunities that they offered me through almost three decades and for the amiable atmosphere that I always found among them.

Eichwalde, March 2022   Detlef Klimm

---

1 Dietrich Schultze (born 28 September 1937).
2 Peter Paufler (born 18 February 1940).

Contents

1 Thermodynamic basis

1.1 Specific heat capacity

1.1.1 Phenomenological description

One of the first reports on experiments, which we would now call "thermal analysis", was given by Le Chatelier (1887),[1] who described the heating of different clay minerals. In his experiments, the sample was heated with constant electric power $W = U_h \cdot I_h$ (U_h and I_h are the voltage and current of the heater) and sample temperature T was recorded by a thermocouple (see Section 1.3.1) at constant time increments. It turned out that the measured heating rate $\dot{T} = dT/dt$ was not constant, because endothermal reactions in the sample (e. g., dehydration processes or phase transitions) resulted in temporally smaller $\dot{T}$. It will be described in Section 2.1 how such a type of heating (or cooling) experiments was developed further to the modern technique of differential thermal analysis (DTA).

The experimental setup of Le Chatelier corresponds to the scheme in Fig. 1.1. If during time t, the amount of heat energy

$$
\begin{aligned}
Q &= W \cdot t \\
&= U_h \cdot I_h \cdot t
\end{aligned}
\tag{1.1}
$$

is transferred to one mole of a sample, then its temperature increases under isochor (volume constant; $\Delta V = 0$) or the more typical isobar (pressure p = const.; often, $p = 1\,\text{bar} = 10^5\,\text{Pa}$) conditions by

$$
\Delta T_V = \frac{\Delta Q}{c_V},
\tag{1.2}
$$

$$
\Delta T_p = \frac{\Delta Q}{c_p},
\tag{1.3}
$$

where c_V and c_p are proportionality parameters called the specific heat capacities of the material under the given conditions, V or p = const. Especially for technical purposes, the amount of substance is often given by the mass $m = nM$, where n is the number of moles, and M is the molar mass. The difference

$$
c_p - c_V = \frac{T}{n}\left(\frac{\partial V}{\partial T}\right)_p\left(\frac{\partial p}{\partial T}\right)_V
\tag{1.4}
$$

---

[1] Henry Louis Le Chatelier (8 October 1850–17 June 1936).

https://doi.org/10.1515/9783110743784-001

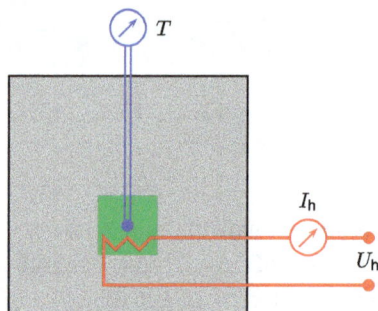

Figure 1.1: The experimental setup used by Le Chatelier (1887) allows the measurement of the specific heat capacity using equation (1.3).

is positive but usually small for condensed phases with small thermal expansion $\partial V/\partial T$ (solids, liquids). For ideal gases, we have $c_p - c_V = R$ (gas constant). Table 1.1 reports some data from the literature (Regen and Brandes, 1979; Paufler, 1986).

Table 1.1: Specific heat capacities near room temperature for several substances. For gases, ideal data are scaled by the gas constant R. Data for solid phases are given in J/(mol K).

Substance	c_p	c_V
1-atomic gas	$\frac{5}{2}R$	$\frac{3}{2}R$
2-atomic gas	$\frac{7}{2}R$	$\frac{5}{2}R$
3-atomic gas, stretched	$\frac{7}{2}R$	$\frac{5}{2}R$
3-atomic gas, tilted	$4R$	$3R$
silver	25.49	24.27
aluminum	24.34	23.03
copper	24.08	23.59
lithium	23.64	22.53
tungsten	24.08	23.59

Figure 1.2 shows calculated heating curves demonstrating the development of the sample temperature T with introduced heating energy Q in the style of the experiments performed by Le Chatelier (1887). For 1 mol = 72.63 g pure germanium, the temperature rises almost linearly up to the melting (fusion) temperature $T_f = 937\,°C$ and remains constant there until the heat of fusion (or "latent heat") $\Delta H_f = 36\,945\,J$ is brought in. A solid solution (or "mixed crystal") with composition 0.9Ge + 0.1Si (1 mol = 68.18 g) starts melting at a higher temperature and, in contrast to pure Ge, T remains not constant. Instead, T rises smoothly over an extended temperature range up to $\approx 1050\,°C$, where melting is finished.

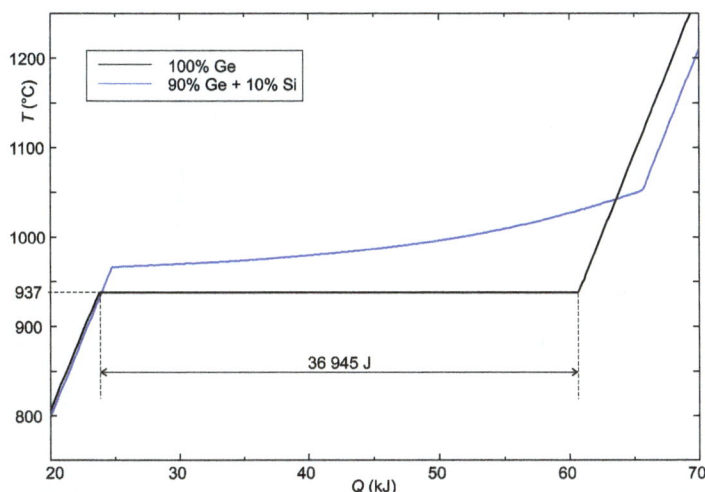

Figure 1.2: Calculated temperature development of 1 mol pure germanium or of a solid solution (cf. Section 1.5.1.2) of 0.9 mol Ge + 0.1 mol Si upon heating.

How long takes the melting of 1 mol Ge if a constant heating power of 100 W is used?

Petit and Dulong (1819)[2,3] reported that for many solid chemical elements, the specific heat capacity approaches 25 J/(mol K) (1.5) at room temperature (Fig. 1.3). For very low temperature, $c_p(T)$ follows a T^3 law (see Fig. 1.8) and vanishes at absolute zero (1.6). Neumann and Kopp[4,5] found that $c_p(T)$ of a chemical compound $A_m B_n$ is nearly the sum of the $c_p(T)$ of its composing chemical elements (1.7).

$$\lim_{T \to 298\,\text{K}} c_p = 3R \approx 25\,\text{J/mol K} \quad \text{(Dulong–Petit)}, \tag{1.5}$$

$$\lim_{T \to 0} c_p = 0, \tag{1.6}$$

$$c_p^{(A_m B_n)}(T) \approx m c_p^{(A)}(T) + n c_p^{(B)}(T) \quad \text{(Neumann–Kopp)}. \tag{1.7}$$

Experimental $c_p(T)$ data are the basis from which "chemical potentials" like H, S, and G (cf. Section 1.2) can be derived and are of great importance for this reason. In Section 2.3.2.2, we will describe how differential scanning calorimetry (DSC), a method

---

2 Alexis Thérèse Petit (2 October 1791–21 June 1820).

3 Pierre Louis Dulong (12 February 1785–19 July 1838).

4 Franz Ernst Neumann (11 September 1798–23 May 1895).

5 Hermann Franz Moritz Kopp (30 October 1817–20 February 1892).

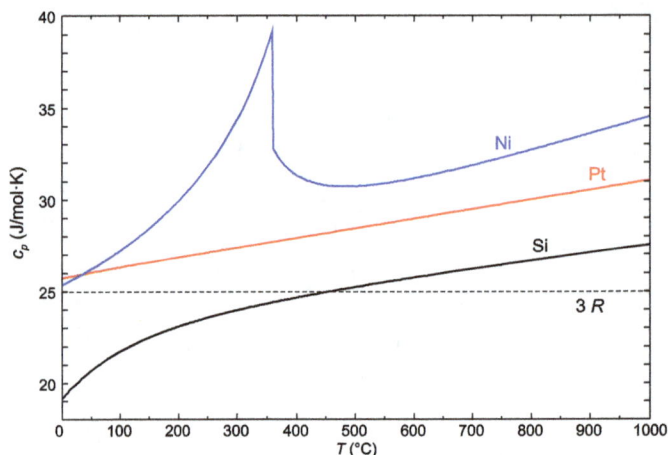

Figure 1.3: $c_p(T)$ functions for 3 chemical elements (data from FactSage 8.0, 2020) compared to the Dulong–Petit law (1.5).

of thermal analysis, can be used to measure $c_p(T)$. For other, partially more accurate methods, the reader is referred, e. g., to Höhne et al. (1996), Navrotsky (2014).[6]

For nearly all chemical elements and for many binary compounds like oxides, sulfides, and halides accurate data are available from printed references, online resources, or other databases, e. g., Barin (1995), NIST (2020), FactSage 8.0 (2020). If for other substances, measurements are not possible, then their $c_p(T)$ functions can be estimated with the Neumann–Kopp rule (1.7) from the components. In Fig. 1.4, this is performed for the chalcopyrite-type compound CuInS$_2$. The black curve shows experimental data obtained by Neumann et al. (1987) from dynamic calorimetry measurements in a Setaram heat flow calorimeter; the authors claim an experimental error of ± 1 J/(mol K). The blue curve is the weighed sum of the contributions $c_p^{Cu}(T) + c_p^{In}(T) + 2 \cdot c_p^{S}(T)$, which for low T is in very good agreement with the experimental data. Difficulties arise at all kinds of phase transitions in one of the phases, here the α/β transition of sulfur and the melting point of indium. Moreover, it is obvious that the slope of the blue curve is significantly too high. This difference can be understood because vibrations of atoms are the most important contribution to c_p, which will be discussed in Section 1.1.3.1. However, the nature of the bonds in metals (Cu, In) and molecular crystals (S) is much different from the bonds with covalent and ionic contributions in a crystal like CuInS$_2$. Thus it is sometimes more appropriate to use fairly similar substances as components for the Neumann–Kopp calculation. The red curve in Fig. 1.4 demonstrates this and is the sum $c_p^{CuS}(T) + c_p^{InS}(T)$ from FactSage 8.0 (2020) data. This curve always presents slightly smaller values than the black one, but the

---

6 Alexandra Navrotsky (born 20 June 1943).

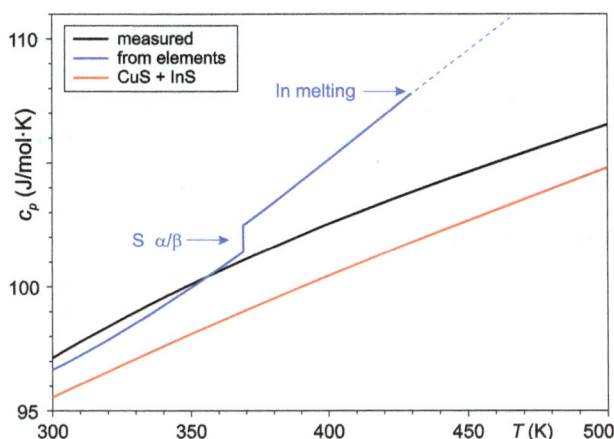

Figure 1.4: $c_p(T)$ data for $CuInS_2$ measured by Neumann et al. (1987) compared to data obtained by adding the FactSage 8.0 (2020) data for the chemical elements or for CuS + InS, according to Neumann–Kopp (1.7).

difference is not much larger than the experimental error. Besides, the slope is very similar. An even better calculation of $c_p(T)$ functions is possible if data for another phase with very high structural similarity are available; an example can be found in Section 3.3.5.

For most substances, the function $c_p(T)$ rises monotonously with T; the slope is steep at low T and becomes flatter near room temperature. Jumps occur at first-order phase transitions, and λ-shaped maxima are typical for second-order transitions. Figure 1.3 shows this for nickel, where c_p has a local maximum at $T_C = 354\,°C$. Such singular temperatures are called Curie[7] points and are often connected with a transformation of a ferromagnetic (or ferroelectric) phase or structure below T_C to a paramagnetic or paraelectric phase above T_C. The following phenomena can be observed for different types of phase transitions (see also Section 1.4.2):

1st-order transition: A certain amount of latent heat energy is required to perform the transition from the low-T to the high-T phase. This latent heat is released again during the back transition from high-T to low-T. The transition occurs sharply at a specific T_t, the two phases are in equilibrium there, and c_p is discontinuous at T_t. Examples: melting, evaporation, α (orthorhombic) $\leftrightarrow$ β (monoclinic) transition of sulfur at $96\,°C$, α (bcc, body-centered cubic) $\leftrightarrow$ γ (fcc, face-centered cubic) transition of iron at $912\,°C$.

2nd-order transition: No latent heat is required to perform the transition, and the transition is continuous. This means that some order parameter starts to change continuously in the low-T phase and reaches its final (high-T) value at the tran-

---

7 Pierre Curie (15 May 1859–19 April 1906).

sition point (often called critical point) T_c. Examples: ferroelectric and ferromagnetic transition, such as Ni in Fig. 1.3, "critical points" where liquids and gases cannot be distinguished. It is sometimes difficult to determine a phase transition to be clearly of 2nd order, because small heats of transition might be overseen. Examples are, e. g., NH_4Cl ($\Delta H = 3950\,J/mol$), $BaTiO_3$ ($\Delta H = 197\,J/mol$), and SiO_2 (α/β quartz, $\Delta H = 499\,J/mol$).

Glass transition: Glasses are metastable phases where ordering of atoms or molecules appears only between close neighbors and wide range ordering (like in crystals) is not present. Typically, glasses show a glass transition point or range T_g, and in this range, c_p rises by a certain amount Δc_p. Glasses are more brittle below and more rubber-like above T_g. In DSC heating curves (see Section 2.1 and Fig. 1.5) a larger c_p leads to a shift of the curve in the endothermal direction. Different points on the curve may be used to define the glass transition, among them the "fictive temperature" $T_f \approx T_{mid}$. (T_f is defined by the size of areas below the curve, and T_{mid} is the middle of the transition range.)

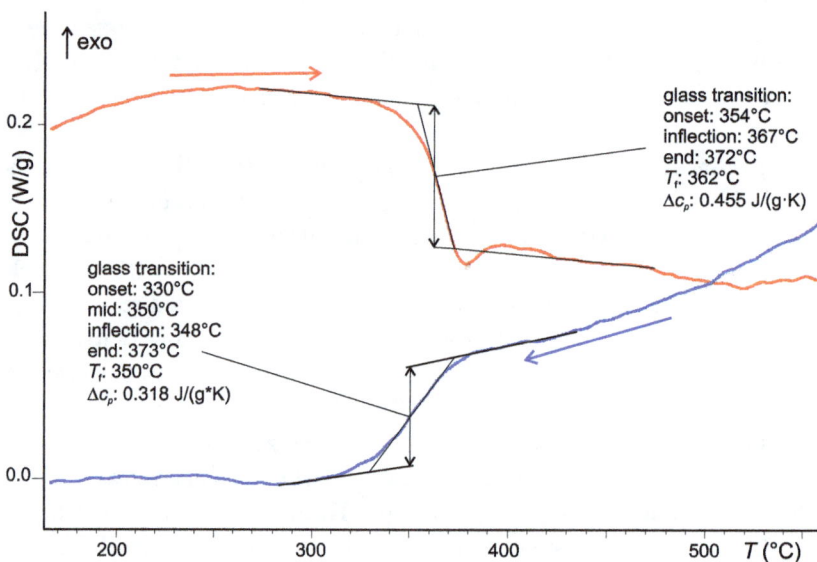

Figure 1.5: DSC measurements (top: heating, bottom: cooling) with 10 K/min of 17.62 mg Na–Gd phosphate glass ($NaPO_3/Gd(PO_3)_3$ with 60% Gd phosphate), cf. Nitsch et al. (2005).

1.1.2 Atomistic description

A thermodynamic system can be described as ensemble of atoms bearing internal energy $U = U_{pot} + U_{kin}$ that may be divided into a potential part (e. g., binding energy) and a kinetic part (different types of oscillations). In the time average, for harmonic oscillators, we have $U_{\text{pot}} = U_{\text{kin}}$, and it can be shown that $U_{\text{kin}} = k_B T/2$ per degree of freedom (where k_B is the Boltzmann[8] constant). Consequently, we have the energy contribution $U' = k_B T$ per degree of freedom. Every atom has 3 degrees of freedom, and in the limit case (where all oscillations are allowed),

$$U = 3N_A k_B T \tag{1.8}$$

is the total molar inner energy of the system (where N_A is the Avogadro[9] constant). c_V corresponds to the amount of energy required to change the temperature of the system (1.2),

$$c_V = \left(\frac{\partial U}{\partial T}\right)_V = 3N_A k_B \approx 25 \frac{\text{J}}{\text{mol K}}, \tag{1.9}$$

and gives in this limit case the value of Dulong–Petit (1.5).

The energies of the $3N$ vibration modes allowed in 1 mole of a substance are different, and the number of modes that can be excited becomes smaller for lower T. A detailed analysis (see textbooks for solid-state physics) results in the T^3 dependence of $c_V(T)$ mentioned above and in $c_V(0\,\text{K}) = 0$ (1.6). In the following, we will give only a short introduction into such a treatment for a one-dimensional hypothetical solid.

The central positions of atoms inside the crystal lattice are defined by the crystal structure, and at $T > 0$ the atoms vibrate around these central positions. In a good approximation, atoms can be considered as mass points. Figure 1.6 shows this for the simple case of a one-dimensional chain of mass points with mass m and distance a. The forces between neighboring atoms are symbolized by springs with spring constant α. Under the assumption of undamped free oscillations for the nth atom, the equation of motion is given by

$$\begin{aligned} m\ddot{x}_n &= -\alpha(x_n - x_{n+1}) - \alpha(x_n - x_{n-1}) \\ &= \alpha(x_{n-1} + x_{n+1} - 2x_n) \end{aligned} \tag{1.10}$$

and can be solved by the ansatz

$$x_n = A\,e^{-i(\omega t - nqa)} \tag{1.11}$$

---

8 Ludwig Eduard Boltzmann (20 February 1844–5 September 1906).

9 Lorenzo Romano Amedeo Carlo Avogadro (9 August 1776–9 July 1856).

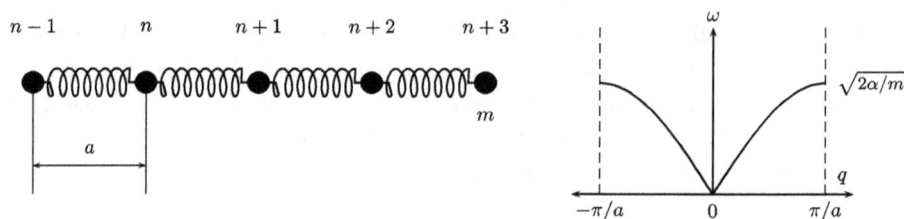

Figure 1.6: Left: Linear monoatomic vibration chain. Right: Resulting dispersion relation (1.12).

describing an elastic wave with circular frequency ω, where the phase factor differs from atom to atom by iqa. We find

$$\dot{x}_n = A\,e^{-i(\omega t - nqa)}\,\omega = -x_n i\omega,$$
$$\ddot{x}_n = A\,e^{-i(\omega t - nqa)}\,\omega^2 = -x_n \omega^2,$$
$$x_{n-1} = x_n\,e^{iqa},$$
$$x_{n+1} = x_n\,e^{-iqa},$$

and, finally, with Euler's[10] formula $e^{ix} = \cos x + i\sin x$,

$$\omega = \sqrt{\frac{2\alpha}{m}(1 - \cos qa)}, \tag{1.12}$$

which is the "dispersion relation" of the vibrational chain showing the dependence $\omega(q)$. A graphical presentation of this function is shown in Fig. 1.6.

The wavenumber q of a vibration state corresponds to the wavelength $\lambda = 2\pi/q$, and possible values for λ range from $\lambda = 2a$ ($q = \pi/a$) to $\lambda = \infty$ ($q = 0$); the latter is a vibration of the rigid lattice. Near $q = 0$, ω is proportional to q, and the slope $\partial\omega/\partial q = v$ is the speed of sound in the lattice.

The number of possible vibration modes is equal to the number of atoms N in the (one-dimensional) chain; in other words, each atom produces exactly one "eigenfrequency" or vibrational mode. A detailed analysis of the three-dimensional case results in $3N$ vibration modes and finally in equations (1.8) and (1.9). The vibrational energy of these modes is a function of q, and at temperature T, only modes with vibrational energy $\leq k_B T$ can be excited. Consequently, the number of possible vibration modes (or "phonons") rises with T, and this is also the origin of the $c_p(T) \approx c_V(T)$ functions, which are typically found by experiment.

It should be noted that the chain in Fig. 1.6 represents the simplest case where only one type of atoms is bound in one direction. In a real crystal, bounds in three directions are formed, and usually more than one sort of atoms are present. Then dispersion relations are more complicated, as compared to Fig. 1.6. The type of modes

---

10 Leonhard Euler (15 April 1707–18 September 1783).

shown there with $\omega(q = 0) = 0$ is an "acoustic mode". If atoms with different masses m, m' are constructing the crystal, then additionally "optical modes" are formed with $\omega(q = 0) \neq 0$.

1.1.3 Contributions to c_V

1.1.3.1 Lattice vibrations: Debye theory

Equation (1.8) gave the inner energy for the limit case that all possible vibration modes (phonons) are excited. More generally, it is given by

$$U(T) = \int_0^\infty g(\omega)x(\omega, T)\, d\omega, \tag{1.13}$$

where the spectral distribution function $g(\omega)\, d\omega$ gives the number of eigenfrequencies between ω and $\omega + d\omega$, and x is the vibration amplitude. Debye[11] assumed an approximation for

$$g(\omega) = \begin{cases} 9N\omega^2/\omega_D^3, & 0 \leq \omega \leq \omega_D, \\ 0, & \omega > \omega_D, \end{cases} \tag{1.14}$$

which means that only vibration states up to some arbitrary Debye frequency $\omega_D = 2\pi\nu_D$ are considered. The Debye temperature $\Theta_D = h\nu_D/k_B$ separates the regions of different functions

$$c_V(T) = \begin{cases} \frac{12\pi^4 R}{5}\left(\frac{T}{\Theta_D}\right)^3, & T \ll \Theta_D, \\ 3R[1 - \frac{1}{20}\left(\frac{\Theta_D}{T}\right)^2], & T \gg \Theta_D, \end{cases} \tag{1.15}$$

and the T^3 dependency for low T and the limit $3R$ for high T are well reproduced, and Θ_D is the limit where all possible vibration states (phonons) in a solid are "just" occupied. It is higher for materials with strong (rigid) bonds and lower if weak (soft) bonds prevail. Some examples (data mainly from the textbook by Kittel, 2004[12]) are given in Table 1.2.

Table 1.2: Debye temperatures for some materials.

Substance	Se	Na	Zn	Au	Ag	Cu	Si	C (graphite/diamond)	α-Al$_2$O$_3$
Θ_D (K)	90	158	327	165	225	345	645	413/$\approx$2000	$\approx$1050

---

[11] Peter Debye (24 March 1884–2 November 1966).
[12] Charles Kittel (18 July 1916–15 May 2019).

Equation (1.15) can be used to derive the Debye temperature of a material if c_p is plotted for $T \ll \Theta_D$ as a function of T^3. By this method Haussühl et al. (2021) obtained $\Theta_D = 676\,\text{K}$ for $LiAlO_2$.

1.1.3.2 Other contributions

Not only phonons, but also other particles or quasi-particles inside solids can carry momentum and contribute this way to c_V. Conductors, and especially metals, bear a high density of electrons resulting in a specific heat contribution

$$c_V^e = \gamma T, \tag{1.16}$$

where $\gamma = \pi^2 k_B^2 / 2\epsilon_F$ holds for free electrons (ϵ_F is the Fermi[13] energy). The function c_V^e is linear in T and relevant especially for low T. Also, magnetic moments and moving atoms or ions can substantially contribute to c_V. Examples are AgI, ZrO_2, and CaF_2, where the cations or anions become mobile below the melting temperature and contribute not only to electrical conductivity ("fast ion conductors"), but also to c_V (Mehrer, 2007).

? Why c_V^e can be relevant especially for low T?

1.2 Derived thermodynamic potentials

1.2.1 Enthalpy

The internal energy U introduced in (1.8) and (1.13) describes the potential and vibrational energy contained in the material itself. In a real system the material has a certain volume V and is exposed to a pressure p. It should be noted that for systems where surface energy can be neglected (phases sufficiently large and phase boundaries with low curvature), p and T are constant over the whole system. The quantity

$$H = U + pV \tag{1.17}$$

is called the enthalpy and is the sum of the internal energy with the amount of work ("volume work") that must be performed to create the phase volume V against the system pressure p. Such as U, H is a thermodynamic potential (= state function) because it depends only on the actual status of the system and not on the way how this status was reached.

---

13 Enrico Fermi (29 September 1901–28 November 1954).

It is usually difficult, if not impossible, to measure the total amount of $\hat{H}$ stored inside a phase, because the energy balance of all contributions to U is often unknown. Contrarily, the enthalpy change of a system from an initial state i to a final state f,

$$\Delta H = H_f - H_i, \qquad (1.18)$$

can usually be measured because $\Delta H = Q$ (the heat added to the system, provided that $p = \text{const.}$ and no other work except volume expansion work is done by the system).

To circumvent the problem of absolute H measurements, it is useful to determine "standard conditions" and a set of basic substances where the H values are fixed at these standard conditions. For this purpose, the US National Bureau of Standards defined $T_0 = 25\,°C = 298\,K$ and $p_0 = 1 \times 10^5\,Pa = 1\,bar$ as the "standard ambient temperature and pressure" (see, e. g., Chase et al., 1982). Under these conditions, $H_i = H_0 = 0$ is defined for every chemical element in the phase state that is stable under these conditions (Fig. 1.7). The function $H(T)$ is usually smooth except at first-order phase transitions, where it jumps by an amount called the heat of fusion (ΔH_f), heat of vaporization (ΔH_v), or in general the heat of transition (ΔH_t). Some first-order transitions for sulfur can be seen in Fig. 1.7. In contrast to this, Fig. 1.8 shows "absolute" $\hat{H}(T)$ data for metallic Li below room temperature.

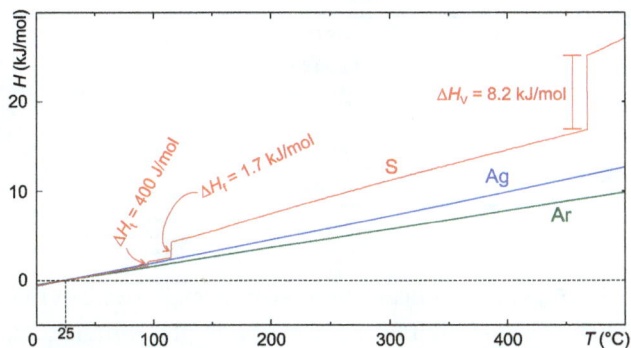

Figure 1.7: $H(T)$ functions for three chemical elements at 1 bar. Sulfur undergoes a monoclinic $\leftrightarrow$ orthorhombic phase transition at 95 °C (transition enthalpy ΔH_t), melts at 115 °C (fusion enthalpy ΔH_f), and vaporizes at 469 °C (vaporization enthalpy ΔH_v). Silver and argon show no transitions in this T range.

The enthalpy of most chemical compounds is strongly negative, because just the release of (binding) energy is the main reason why chemical compounds are formed from the elements. Some examples can be found in Table 1.3. Compounds with $H > 0$ are not stable and are either entropically stabilized (see Section 1.2.2) or exist only metastable, because their decomposition is hindered by kinetic reasons. The latter case is more typical and is found in Table 1.3 for two chlorine oxides, which are ex-

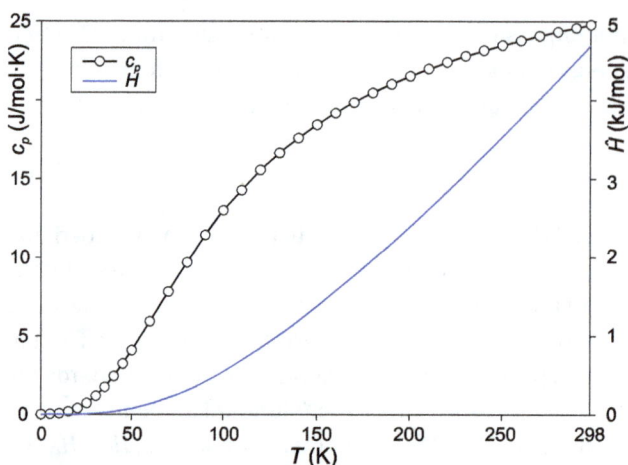

Figure 1.8: Measured $c_p(T)$ data for lithium below room temperature (Alcock et al., 1994). $\hat{H} = H - H(0\,K)$ is the "real" enthalpy that starts from 0 at absolute zero.

Table 1.3: Enthalpy H of several substances under standard conditions (25 °C, 1 bar). Data from Fact-Sage 8.0 (2020).

Substance	H (kJ/mol)	Substance	H (kJ/mol)
Na (sol)	0	Cl_2O (gas)	+87.86
Cl_2 (gas)	0	ClO_2 (gas)	+104.60
O_2 (gas)	0	AgN_3 (sol)	+308.78
NaCl (sol)	−411.12	Au_2O_3 (sol)	−3.35
NaCl (liq)	−394.96	H_2O (liq)	−285.83
Na_2O (sol)	−417.98		

tremely reactive, and silver azide AgN_3 is technically used even as explosive. For this substance, Schmidt et al. (2007) observed by DSC an irreversible phase transition, which peaks at 199.4 °C to a high-T phase, which is another indication of the unstable nature of the substance.

Most compounds of "noble" metals, like Au_2O_3 in Table 1.3, possess very small enthalpies. This is the reason why already a small energetic activation, e. g., by heating, usually decomposes such compounds—in the case of Au_2O_3 at 150...160 °C. Silver forms compounds that are slightly more stable but are so sensitive that silver halides can be partially decomposed by the energy of light. This was invented by Schulze (1719)[14] and gave the basis for the first (still fleeting) photograms. Silver(I) oxide Ag_2O is stable only at low $T < 200$ °C under oxidizing conditions (Fig. C.31).

---

14 Johann Heinrich Schulze (12 May 1687–10 October 1744).

The enthalpy of substances for other temperatures, different from the standard value for $T_0 = 25\,°C$, can be calculated by

$$H(T) = H_0 + \int_{T_0}^{T} c_p(T)\,\mathrm{d}\,T \tag{1.19}$$

from experimental $c_p(T)$ data. If we start with $\hat{H}(0\,\mathrm{K}) = 0$, then (1.19) represents the area under the $c_p(T)$ curve in Fig. 1.8. $c_p(T)$ rises with T^3 only for low $T \leq 50\,\mathrm{K}$ and approaches quickly the Dulong–Petit value 25 J/(mol K) (1.5) and then slowly changes; consequently, for many substances, $H(T)$ is an almost linear function at high T.

Enthalpy H, entropy S (see Section 1.2.2), and Gibbs[15] free energy G (see Section 1.2.3) are thermo- **!**
dynamic potentials (state functions). According to Hess'[16] law, the value of such a potential does not
depend on the path of the reaction, like depicted in Fig. 1.9. With Hess' law the dependence of reaction enthalpies on T can be calculated, and also transformation enthalpies under nonequilibrium
conditions.

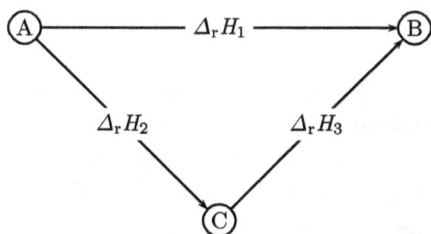

Figure 1.9: Hess' law: $\Delta_r H_1 = \Delta_r H_2 + \Delta_r H_2$.

Hess' law is the basis for the Born[17]–Haber[18] cycle and can be used to calculate lattice energies of compounds, which otherwise are not available. This lattice energy corresponds to the inner energy U in equations (1.8) and (1.19). Figure 1.10 demonstrates this for magnesium oxide: If the formation enthalpy $\Delta H_f = -602\,\mathrm{kJ/mol}$ is known, e. g., from calorimetric measurements, U^{lattice} can be calculated from this circle. Data for the four energies or enthalpies from the left side (blue labels) can be found else- where in the literature, and from these values we obtain $U^{\mathrm{lattice}} = -3882\,\mathrm{kJ/mol}$.

---

15 Josiah Willard Gibbs (11 February 1839–28 April 1903).

16 Germain Henri Hess (7 August 1802–30 November 1850).

17 Max Born (11 December 1882–5 January 1970).

18 Fritz Haber (9 December 1868–29 January 1934).

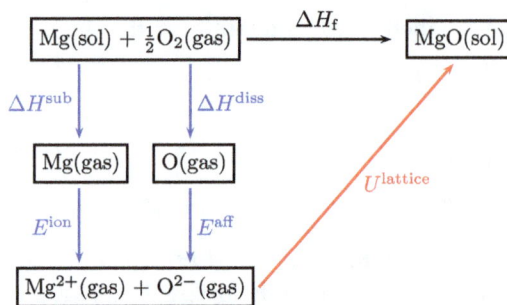

Figure 1.10: Born–Haber cycle for MgO: $\Delta H_f = \Delta H^{sub} + E^{ion} + \Delta H^{diss} + E^{aff} + U^{lattice}$. ($\Delta H^{sub} = 148\,kJ/mol$ – sublimation enthalpy, $E^{ion} = 2188\,kJ/mol$ – ionization energy, $\Delta H^{diss} = 249\,kJ/mol$ – dissociation enthalpy, $E^{aff} = 695\,kJ/mol$ – electron affinity.)

? With some care, very pure water can be "supercooled" below 0 °C and then crystallized. Calculate the heat of crystallization $\Delta H_f^{(-10)}$ at −10 °C using the following data: $\Delta H_f^{(0)} = 6001\,J/mol$, $c_p^{ice} = 36\,J/(mol\,K)$, $c_p^{liq} = 76\,J/(mol\,K)$.

1.2.2 Entropy

The term entropy describes another thermodynamic potential, which can be defined either statistically or from the thermodynamic viewpoint. The statistical interpretation is related to the Shannon[19] entropy used in information theory and corresponds to the average information density in a system of symbols (or atoms). In the statistical interpretation the entropy S is a measure of "uncertainty" of a given state; in other words, it is a measure of the number of equivalent arrangements of parts (e. g., atoms) setting up the system. This is shown in Fig. 1.11 (a) for the case of a planar lattice where all possible sites are occupied by atoms. Another equivalent possibility for the atoms does not exist, and the probability for the shown state is $P = 1$. Not so in Fig. 1.11 (b), where two atoms are missing: 48 options exist for selecting the first missing atom, and 47 options remain for the second missing atom, but it is indistinguishable which atom was missing first. This means that $48 \times 47/2 = 1128$ equivalent microstates exist, which result in the same macrostate "plane lattice" with 48 sites. If P_i is the probability of the ith microstate, then

$$S = -k_B \sum_i P_i \ln P_i \tag{1.20}$$

is the entropy of the corresponding macrostate.

---

19 Claude Elwood Shannon (30 April 1916–24 February 2001).

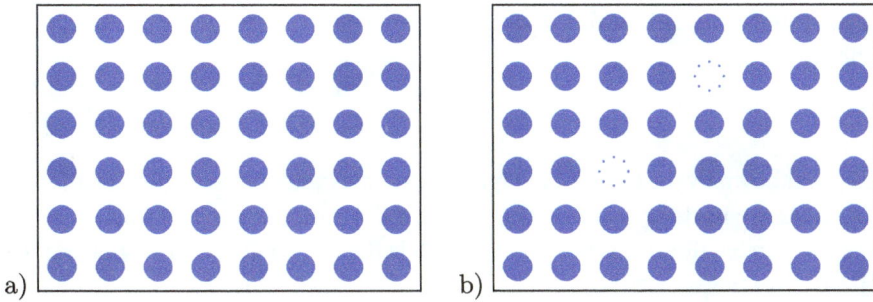

Figure 1.11: Statistical interpretation of entropy: (a) 1 possible option to distribute $6 \times 8 = 48 = n$ atoms on n lattice sites. (b) shows one of $48 \times 47/2 = 1128$ equivalent options to arrange $f = 2$ faults (e. g., unoccupied sites or vacancies) on 48 lattice sites.

For the arrangement from Fig. 1.11, this is demonstrated in Fig. 1.12. The system has N atom sites that are either "faulty" (f sites) or occupied ($N - f$ sites). The number of possible arrangements of faulty sites (= number of microstates) is

$$\frac{1}{P} = \Omega = \frac{N!}{(N-f)!f!} \tag{1.21}$$

and grows drastically with f to a maximum value, which is obtained for $f = N/2$. The result of (1.21) for $N = 48$ is shown by the bar plot in Fig. 1.12.

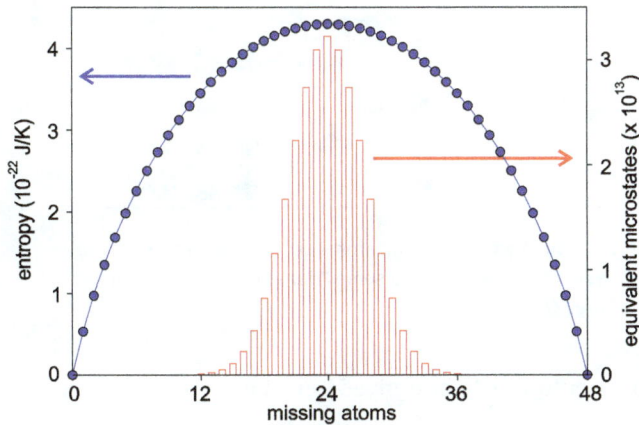

Figure 1.12: Derivation of entropy S for the ensemble shown in Fig. 1.11 with different numbers of missing atoms. The number Ω of equivalent microstates is maximum if 50% of the sites are faulty.

In closed systems in equilibrium, P is often identical for all microstates, and (1.20) simplifies to

$$S = k_B \ln \Omega,$$ (1.22)

where $S(f)$ is shown by the line plot in Fig. 1.12. In a real system with $f > 0$ and at $T > 0$, all faulty sites are initially ordered ($\Omega(t = 0) = 1$); this ordering drops with time t. It will be shown in Section 1.2.3 that increase of S is the driving force for increasing the "uncertainty" of the system.

The thermodynamic interpretation of S does not depend on the atomistic nature of matter but can be related to it. Figure 1.13 (a) shows an ideal lattice, where the atoms are connected by chemical bonds. We can assume that this solid is formed because the creation of every bond reduces the internal energy of the material (compared to sole atoms) by a certain amount E. In Fig. 1.13 (b), some vacancies are introduced, which increase the disorder, or the "uncertainty", of the system and increase its entropy S. If for this purpose, n bonds had to be broken, then the energy $Q = n \cdot E$ was used. Assuming that this process is performed under equilibrium conditions, it is reversible, and the relation

$$\Delta S = \frac{Q_{rev}}{T}$$ (1.23)

describes the entropy change of the system.

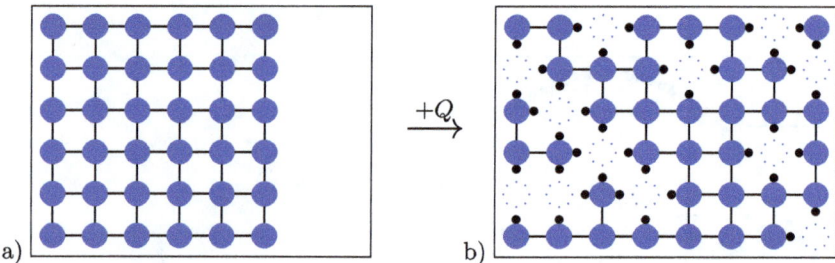

Figure 1.13: Thermodynamic interpretation of entropy: (a) all atoms occupy lattice sites and form ideal bonds. (b) faults (here vacancies) are introduced, and some "dangling bonds" are created. For this process, a certain amount of energy Q was used.

In analogy to equation (1.17) the entropy can be derived from the specific heat capacity by

$$S(T) = S_0 + \int_{T_0}^{T} \frac{c_p(T)}{T} \, dT,$$ (1.24)

and we have $S_0(0\,K) = 0$.

1.2.3 Gibbs-free energy

1.2.3.1 Pure substances

The enthalpy H introduced in Section 1.2.1 is a measure of the amount of energy added (or extracted) from a system. However, this process is usually not reversible: Even if lattice vibrations are considered undamped (contribution U in equation (1.17)) and if the volume work can be restored (contribution $p \cdot V$ in equation (1.17)), the entropy S of the system increases upon heating. Once the disorder of a system becomes larger, this is in general an irreversible process.

It is useful to define the "free energy" (or Gibbs energy)

$$G = H - TS = U + pV - TS, \qquad (1.25)$$

which is the amount of energy that can be reversibly added or extracted from the system. As H and S, G is a thermodynamic potential. This means that for a given state (x_i, T, p) of a system, G does not depend on the way this state was reached.

In his seminal paper on heterogeneous equilibria, Gibbs (1874–1878) derived this thermodynamic potential by combining the principles of maximum entropy: "For a closed system with fixed internal energy, the entropy is maximized at equillibrium" and of minimum energy "For a closed system with fixed entropy, the total energy is minimized at equillibrium".

For elevated T (room temperature or beyond), H changes only weakly (see, e. g., Fig. 1.7), and the same holds for S. From (1.25) it is obvious that $G(T)$ is a function that drops nearly linearly for most phases and systems. Figure 1.14 demonstrates this for the three aggregation states of sodium chloride. The solid phase develops the strongest binding forces between atoms, this way reducing H and resulting in the most negative $G^{\text{sol}}[0\,^\circ\text{C}] = -430.877\,\text{kJ/mol}$. The energy gain by bonding is less significant in the liquid ($G^{\text{liq}}[0\,^\circ\text{C}] = -415.818\,\text{kJ/mol}$) or even smaller in the gas with much weaker attractive forces between atoms ($G^{\text{gas}}[0\,^\circ\text{C}] = -244.230\,\text{kJ/mol}$). On the other hand, the degree of disorder and, consequently, the slope $-S$ of the G functions become larger in this order. The result is that solid, liquid, and gas phases have the lowest G and become stable one after the other with larger T.

For all solid curves in Fig. 1.14, we assumed that $p = 1\,\text{bar}$. The influence of p on S and H is usually small for condensed phases (solids, liquids). In contrast, gases show high compressibility, which leads to a high dependency of atomic interactions on pressure. For lower p, $G^{\text{gas}}(T)$ shifts to bottom left, leading to an intersection with $G^{\text{liq}}(T)$ at lower T, which means that T_{boil} drops with p. If for sufficiently small p, the intersection is below T_{f}, then the liquid phase is never stable, and the substance undergoes sublimation, which is a first-order phase transition from the solid to the gas, and vice versa. This is shown by the dashed curve in Fig. 1.14.

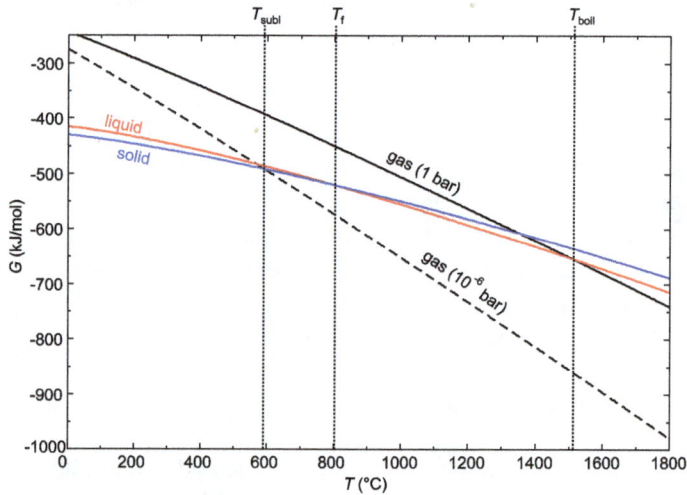

Figure 1.14: G as a function of T for the solid, liquid, and gaseous phases of sodium chloride (NaCl); $G(T)$ for the gas is drawn for two pressures $p = 1$ bar and $p = 10^{-6}$ bar. T_f – fusion point, T_{boil} – boiling point, T_{subl} – sublimation point.

1.2.3.2 Mixture phases

The Gibbs energy of a mixed phase Φ can be expressed as the sum

$$G^\Phi = G^0 + G^{id} + G^{ex}, \tag{1.26}$$

where G^0 is the scaled contribution of the pure components. In systems with two components A and B, this is the straight line

$$G^0 = xG^B + (1 - x)G^A \tag{1.27}$$

connecting the values for the pure components, where x is the molar fraction of component B (see Fig. 2.45). If A and B can be mixed in one phase (e. g., in a solid or liquid solution), then the disorder of this phase grows, and this enhances its entropy S. This is basically a statistic effect, which is independent of the chemical nature of the corresponding atoms. Nevertheless, increasing S results with (1.25) in a reduction of G for $T > 0$ and hence in a stabilization of the phase. This stabilizing Gibbs energy contribution G^{id} to G^Φ (1.26) appears without energetic interaction of the species ($\Delta H^{id} = 0$, hence obeying Raoult's[20] law),

$$G^{id} = -RT \sum_i^C x_i \ln x_i \quad (C \text{ components}) \tag{1.28}$$

$$= -RT[x \ln x + (1 - x) \ln(1 - x)] \quad (2 \text{ components}), \tag{1.29}$$

---

20 François Marie Raoult (10 May 1830–1 April 1901).

and lowers the Gibbs energy of the phase for intermediate compositions; the effect is maximal if the concentrations of all components are equal ($x = 0.5$ for systems with two components). Also, for mixture phases with high number of components ($C \geq 5$), the maximum G^{id} is reached for identical concentrations for all components, and unexpected metallic or compound mixture phases, called "high-entropy alloys", were reported (HEAs; see Section 1.5.3.5).

Simplify equation (1.28) for the case of C components with identical composition. How large is the impact on G^{Φ} for $C = 6$ at $T = 1000$ K?

The contribution G^{ex} summarizes deviations from the ideal behavior, which result, e. g., from attraction or repulsion or from complicated shapes of the interacting species. In cases where the components are very similar and for very high temperature, G^{ex} can sometimes be neglected, and the phase behaves ideal. If a mixed liquid phase and a mixed solid phase are in equilibrium, then a "mixed crystal" system is formed, which will be described in Section 1.5.1.2.

1.3 Thermoelectricity, thermocouples

1.3.1 Phenomenological description

The transport of heat and electrons in solids is coupled, and these coupling effects are called thermoelectric effects, among them, Seebeck[21] and Peltier[22] effects. Both effects are reciprocal. Compared to these effects, Thomson[23] and Bridgman[24] effects are usually small. The Seebeck effect gives a physical basis for thermocouples.

Seebeck: Two different metals A and B (or other conductors) are electrically connected in a way schematically shown in Fig. 1.15, and the connections are held at different temperatures. Then a thermovoltage $\Delta U^{AB} = \Sigma^{AB} \Delta T$ appears between both connections. A detailed treatment shows that Σ is a second-rank tensor. However, the tensor description is necessary only for most anisotropic materials like crystals (cf. the two different values for Sb crystals in Table 1.5) and textures, but not for isotrope polycrystals like thermocouple wires (Nye, 1957).[25]

Peltier: In a setup similar to Fig. 1.15 an electric current is created by applying a voltage between poles (1) and (3). Then, in addition to the Joule[26] heating RI^2 (where R

---

21 Thomas Johann Seebeck (29 March 1770–10 December 1831).

22 Jean Charles Athanase Peltier (22 February 1785–27 October 1845).

23 William Thomson, 1. Baron Kelvin (26 June 1824–17 December 1907).

24 Percy Williams Bridgman (21 April 1882–20 August 1961).

25 John Nye (26 February 1923–8 January 2019).

26 James Prescott Joule (24 December 1818–11 October 1889).

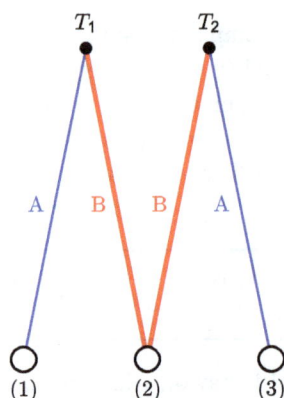

Figure 1.15: Measurement of the thermovoltage between two different metals A and B forming a thermocouple pair. The two welding points between the wires are held at different T_1, T_2. All other points are held at some constant T. (1)–(2) is the thermovoltage related to T_1; (1)–(3) is related to the difference $T_1 - T_2$, and hence the *differential thermal analysis* (DTA) signal; see Fig. 2.1 (a).

is the electrical resistivity), which is created everywhere along the different wires, thermoelectric heating $\pm\Pi^{AB}I$ or cooling appears at the welding points. The Peltier and Seebeck coefficients are related by $\Pi^{AB} = T\Sigma^{AB}$, which implies that Π is also a second-rank tensor.

Thomson: Additional heating and cooling $\pm\tau^A I \partial T/\partial x$ also appears if the system of different wires in Fig. 1.15 is replaced by one single wire A and, at the same time, a temperature gradient is opposed to it. The Thomson effect describes the temperature dependence of the Seebeck effect, which can be expressed as $\partial\Sigma^{AB}/\partial T = -\tau^A(T)/T + \tau^B(T)/T$.

Bridgman: Additional heating and cooling appears in one material if the current density inside the wire changes with x. This may result from nonconstant cross-section, but also due to variable crystallographic orientation inside the wire.

If the measurement of ΔU^{AB} in Fig. 1.15 is performed without current flow, e. g., by compensation, then it depends only on the temperature difference ΔT, and Σ^{AB} is a materials parameter of the combination A–B. If, moreover, one of the welding points is placed in a thermostat (e. g., $T_1 = $ const. $= 0\,°C$ in ice/water), then the thermovoltage depends only on the temperature of the other welding point and can be used to measure T_2. This is technically used by thermocouples (Table 1.4), which are available for a wide range of temperatures and chemical environments. For very low temperatures, however, thermocouples cannot be used as $\Sigma^{AB} \to 0$ for $T \to 0$.

! Thermocouples are based on the Seebeck effect and represent an important device for temperature measurements in thermal analysis.

Table 1.4: Thermovoltage (in mV, DIN EN 60584-1) of several common thermocouples, positive pole first. Type J = "iron/constantan". Type D is occasionally called "type W".

T (°C)	Type J Fe/Cu55Ni45	Type S Pt90Rh10/Pt	Type B Pt70Rh30/Pt94Rh6	Type D W97Re3/W75Re25
−200	−8.15	–	–	–
−100	−4.60	–	–	–
0	0.00	0.0000	0.0000	0.000
100	5.37	0.6459	0.0332	1.145
200	10.95	1.4408	0.1783	2.602
300	16.55	2.3230	0.4306	4.286
400	22.15	3.2594	0.7865	6.129
500	27.84	4.2333	1.2418	8.076
600	33.66	5.2387	1.7919	10.085
800	46.23	7.3445	3.1536	14.170
1000	–	9.5870	4.8343	18.226
1200	–	11.9510	6.7864	22.142
1400	–	14.3723	8.9562	25.875
1600	–	16.7768	11.2630	29.403
1800	–	–	13.5913	32.702
2000	–	–	–	35.707
2200	–	–	–	38.289
2400	–	–	–	40.223

It should be noted that the thermoelectric properties depend on crystal defects, such as impurities or large dislocation densities. Consequently, thermocouples must be handled with care, and strong bending of the wires must be avoided. For thermocouples made of platinum (type S or B), oxidizing conditions are optimal. If heated for long time under strongly reducing conditions, alloying with C, S, As, or volatile metals (Ga, Bi) may occur ("platinum poisons"). This can affect the measured ΔU^{AB} and finally destroy the thermowire.

The Peltier effect is used mainly for refrigerators without moving parts for application in households, laboratory, and industry. "Camping coolers" that can be run by the 12 V DC power supply of cars, cooling stages for microscopes, and coolers for SDDs (silicon drift detectors) in EDX are examples. As the Peltier and Seebeck effects are reciprocal, the same materials show high efficiency. It should be noted, however, that the Seebeck effect is mainly used with thermocouples for measuring T, where the total height of the signal is not so important, and mainly long-term stability is required. In contrast, for Peltier coolers, high efficiency is inevitable to overcome the Joule heating (which is always present) by the thermoelectric cooling effect. Some data can be found in Table 1.5.

Table 1.5: Absolute Seebeck coefficients for several materials near room temperature, mainly after Paufler (1986).

Material		Σ^A (μV K^{-1})	Material	Σ^A (μV K^{-1})	Material	Σ^A (μV K^{-1})
Se		+1000	Au	+1.1	n-Bi$_2$Te$_3$	−230
Sb	$\|\vec{c}$	+20.6	Pt	−4.4	Pb$_{15}$Ge$_{37}$Se$_{58}$	−1990
	$\perp\vec{c}$	+46.8	Ni	−19.9		

1.3.2 Atomistic description

As indicated before, all thermoelectric effects can be related to the Seebeck coefficient Σ and its dependence on material, temperature, and crystallographic orientation. The different dependence of the electronic structure for different materials is responsible for these effects. We find

$$\Sigma = -\frac{\pi^2}{3}\frac{k_B^2}{e}T\left(\frac{\partial \ln \sigma_{\text{elect.}}}{\partial \epsilon}\right)_{\epsilon=\epsilon_F}, \tag{1.30}$$

where the last term is the dependence of the electrical conductivity $\sigma_{\text{elect.}}$ on the energy of electrons at the Fermi[27] energy ϵ_F.

1.4 Phases

A phase is a volume filled homogeneously with matter in dimensions that are large compared to interatomic distances. The latter restriction is necessary because atoms of different chemical elements are of course different; nevertheless, chemical compounds can form phases. In every system where gases exist (including thermal analyzers), these gases form usually one phase only, because under not too extreme conditions (see, e. g., Senesi et al., 1997), all gases intermix in arbitrary ratio. This is not always so for the liquid aggregation state of different substances. Substances of similar chemical nature can often (but not always) be mixed in their liquid (molten) states:
- Many molten metals form one liquid metallic phase, but, e. g., liquid Pb shows with many other liquid metals a "miscibility gap" with two separate liquid phases (Teppo et al., 1991).
- Water mixes in arbitrary ratio with lighter alcohols like CH$_3$OH (methanol) and C$_2$H$_5$OH (ethanol), but from 1-butanol upwards the mutual miscibility is increasingly limited at room temperature (Barton, 1984).
- Ionic compounds are often miscible in the liquid state if their anions are identical or similar.

---

27 Enrico Fermi (29 September 1901–28 November 1954).

If the liquid aggregation states of compounds cannot be mixed, then they form two or even more liquid phases, e. g., water and oil.

The miscibility of different solid substances depends on even more conditions. Besides a sufficient degree of chemical similarity (like given above), some crystallographic aspects are important, which will be described in the next subsection.

1.4.1 Some crystallography

A vast majority of solids appears in crystalline state. Crystals are not necessarily large (and often beautiful) specimen that can be found in mineralogical collections worldwide or as gemstones, in technical devices like computers, lasers, light emitting diodes (LEDs), or some sensors. Nearly every stone, every nail in the hardware store, and even parts of our human bodies are crystalline at least on a microscopic scale. Besides crystals, two other types of solids exist:

Glasses are solids that exist metastably in nonequilibrium. This leads to some specific features like glass transitions, where thermal analysis curves show characteristic effects, like those shown in Fig. 1.5. Glasses have no sharp melting point; instead, they become softer upon heating beyond the glass transition until they finally form a melt, usually, with high viscosity. From the structural point of view, glasses are short-range ordered over a range of a few interatomic distances but are disordered over longer distances.

Quasicrystals were discovered by Shechtman et al. (1984)[28] and represent ordered solids on short and long distances. It should be mentioned, however, that the existence of structures possessing as well short-range as long-range order, but no translational symmetry in three dimensions (what accurately meets the definition of quasicrystals, too), was very detailed dealt with already more than 20 years earlier by Dornberger-Schiff (1964),[29] who named them OD-structures (order-disorder-structures), and by Bohm (1967, 1968),[30] who named them metacrystals. The thermodynamic behavior of these still somewhat exotic materials is very similar to "normal" crystals (see below); but from the structural point of view, they are significantly different, because quasicrystals have no translation symmetry. This means that the atomic arrangement of quasicrystals cannot be built up by stacking smaller structural units ("unit cells") periodically in three dimensions like shown for a crystal in Fig. 1.16. This is an exclusive property of crystals.

---

28 Daniel "Dan" Shechtman (born 24 January 1941).

29 Katharina Boll-Dornberger, or Käthe Dornberger-Schiff (2 November 1909–27 July 1981).

30 Joachim Bohm (born 25 March 1935).

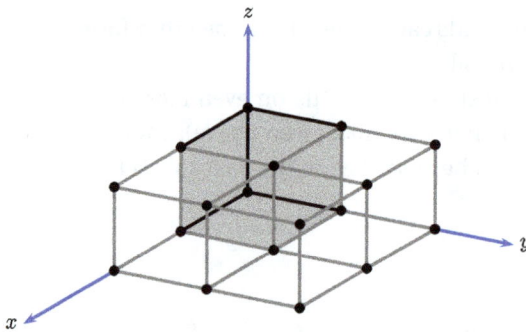

Figure 1.16: $2 \times 2 \times 1$ stacking of orthorhombic unit cells. The dimensions along the x, y, z axes in the orthorhombic system (Table 1.6) are the lattice parameters $a \neq b \neq c$. In the orthorhombic system the angles between axes are $\angle(b, c) = \alpha = \angle(c, a) = \beta = \angle(a, b) = \gamma = 90°$.

It is beyond the scope of this book to give here an extensive introduction into the science of crystallography. For this, the reader is referred to other print or web resources, e. g., Kleber et al. (2021),[31] Borchardt-Ott (2012),[32] Julian (2015), IUCr (2020). However, it is important to know that crystals are stackings of identical unit cells in all three dimensions of space. Figure 1.16 shows this schematically for a stacking in two dimensions; stacking in the third was not drawn for better recognizability. Every unit cell (e. g., the gray cell in the origin of the coordinate system in Fig. 1.16) contains a set of atoms at specific coordinates inside the cell. These atoms must occur in the stoichiometric ratio of the substance, because all cells are identical: a copper unit cell contains only Cu atoms, a sodium chloride unit cell contains Na^+ and Cl^- as 1 : 1, and a cane sugar unit cell contains carbon, hydrogen, and oxygen in the ratio 12 : 22 : 11, because the chemical sum formula is $C_{12}H_{22}O_{11}$. The atomic positions inside the cell, together with the geometrical dimensions of the unit cell $a, b, c, \alpha, \beta, \gamma$, describe the crystal structure of the substance. Crystal structures are often determined by diffraction of X-ray radiation.

The most general shape of a unit cell is the parallelepiped. Such volume is limited by three pairs of identical parallelograms with arbitrary sides a, b; b, c; and c, a. 3D stacking of parallelepipeds gives a triclinic lattice, which can be described by triclinic coordinates. A triclinic lattice has either no symmetry elements, except the identity 1, or it has besides 1 the inversion center $\bar{1}$ (spoken "bar one"), which combines an atom A at coordinates x_A, y_A, z_A with an identical atom at coordinates $-x_A, -y_A, -z_A$. The translations are symmetry elements of every lattice too. More options do not exist, and these sets of possible combinations of symmetry elements are called a space group. Hence we can conclude that only two space groups ($P1$ and $P\bar{1}$) are compatible

---

31 Wilhelm "Will" Kleber (15 December 1906–27 August 1970).

32 Walter Borchardt-Ott (16 November 1933–19 January 2012).

with a triclinic lattice. The letter P stands for a "primitive" lattice, because it has lattice points only at the corners of the unit cells, like the orthorhombic lattice in Fig. 1.16. Besides P, in general, the following lattices exist: F (face centered = one additional lattice point on every face center); I (german *innenzentriert* = body centered = one additional lattice point in the center of the unit cell); C, A (lattice points on the faces where z or x points out; R (rhombohedral lattice, see caption of Table 1.6).

Table 1.6: The seven crystallographic coordinate systems ("crystal systems") and the corresponding numbers of space groups. The total number of space groups is 230. The hexagonal lattice can alternatively be described in rhombohedral coordinates with $a = b = c$ and $\alpha = \beta = \gamma \neq 90°$.

Crystal system	Lattice parameters	Space groups
triclinic	$a \neq b \neq c; \alpha \neq \beta \neq \gamma$	2
monoclinic	$a \neq b \neq c; \alpha = \gamma = 90°; \beta \neq 90°$	13
orthorhombic	$a \neq b \neq c; \alpha = \beta = \gamma = 90°$	59
tetragonal	$a = b \neq c; \alpha = \beta = \gamma = 90°$	68
trigonal	$a = b \neq c; \alpha = \gamma = 90°; \beta = 120°$	25
hexagonal	$a = b \neq c; \alpha = \gamma = 90°; \beta = 60°$	27
cubic	$a = b = c; \alpha = \beta = \gamma = 90°$	36

A space group symbol in the nomenclature of the International Union of Crystallography IUCr, as defined by Hahn et al. (2016),[33] consists in its first position of an italic letter (P, I, F, A, C, R) describing the lattice type. Then for up to three symmetrically independent directions (so-called viewing directions), symmetry elements are given: 6, 4, 3, 2, 1 for rotation axes, where 1 is the identity; $\bar{6}, \bar{4}, \bar{3}, \bar{2} \equiv m, \bar{1}$ are combinations of a rotation with an inversion (rotoinversion), where m is a mirror plane, and $\bar{1}$ is the inversion center. Symbols like 3_2 stand for screw axes, and letters like c, b, a, d, n for glide planes. Space group symbols like $P\frac{2}{m} \equiv P2/m$ mean a 2-fold axis 2 standing orthogonal on a mirror plane m.

Substances with triclinic crystal lattices are found seldom in nature, and lattices with higher symmetry are more common. Organic solids often have monoclinic or orthorhombic lattices. Very simple inorganics (many metals, simple oxides, or halides) are often cubic. However, this is only a general trend: for example, Zn is hexagonal, Sn is tetragonal, and Ga is orthorhombic. With the number of symmetry elements, also the number of possible combinations of these symmetry elements, and hence the number of space groups, grows significantly. In total, 230 space groups where shown to exist, and each of them is compatible with one of seven lattices given in Table 1.6.

In the beginning of this section, some preconditions for the miscibility of liquids were given, and it was mentioned that the requirements for solids are even harder:

---

33 Theo Hahn (3 January 1928–12 February 2016).

> **!** It is a necessary but not sufficient condition for complete miscibility of different crystalline solids that they have identical crystal structure, which implies that they belong to the same space group. Another requirement is that their lattice parameters may not be too different, which typically means that the lattice constants should not differ by more than $\approx 10 \ldots 15\,\%$.

A small selection of the thousands known crystal structures is shown in Figs. 1.17 and 1.18, and in the caption, for each structure, the space group and the required lattice parameters are given. All structures from Fig. 1.17 are cubic, and from Table 1.6 we see that only one lattice parameter (the edge length a of the unit cell) is independent. The tungsten structure in Fig. 1.17 (a), already solved by Davey (1925),[34] is identical with the body centered cubic (bcc) lattice, but copper in Fig. 1.17 (b), taken from the famous collection by Wyckoff (1963),[35] corresponds to the face centered cubic (f. c. c.) lattice. This difference of crystal structures does not allow the formation of an unlimited series of mixed crystals between them, and hence in the solid state, W and Cu will form separate phases if they coexist in a system. (It should be mentioned, however, that a minor "rim" mutual solubility is not forbidden, see Section 1.5.1.2).

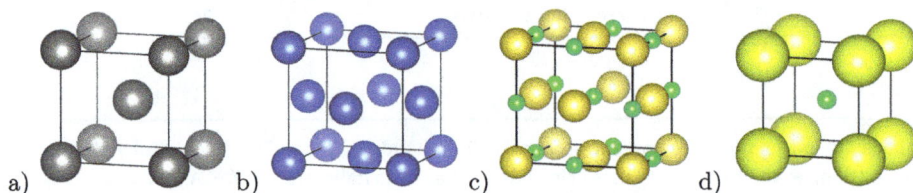

Figure 1.17: Four simple cubic crystal structures. (a) W (Davey, 1925, $Im\bar{3}m$, a = 3.158 Å), (b) Cu (Wyckoff, 1963, $Fm\bar{3}m$, a = 3.61496 Å), (c) NaCl (Na$^+$ yellow, Cl$^-$ green, Abrahams and Bernstein, 1965, $Fm\bar{3}m$, a = 5.62 Å), (d) CsCl (Cs$^+$, yellow, Cl$^-$ green, Wyckoff, 1963, $Pm\bar{3}m$, a = 4.123 Å).

Most alkali halides and also some oxides (e. g., MgO), sulfides (e. g., BaS), and nitrides (e. g., TiN) crystallize in the NaCl ("halite", Abrahams and Bernstein, 1965[36]) structure, which is shown in Fig. 1.17 (c). In this structure, every cation is octahedrally coordinated to six anions, and vice versa. (The coordination polyhedra are not shown in Fig. 1.17.) CsCl is an exception and crystallizes in a crystal structure shown in Fig. 1.17 (d) and received its name from this substance. Here the coordination polyhedron is a cube (also not shown).

Figure 1.18 shows two different structures of the compound $CaCO_3$. Both modifications can be found as minerals in nature, but only the calcite structure is thermodynamically stable under ambient conditions. The aragonite modification has a higher

---

34 Wheeler Pedlar Davey (19 March 1886–12 October 1959).

35 Ralph Walter Graystone Wyckoff (9 August 1897–3 November 1994).

36 Sidney Cyril Abrahams (28 May 1924–9 February 2021).

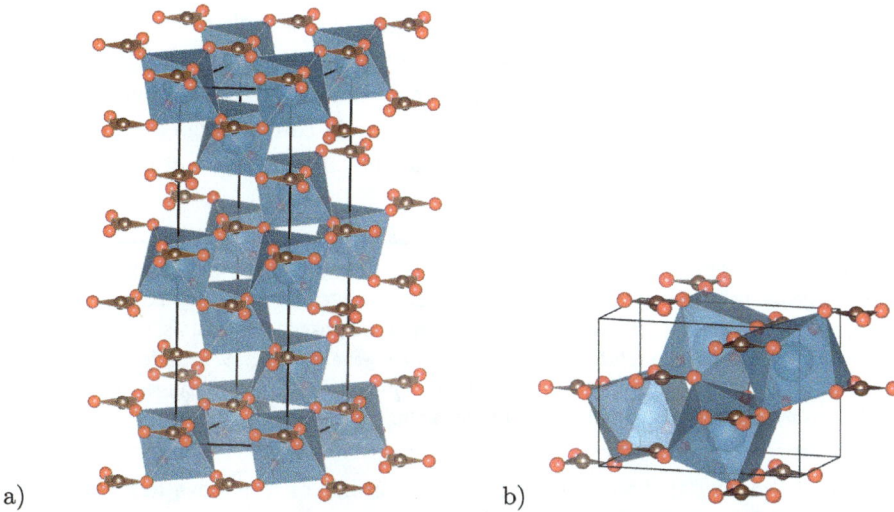

Figure 1.18: Crystal structures of two modifications of $CaCO_3$ in polyeder projection (Ca^{2+} blue, C^{4+} brown, O^{2-} red, Ca^{2+} is octahedrally coordinated by six O^{2-}, C^{4-} is planar coordinated by three O^{2-} to CO_3^{2-} anions. (a) Calcite (Graf, 1961, $R\bar{3}c, a = 4.99\,\text{Å}, c = 17.0615\,\text{Å}$). (b) Aragonite (Villiers, 1971, $Pncm, a = 4.9614\,\text{Å}, b = 7.9671\,\text{Å}, c = 5.7404\,\text{Å}$.)

Gibbs energy than the calcite modification (Bäckström, 1925),[37] but the phase transition rate to the stable calcite modification is so low that aragonite can be considered metastable. In contrast to Fig. 1.17, the structures in Fig. 1.18 where drawn with coordination polyhedra to make their structural similarity more obvious. However, calcite contains planar CO_3^{2-} (carbonate) anions, and Ca^{2+} in octahedral (6-fold) coordination, whereas the Ca^{2+} coordination is 9-fold in aragonite.

1.4.2 Phase transitions

1.4.2.1 Structural aspects

A phase transition is often associated with a significant change of the atomistic structure of matter, where usually its symmetry also changes. We have to compare and discuss the crystal structures of the phases involved, which is performed in the framework of crystallography and their terminology. Here we give only a very brief overview; for details, the reader is referred to the textbooks on crystallography mentioned in Section 1.4.1.

The symmetry of a crystal can be described by its space group. 230 space groups in three dimensions are described in the "International Tables for Crystallography" (Hahn et al., 2016). Each of these groups can be attributed to one of the seven

---

37 Hans Lemmich Juel Bäckström (18 December 1896–1 August 1977).

coordinate systems given in Table 1.6. For the subject of this book, transitions between solid phases are often relevant, e. g., between the aragonite and calcite structures of $CaCO_3$ (Fig. 1.18). Also, melting, the transition between a solid (crystalline) phase and a liquid, is a phase transition. However, also a liquid can undergo a phase transition to another liquid phase, e. g., the nematic/smectic transition of liquid crystals (Lavrentovich and Terent'ev, 1986).

---

i Crystals may contain different types of symmetry elements, among them, 6-, 4-, 3-, and 2-fold rotation axes, which generate rotations with angles of 60°, 90°, 120°, or 180°, respectively, and their multiples. Besides, rotoinversion axes, screw axes, and glide planes are possible in crystals. In quasicrystals, however, also "noncrystallographic" axes can occur, which generate, e. g., 5- or 10-fold symmetry. In liquids and in ceramic and metallic textures, ∞-fold rotation axes are also observed, which means that rotations by infinitesimally small angles are possible.

A change of the crystal structure can mean that atoms or ions have to move by diffusional steps over significant distances. Cesium chloride is an example: this substance undergoes at $T_t = 470\,°C$ a transition from the CsCl crystal structure (space group $Pm\bar{3}m$) to the NaCl structure (space group $Fm\bar{3}m$), which is stable at $T > T_t$ (see Fig. 1.17). During this large structural change, even the coordination of the ions is changed, which means almost a reconstruction of the crystal structure, and this transformation is called a reconstructive phase transition, following the classification of Buerger (1951).[38] Reconstructive transitions are accompanied by a change of enthalpy; for the CsCl transition, this is $\Delta H_t = 3765.6\,J/mol$.

The substance SiO_2 can appear in a wide variety of modifications, and Fig. 1.19 shows two of them: α-quartz is stable at ambient conditions and is the most abundant modification in nature. Already below $T_t = 573\,°C = 846\,K$ the $[SiO_4]$ tetrahedra start to turn and stretch slightly, and this process is finished at T_t. The atomic coordinates in the resulting β-quartz modification Fig. 1.19 (b) are only slightly different from those of the α ("low quartz") modification shown in the left drawing. It is called a displacive transition, because the transition proceeds by slight displacements of atoms without breaking bonds between them. Christy (1993) points out that the distinction between displacive and reconstructive phase transitions is somewhat subjective and proposes a classification according the symmetry relationships between the modification; then for a "type I" transition, the space group of one modification should be a subgroup of the space group of the other modification. For the $\alpha - \beta$ transition of quartz, this is the case, because the low-T space group $P3_221$ is a subgroup of $P6_222$.

For the $CaCO_3$ modifications aragonite and calcite shown in Fig. 1.18, structural relationships are not as obvious as for quartz. In both modifications, planar CO_3^{2-} complex ions are connected via Ca^{2+} ions; however, the aragonite structure is denser packed (the density of calcite is $2.711\,g/cm^3$, aragonite $2.930\,g/cm^3$), and every Ca^{2+} is

---

38 Martin Julian Buerger (8 April 1903–26 February 1986).

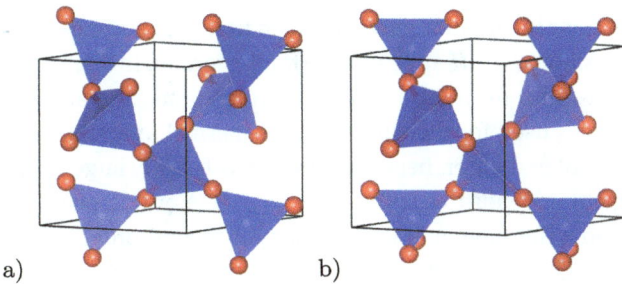

Figure 1.19: Crystal structures of two modifications of SiO_2 in polyeder projection (data from Kihara, 1990, Si^{4+} blue, O^{2-} red, (a) α-quartz at $T = 298$ K ($P3_221, a = 4.9137$ Å, $c = 5.4047$ Å). (b) β-quartz at $T = 848$ K ($P6_222, a = 4.9965$ Å, $c = 5.4570$ Å.)

coordinated to 9 O^{2-} for aragonite, rather than to 6 O^{2-} for calcite (Reuß, 2003). However, again small rotations and deformations of the coordination polyhedra can convert one modification into the other. In contrast to quartz, the symmetry is significantly changed here between rhombohedral (trigonal, space group $R\bar{3}c$) and orthorhombic (space group $Pmcn$).

If a solid substance undergoes a phase transition and changes its crystal structure to a new modification, then it becomes a new phase. In equilibrium, sometimes two (or at a triple point, even three) modifications may coexist; then each modification forms a separate phase.

As a rule of thumb, the modification that is stable at higher T has often a higher symmetry. If the subsequent modifications are named by greek letters, then one often starts with α for the modification that is stable immediately below the melting point, and with lower T, $\beta, \gamma, \ldots$ are following. However, exceptions occur, and, e. g., α-quartz is more stable at lower T than β-quartz.

1.4.2.2 Thermodynamic aspects

An alternative classification of phase transitions was introduced by Ehrenfest[39] (Jaeger, 1998) and relies on the behavior of derivatives of thermodynamic potentials.

In the Ehrenfest classification, a phase transition of order n shows a discontinuity in the nth derivative of a thermodynamic potential. This means that for a first-order transition, from equation (1.25) we obtain $(\partial G/\partial p)_T = V$ and $(\partial G/\partial T)_p = -S$, and hence V and S undergo abrupt changes at a first-order transition (also, $H = \partial(G/T)/\partial(1/T)$ is discontinuous). For a second-order transition, the first derivatives of G change continuously, but the second change abrupt, e. g., $(d H/ d T)_p = c_p$, see equation (1.19). Transitions of third or higher order are rarely reported, and experimental evidence is tenuous (Kumar, 2003; Janke et al., 2006).

---

39 Paul Ehrenfest (18 January 1880–25 September 1933).

For thermal analysis, the most significant difference between first- and second-order transitions is that a heat of transition $Q = \Delta H$ appears only for the former. In Fig. 1.7, all three steps in the $H(T)$ curve of sulfur result from first-order transitions, and also the transitions of selected substances used for the calibration of thermal analyzers (see Table 2.2) are without exception of first order, because only a significantly large latent heat (transformation, often melting) leads to a sufficiently large DTA or DSC peak.

Landau (1936)[40] introduced a thermodynamic approach for second-order transitions, where an order parameter Q results in the excess Gibbs energy term

$$G = H - TS = \frac{1}{2}A(T - T_c)Q^2 + \frac{1}{4}BQ^4 + \frac{1}{6}Q^6 \qquad (1.31)$$

(A, T_c, B, C are constants), which describes the behavior of measurable physical quantities M such as specific heat capacity, polarization, or atomic positions near second-order phase transitions by expressions of the type

$$M \propto |T_c - T|^{-\alpha}, \qquad (1.32)$$

where T_c is the transition (sometimes called "critical" or Curie[41]) temperature, and $\alpha \approx \frac{1}{2}$ is the "critical exponent". Second-order transitions proceed by different mechanisms: Small movements of atoms and turns of coordination polyhedra are one option ("displacement type"), but some compounds like the 1 : 1 CuZn alloy (β/β' brass) show "order–disorder" transitions between a body-centered disordered cubic structure (space group $Im\bar{3}m$ at high T) and a CsCl-type ordered $Pm\bar{3}m$ structure at low T. In both structures the atoms sit on identical sites in the crystal lattice, but the distribution of Cu and Zn over these sites is different (Yeomans, 1992). The c_p anomaly of nickel in Fig. 1.3 results from a ferromagnetic/paramagnetic second-order transition of this metal, which was also observed by DSC measurements (Lopeandía et al., 2008) and neutron diffraction (Drabkin et al., 1969).

The transitions between a ferroelectric phase (at low T) and a paraelectric phase (with higher symmetry, at high T) are usually of the second order; an example with high technical relevance is lithium niobate $Li_{1-x}NbO_3$, which crystallizes from the melt in the paraelectric $R\bar{3}c$ phase and transforms below the Curie temperature $T_C \approx$ 1020 °C to a ferroelectric $R3c$ phase (T_C depends on x, Carruthers et al., 1971). Dohnke et al. (2004) observed this transition by weak DSC effects. For a comprehensive introduction to the Landau theory, the reader is referred, e. g., to Salje (1990).[42]

From the structural point of view, first-order transitions are often reconstructive and proceed at one specific transition temperature T_t. At T_t, low-T and high-T phases coexist in equilibrium, but immediately below or above T_t, one of the coexisting

40 Lev Davidovich Landau (9 January 1908–1 April 1968).

41 Pierre Curie (15 May 1859–19 April 1906).

42 Ekhard Salje (born 26 October 1946).

phases disappears. The melting of ice to water at $T_f = 0\,°C$ is a good example. $H(T)$ jumps at T_t from a value for the low-T phase to a different value for the high-T phase discontinuously, and hence transitions of first order are often called "discontinuous transitions". However, it should be taken into account that this is strictly true only for pure substances (one component). Some phases like mixed crystals, which will be discussed in Section 1.5.1.2, show melting over a finite T range. Also, this melting needs a certain heat of fusion ΔH and is therefore clearly of first order, but the term "discontinuous transition" could be misinterpreted, because the melting process is not restricted to a fixed value T_f.

Second-order transitions are from the structural point of view often displacive. They are smeared over a transition range of typically several 10 K or even more, and they are (at least in theory) not accompanied by a latent heat of transformation. It was already mentioned at the end of Section 1.1.1 that the discrimination if a transition is in fact of second order or "weakly" first order is often not straightforward. Not only the enthalpy increment necessary to perform a first-order transition, but also the $c_p(T)$ maximum (λ; see the Ni curve in Fig. 1.3) connected with a second-order transition results in an increased demand of energy during heating cycles in DTA or DSC measurements. As described in Sections 2.1 and 2.3, this can be interpreted as "weakly" first-order transitions. In the FactSage 8.0 (2020) databases, we find, e. g., $\Delta H_t = 498.9\,J/mol$ for the displacive transitions α-quartz $\rightarrow$ β-quartz (Fig. 1.19) and $\Delta H_t = 183.3\,J/mol$ for the displacive transition aragonite $\rightarrow$ calcite (Fig. 1.18), which are very low compared to a reconstructive transition between solid phases (CsCl, Figs. 1.17 (c) and (d), $\Delta H_t = 3.7656\,kJ/mol$) or melting (for CsCl, $\Delta H_f = 15.8992\,kJ/mol$).

At the melting point of NaCl ($T_f = 800.65\,°C = 1073.8\,K$, Fig. 1.14), the entropy of the solid phase **[?]** is $S_{NaCl}^{sol} = 143.655\,J/(mol\,K)$, and the entropy of the liquid phase has the higher value $S_{NaCl}^{liq} = 169.878\,J/(mol\,K)$. How large is the heat of fusion (melting enthalpy ΔH_f)? Hint: Use equation (1.25).

1.5 Phase diagrams

In this chapter, we present the most prevailing types (topologies, geometries) of $x - T$ phase diagrams, where x stands for the concentration, and T stands for the temperature. Other possible coordinates for phase diagrams (pressure p, magnetic or electric fields $\vec{H}, \vec{E}, ...$) are not so relevant here because the variation of T for different sample compositions x is a basic principle of thermal analysis.

Whenever possible, concentrations should be given as molar fractions x of the components. More- **[i]** over, it is sometimes useful to choose components in such a way that their chemical formula units contain a similar number of atoms or ions. For instance, instead of $SrO-Ga_2O_3$ (2–5 atoms), we can choose $SrO-\frac{1}{2}Ga_2O_3$ (2–2.5 atoms) as components. The reason is that the slope of phase boundaries is substantially determined by the number of microstates of the system and hence by its entropy (see Section 1.2.2), and this scales with the number of interacting species.

1.5.1 Binary phase diagrams without intermediate compound

1.5.1.1 Lever rule

? Why three phases cannot coexist in a phase field (area) of an isobar binary phase diagram?.

Binary phase diagrams in concentration–temperature $(x - T)$ coordinates contain phase fields (areas) with one or two coexisting phases. In Fig. 1.20 the gray fields contain only the phase Φ_1 or Φ_2. Within both fields, every point designates specific thermodynamic conditions (a concentration of the corresponding phase x and its temperature T) where this phase is stable.

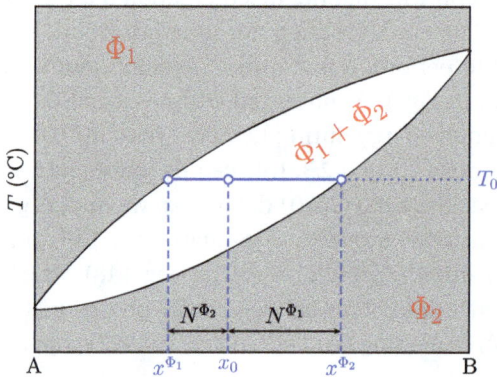

Figure 1.20: The lever rule for a binary system B–A. The gray regions are phase fields where only one phase Φ_i is stable. (Often Φ_1 is the melt, and Φ_2 is a solid solution.) The compositions x^{Φ_1} and x^{Φ_2} and their amounts N^{Φ_1} and N^{Φ_2} of both phases that are in equilibrium for an overall (system) composition x_0 can be derived from the "lever" or "tie line" at the corresponding temperature T_0.

It is not so for the white phase field $\Phi_1 + \Phi_2$ around the diagonal of the phase diagram. There every point, say (x_0, T_0), describes only the state of the whole system, which is composed of two phases in this field. The system must be isothermal, because phase diagrams are constructed for thermodynamic equilibrium. Hence an isotherm is represented by a horizontal line in the diagram. The blue horizontal line in Fig. 1.20 is such an isotherm; it connects the states of Φ_1 and Φ_2, which are in equilibrium at the temperature T_0 with concentrations x^{Φ_1} and x^{Φ_2}, respectively. The full blue line is called a "tie line". Figure 1.20 describes a "mixed crystal" system. Such systems will be described in Section 1.5.1.2, but the exact kind of the system is not relevant for the discussion here; it is only important that the 2-phase-field is neighbor to 1-phase-fields.

---

i A tie line is an isotherm in a phase diagram that connects equilibrium states.

The tie line in Fig. 1.20 allows us to determine not only the concentrations x^{Φ_1} and x^{Φ_2} for both equilibrium phases, but also the corresponding amounts N^{Φ_1} and N^{Φ_2}. The lever rule

$$\frac{N^{\Phi_1}}{N^{\Phi_2}} = \frac{x^{\Phi_2} - x_0}{x_0 - x^{\Phi_1}} \tag{1.33}$$

says that the tie line can be considered as a mechanical lever suspended at x_0. The two loads act on both ends of the lever and are inverse proportional to the corresponding amounts of the phases. In Fig. 1.20 the point x_0 is slightly left from the center of the 2-phase-field, closer to the Φ_1 field. Consequently, the ratio N^{Φ_1}/N^{Φ_2} is the length of the (longer) right lever arm divided by the (shorter) left arm.

We will show in the following sections that different types of horizontal (isotherm) phase boundaries may occur. Such isotherms include
- eutectic lines (Section 1.5.1.3)
- peritectic lines (Section 1.5.2.2)
- monotectics (Section 1.5.1.4)
- phase transitions (Section 1.4.2)

and can result from the properties of one specific phase (e. g., phase transitions) or from the interaction of two or more phases (e. g., eutectics). However, the same process always appears at the same temperature. Then the enthalpy change of the system is only dependent on the amount of material that undergoes the transition. Plots of the related DTA or DSC peak area $A(x_0)$ are linear, because the lever rule (1.33) is a linear function of the sample composition x_0. Such plots were already introduced by Tammann (1905)[43] and can be used, e. g., for the determination of the composition where an effect has its maximum. In Section 3.3, we show this for the determination of the "eutectic point", but other applications are possible too.

The idea behind the lever rule can be extended also to systems with three or more components. This is the concept of "barycentric coordinates" introduced by Möbius (1827).[44] For three components A, B, C, this is demonstrated by the mechanical model in Fig. 1.21, where the blue dot represents the overall composition of the system $x_0^{(A)}, x_0^{(B)}, x_0^{(C)}$. The concentration values are not necessarily scaled with respect to pure A, B, C. Instead, the three phases Φ_i ($i = 1, \ldots, 3$) are used as a basis. The aggregation state of Φ_i is not relevant; variable compositions (e. g., melts) are also possible. For details, the reader is referred, e. g., to Vince (2006), Wolfram Research (2020).[45]

---

43 Gustav Heinrich Johann Apollon Tammann (16 May 1861–17 December 1938).

44 August Ferdinand Möbius (17 November 1790–26 September 1868).

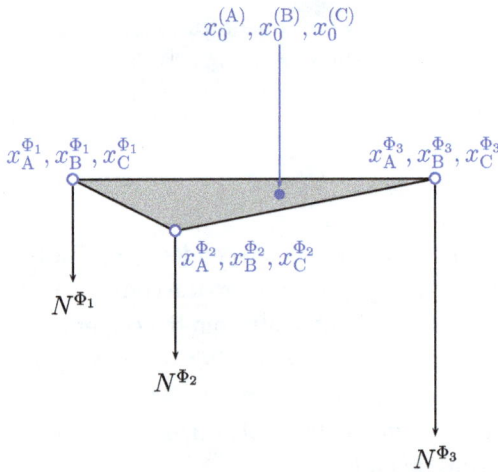

Figure 1.21: The barycentric theorem as extension of the lever rule (Fig. 1.20) to ternary systems. If the overall (system) composition $x_0^{(A)}, x_0^{(B)}, x_0^{(C)}$ evokes from 3 phases, then their amounts N^{Φ_i} ($i = 1, \dots, 3$) balance the gray concentration triangle. The edges of the triangle are tie lines.

An analytic description of the barycentric theorem is given by

$$N^{\Phi_1}\left(x_0^{(k)} - x_k^{\Phi_1}\right) + N^{\Phi_2}\left(x_0^{(k)} - x_k^{\Phi_2}\right) + N^{\Phi_3}\left(x_0^{(k)} - x_k^{\Phi_3}\right) = 0 \tag{1.34}$$

(k = A, B, C; Paufler, 1986) and can be extended to an arbitrary number of phases and components.

1.5.1.2 Mixed crystals

If the liquid and solid phases of two components A and B can mix in arbitrary ratio, then the binary phase diagram is similar to Fig. 1.22 (a). At high T, only one phase field *liquid* exists with variable composition ranging from pure A ($x = 0$) to pure B ($x = 1$). Also, at low T a phase field *solid* exists with only one phase, which is a "solid solution" or synonymous "mixed crystal". The variable composition of this phase can be expressed in different ways, e. g., as $A_{1-x}B_x$. If the real composition is not relevant or not known, then spellings like (A,B) are sometimes used, or A:B for very small concentrations (doping) of, e. g., A in B.

In the *liquid + solid* field, both phases coexist, and the equilibrium concentrations x^{liq} and x^{sol} of the liquid and solid mixture phases for every temperature between the melting points of the pure components can be obtained if a horizontal "tie line" is drawn at this temperature. Figure 1.22 (b) shows two options, which can be considered as limit cases for the crystallization of a melt with an arbitrary starting composition x_0:

---

45 Stephen Wolfram (born 29 August 1959).

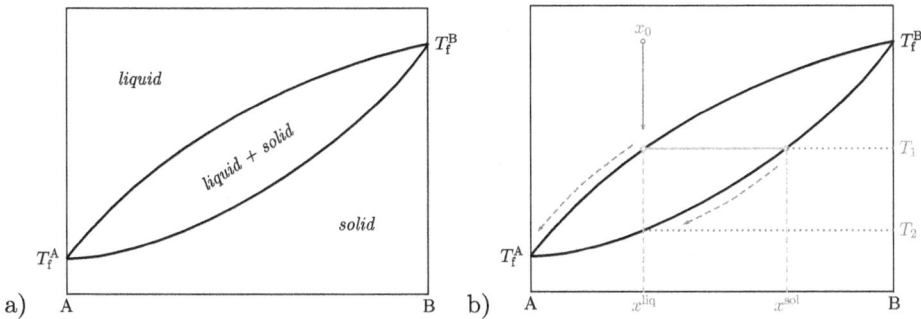

Figure 1.22: Binary phase diagram A–B, where both components show complete miscibility in the liquid and solid states. (a) with labels for the three phase fields; (b) shows possible crystallization paths during cooling of a melt with initial composition x_0.

1. Cooling of the homogeneous melt proceeds down to T_1, where crystallization of a solid solution with composition x^{sol} starts with an infinitesimal amount of the solid. For a lower temperature T' ($T_2 < T' < T_1$, not shown in Fig. 1.22 (b)), the corresponding tie line would result in another $x'^{sol} < x^{sol}$, which is in equilibrium with another $x'^{liq} < x^{liq}$. This process continues until T_2 is reached, and one homogeneous solid solution remains, which has the same composition x_0 like the initial melt. This process has the name
    ⇒ *equilibrium crystallization.*

2. Again, cooling of the homogeneous melt proceeds down to T_1, where an infinitesimal amount of a solid with composition x^{sol} crystallizes. However, if the initially crystallized material cannot equilibrate continuously with the remaining melt, then the melt depletes by the higher melting component (here B), and its composition moves downward the upper phase boundary ("liquidus") toward the pure component A. Accordingly, then the finally crystallizing material solidifies at T_f^A and is almost pure A. This process of component segregation has the name
    ⇒ *Scheil[46]–Gulliver[47] crystallization.*

The crystallization path of real systems is often intermediate between equilibrium and Scheil–Gulliver   **!**
crystallization. Especially for the typical cooling rate $\dot{T} \approx 10$ K/min during thermal analysis, equilibrium often cannot be reached, and the process corresponds more to Scheil–Gulliver.

Above the upper phase boundary in Fig. 1.22, the whole system is liquid, and correspondingly this boundary is called the liquidus (or liquidus line) of the system. Be-

---

46 Erich Scheil (8 December 1897–2 April 1962).
47 Gilbert Henry Gulliver (1878–24 October 1952).

low the lower phase boundary, the whole system is solid, and this is the solidus (or solidus line). Under some simplifying conditions, both lines can be described analytically with the following Schröder[48]–van Laar[49] equations, which are valid in this simple form if no excess Gibbs-free energy contributions are involved ($G^{ex} = 0$, see Section 1.2.3.2 and Schröder, 1893; van Laar, 1908):

$$x^{sol} = \frac{\exp\left[-\frac{\Delta H_f^A}{R}\left(\frac{1}{T} - \frac{1}{T_A}\right)\right] - 1}{\exp\left[-\frac{\Delta H_f^A}{R}\left(\frac{1}{T} - \frac{1}{T_A}\right)\right] - \exp\left[-\frac{\Delta H_f^B}{R}\left(\frac{1}{T} - \frac{1}{T_B}\right)\right]}, \tag{1.35}$$

$$x^{liq} = \exp\left[-\frac{\Delta H_f^B}{R}\left(\frac{1}{T} - \frac{1}{T_B}\right)\right] \times x^{sol} \quad (T_A \lesseqgtr T \lesseqgtr T_B). \tag{1.36}$$

Equations (1.35) and (1.36) describe a lens-shaped 2-phase-field between liquidus and solidus, similar to Fig. 1.22. The lens becomes broader for large ΔH_f^A and ΔH_f^B and is asymmetrical (flatter on one side and steeper on the other side) for very different values of ΔH_f^A and ΔH_f^B. However, both lines are monotonously rising or falling, without intermediate extrema. If, however, such extrema are found experimentally, then a $G^{ex} \neq 0$ term is required for a numerical description of the system. Such terms often appear for the solid phase, and they often give a positive contribution to the Gibbs-free energy of this phase, which destabilizes the solid with respect to the liquid and leads to a local minimum of liquidus and solidus. Uecker et al. (2017)[50] reported such a minimum for the pseudo-binary system $LaLuO_3$–$LaScO_3$ close to $LaLuO_3$, with ca. 10 % $LaScO_3$. With their data, we obtain $G^{ex} \approx 1\,kJ/mol$.

In contrast, pseudo-binary systems between near rare-earth scandates $REScO_3$ were found by Uecker et al. (2013) almost ideal and could be described satisfactorily by the Schröder–van Laar equations (1.35) and (1.36), because the chemical and crystallographic similarity of the components is higher in comparison to $LaLuO_3$–$LaScO_3$.

If in a mixed crystal system, liquidus and solidus have a local extremum, then this must be a common extremum for both curves. This means that it must appear at the same composition x_{aze} and the same temperature T_{aze}. The point x_{aze}, T_{aze} is called the azeotrope point. There crystallization occurs without segregation: liquid and solid in equilibrium have the same composition x_{aze}.

It should be noted that from the thermodynamic point of view, only the type of the phase transition (first-order) is relevant for the description by the Schröder–van Laar equations (1.35) and (1.36), but not the aggregation state of the involved phases. Hence also the evaporation of liquids or first-order transition between solids leads to a similar topology of the phase diagram. Abriata et al. (1982) reported the system Zr–Hf,

48 Iwan Schröder (PhD thesis in St. Petersburg, Russia, 1890).

49 Johannes Jacobus van Laar (10 July 1860–9 December 1938).

50 Reinhard Uecker (born 11 May 1951).

where both components undergo subsequent transitions hexagonal $\rightleftharpoons$ cubic $\rightleftharpoons$ liquid with complete miscibility of the components in all three phases. Correspondingly, the binary phase diagram shows two stacked 2-phase fields "hexagonal + cubic" and "cubic + liquid", which separate "hexagonal" (bottom) from "cubic" (middle) and then "liquid" (top).

Very often, the phases in mixed crystal systems are not ideal, and then the topology of the binary phase diagram can significantly deviate from the ideal lens shape in Fig. 1.22. As an example, $LaLuO_3$–$LaScO_3$ was mentioned above, and Fig. 1.23 shows the system KCl–KI, which also deviates significantly from ideality and is discussed controversially in the literature. This system was reported by Wrzesnewsky (1912) with an azeotrope point at $x \approx 0.45$, and Le Chatelier (1894) found the azeotrope at $x \approx 0.50$. More recently, Sangster and Pelton (1987) reviewed the data on this system and proposed a small (4 %) solubility of KI in KCl and a rather large (45 %, equivalent to $x = 0.55$ in Fig. 1.23) solubility of KCl in KI with an intermediate miscibility gap.

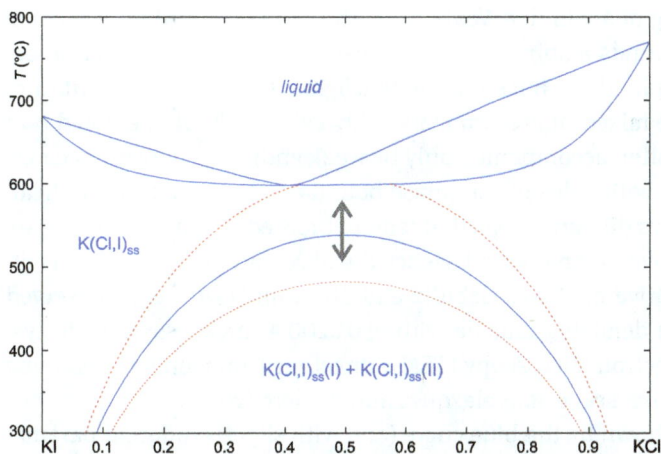

Figure 1.23: Blue lines: calculated (but incorrect!) binary phase diagram KCl–KI with an azeotrope near $x = 0.4$ and a miscibility gap. The red dotted lines indicate hypothetical alternative boundaries of this gap. Compare with the correct version in Fig. 3.25.

These discrepancies can be explained by slight variations of the phase diagram. First, a minimum in the liquidus can be explained by an azeotrope, like shown in Fig. 1.23, or by a eutectic, like shown in Fig. 1.24. The difference is that below the azeotrope, one single solid solution phase is stable, and below the eutectic, a two-phase mixture. However, solution phases are stabilized mainly by the Gibbs-free energy of mixing (equation (1.29)), which is proportional T. Consequently, all solutions become less stable upon cooling, which can result in "demixing". For the KCl–KI system, this means that below the lower dome-shaped phase boundary, an initially homogeneous

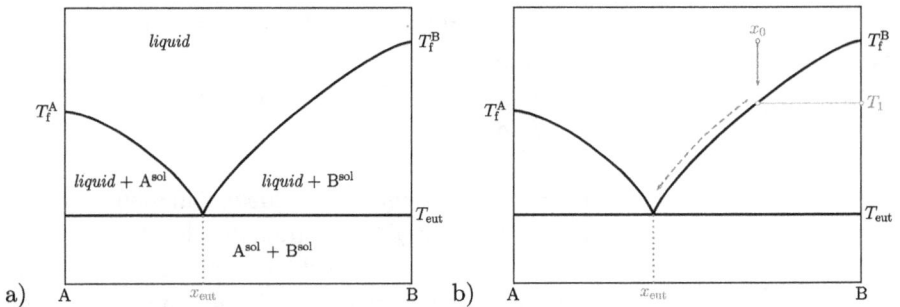

Figure 1.24: Binary phase diagram A–B, where both components show complete miscibility in the liquid state and no miscibility in their solid phases. (a) With labels for the four phase fields; (b) shows the crystallization path during cooling of a melt with initial composition x_0.

$K(Cl,I)_{ss}$ solid solution decomposes into two isostructural solid solutions $K(Cl,I)_{ss}(I)$ + $K(Cl,I)_{ss}(II)$ with different compositions.

Like for the demixing of a van der Waals[51] gas, the demixing equilibrium curve (shown in Fig. 1.23) surrounds a spinodal curve, where demixing is expected to occur spontaneously. The spinodal is not shown in this figure. Between the equilibrium and spinodal curves, one mixed phase can metastably exist. It should be noted that the demixing process is often accompanied only by weak enthalpy changes, because the crystal structures on both sides of the phase boundary are similar or identical. Besides, demixing requires diffusive steps that tend to proceed slowly, which smears the small thermal effect over a long period. Hence the observation of demixing is often performed by alternative methods. Schultz and Stubican (1970), e. g., observed dark/light contrasts from demixing lamella with ≈50...200 Å thickness in oxide systems by Transmission Electron Microscopy (TEM). Direct measurements of demixing by thermal analysis are rare; see, e. g., Velazquez and Romero (2020).

The miscibility gap shown by the blue curve (and with slightly different parameters by the dotted red curves) marks the stability limits. This means that below these lines, one homogeneous $K(Cl,I)_{ss}$ phase is not thermodynamically stable. However, like for monotectic demixing in a liquid phase (see Section 1.5.1.4), phase separation requires overcoming some energy barrier. This is possible inside a "spinodal" curve given by $\partial^2 G/\partial x^2 = 0$. Hillert (1961)[52] has shown that "upward diffusion" can lead to periodic concentration fluctuations there.

In Section 3.3.2, we will show by experimental data that the blue lines in Fig. 1.23 are an incorrect representation of the KCl–KI system, because there unlimited mutual miscibility at least for high temperatures is assumed. It turns out that rather the upper red dotted phase boundary approaches the truth.

---

51 Johannes Diderik van der Waals (23 November 1837–8 March 1923) .

52 Mats Hilding Hillert (born 28 November 1924).

1.5.1.3 Eutectics

In the beginning of Section 1.4, we pointed out that the conditions for mutual miscibility of liquids are significantly less restrictive than for solids: the latin *similia similibus solvuntur* means basically that we have a good chance that the melts of similar chemical compounds can form one single phase. It is not so for solids, where additionally identical crystal structure is required (see note on p. 26).

A eutectic system A–B occurs if A(liq) and B(liq) are miscible, and A(sol) and B(sol) not, or at least not completely. In the easiest case the mutual solubility of B in A(sol) and of A in B(sol) is negligible; then the result is a binary phase diagram shown in Fig. 1.24. In Section 1.2.3.1, we have seen that a pure substance, e. g., A, has its melting point T_f^A there, where the $G^{liq}(T) = H^{liq}(T) - T \cdot S^{liq}(T)$ curve of the liquid phase falls below $G^{sol}(T) = H^{sol}(T) - T \cdot S^{sol}(T)$ of the solid phase. If we add some B to the system, then $G^{sol}(T)$ is not influenced because no solubility occurs. In contrast, $S^{liq}(T, x)$ is significantly increased by the larger disorder resulting from the B particles (1.20). The solid/liquid phase boundary is more and more bent downward, because $S^{liq}(T, x)$ continuously grows with x. The same discussion can be performed starting at pure B, and under ideal conditions (no excess enthalpy or entropy), we can derive the equations

$$x^{liq} = 1 - \exp\left[-\frac{Q_A}{R}\left(\frac{1}{T} - \frac{1}{T_A}\right)\right],$$

(1.37)

$$x^{liq'} = \exp\left[-\frac{Q_B}{R}\left(\frac{1}{T} - \frac{1}{T_B}\right)\right]$$

(1.38)

for both liquidus curves. In analogy to (1.35) and (1.36), also (1.37) and (1.38) are often called the Schröder–van Laar equations. Both curves intersect at the eutectic point (x_{eut}, T_{eut}). If a sample with composition x_{eut} is heated, then it melts completely at the temperature T_{eut}. Thermal analysis will produce a sharp peak that is not different from the melting peak of a pure substance. Samples with $x' < x_{eut}$ start to melt at T_{eut} and first form a melt with composition x_{eut} (which is richer in B) under the release of A(sol). Upon further heating, this A(sol) dissolves continuously in the melt until the liquidus is reached and a clear melt with composition x' is formed. Also, samples with $x'' > x_{eut}$ start to melt at T_{eut}, but under the release of B(sol) until, after passing of the liquidus, a single melt phase with composition x'' is formed.

In Fig. 1.24 (b) the reverse case is shown, where a melt x_0 is cooled until the crystallization of B(sol) starts at T_1. Consequently, the melt depletes in B, and with further cooling, its composition moves toward x_{eut} (gray dashed arrow). This eutectic melt composition is reached at T_{eut}, and then the remaining melt completely crystallizes.

A practical example for the determination of a eutectic phase diagram is given in Section 3.3. We will show there that it is useful to plot the area of the eutectic DTA peak A versus the sample composition x. Already Tammann (1905) showed that the functions $A(x)$ rise linearly from both components to x_{eut}, because for every sample composition at the identical temperature T_{eut}, the first portion of melt with identical composition x_{eut} is formed. Consequently, the required amount of heat, which is proportional to A, depends only on the portion of the sample that melts at T_{eut}. According to the lever rule (Section 1.5.1.1), this portion rises linearly from both pure components (no eutectic melt formed) to x_{eut} (100 % eutectic melt formed); see also Rycerz (2013).

Haseli et al. (2021) investigated the pseudo-binary eutectic system KCl-Na$_2$CO$_3$ by DTA, in combination with scanning electron microscpy (SEM) and energy dispersive X-ray spectroscopy (EDS) of quenched melts. Besides standard measurements with rates of $\dot{T} = \pm 10$ K/min, for the determination of x_{eut}, the authors performed heating/cooling runs with rates down to ± 1 K/min. It was the purpose of these slow rates to gain, with a better thermal resolution of the system, a better separation of the eutectic peak from the liquidus shoulder close to it. As expected, this better resolution was obtained; nevertheless, the Tammann construction proved superior for the determination of the eutectic composition.

We will show later that crystallization takes place often below the equilibrium temperature where it should occur, a kinetic phenomenon called supercooling. Figure 2.21 demonstrates this for pure gold. The degree of supercooling depends on the thermal conditions and on the phase to be crystallized. If near a eutectic point the crystallization of one phase is kinetically prohibited, then nonequilibrium crystallization can occur. Nakamura et al. (2013) observed this in the system LiF–BiF$_3$, where crystallization of the intermediate phase LiBiF$_4$ can be kinetically hindered for BiF$_3$-rich compositions. As a consequence, DTA peaks related to a metastable LiF–BiF$_3$ eutectic were observed.

In such cases, it is useful to extend the liquidus curves of the neighboring phases below T_{eut}. In Fig. 1.25, this was done by overlaying three (FactSage 8.0, 2020) calculations:

1. A full calculation, where all solid phases and the Al$_2$O$_3$/Er$_2$O$_3$ melt (*liquid*) are taken into account.
2. A calculation where the right constituent of the eutectic (α-Al$_2$O$_3$) is made dormant, which means that it is not allowed to be formed. This results in the dashed right extension of the Er$_3$Al$_5$O$_{12}$ liquidus below T_{eut}.
3. A calculation where the left constituent (Er$_3$Al$_5$O$_{12}$) is made dormant. This results in the dashed left extension of the α-Al$_2$O$_3$ liquidus below T_{eut}.

If in an experiment the crystallization of one neighboring phase is kinetically hindered, then the formation of the other phase may occur at these dashed lines. In the shaded, almost triangular region below both dashed lines, self-organized eutectic structures can occur. Orera et al. (2012) gives an overview on the production of such structures by directional solidification of eutectic melts. The preparation and measurements of optical and mechanical properties of Er$_3$Al$_5$O$_{12}$/α-Al$_2$O$_3$ eutectic metamaterials were described by Nakagawa et al. (2005).

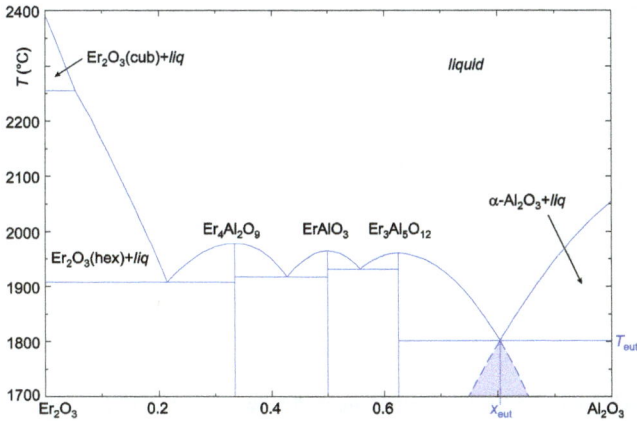

Figure 1.25: Phase diagram Al_2O_3–Er_2O_3 calculated with FactSage 8.0 (2020). According Wu and Pelton (1992), for the eutectic between $Er_3Al_5O_{12}$ and Al_2O_3, we have $T_{eut} = 1802\,°C$ and $x_{eut} = 0.805$. The liquidus curves of these two phases are prolongated below T_{eut}.

1.5.1.4 Monotectics

Not all liquids (melts) can be mixed in arbitrary ratio. Already in the introduction to Section 1.4, we pointed out that substances with very different chemical nature (like water and oil) can hardly be mixed in any condensed phase. However, for inorganics with often ionic or metallic bonding, the chemical similarity may not be so strong like it seems on the first look: Some metals cannot be mixed in molten state, and some "network forming" oxides tend to form complex anions. Silicates and borates, e. g., have often complex anions of the kinds $[SiO_4]^{4-}$, $[Si_2O_7]^{6-}$, $[BO_3]^{3-}$, $[B_2O_5]^{4-}$, and so on. If such complex anions occur in a system together with simple oxides, like CaO, sometimes, a homogeneous melt cannot be formed; rather, a miscibility gap ("demixing") may occur. The CaO–SiO_2 system is an example of high technical relevance, especially in the building materials industry for the production of portland cement; there a miscibility gap occurs for SiO_2-rich melts (Eriksson et al., 1994).[53]

A miscibility gap in the melt leads to monotectic melting, and the topology of a simple monotectic system is shown in Fig. 1.26. In this system, both components are immiscible in the solid state, but complete miscibility in the molten state occurs only above some critical temperature T_c. Below T_c the mutual equilibrium solubility limits are described by the dome-shaped line in the center of the diagram. It should be added that demixing occurs usually not immediately below the solid dome, because the phase separation requires some activation energy. To overcome this energy barrier, the second derivative of the Gibbs energy $\partial^2 G/\partial x^2$ must be negative. The points where $\partial^2 G/\partial x^2 = 0$ are the "spinodals" of the system and are shown as a dashed line inside the monotectic region of Fig. 1.26. Impressive tomographic pictures of Al–In alloys with monotectic demixing were obtained, e. g., by Kaban et al. (2012).

---

53 Gunnar Eriksson (born 7 February 1942).

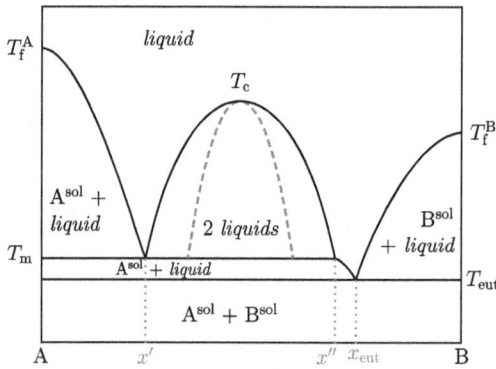

Figure 1.26: Phase diagram A–B with monotectic demixing of the melt between x' and x'' (apex at T_c). The solid "dome" shows the equilibrium boundary of demixing; inside the dashed region spinodal demixing occurs. x' is often called the monotectic composition x_m, T_m is the monotectic temperature, and x_{eut} is the eutectic composition.

For potential laser applications, Klimm et al. (2002) investigated the pseudo-binary phase diagram $YCa_4O(BO_3)_3$–$GdCa_4O(BO_3)_3$. Both materials crystallize in identical monoclinic crystal structures with space group *Cm*, and as a result of their chemical and structural similarity, an almost ideal mixed crystal system is formed. With the measured T_f and ΔH_f for both components, the phase diagram follows almost perfectly the Schröder–van Laar equations (1.35) and (1.36). However, an experimental problem occurred for high concentrations of the Y-phase ($x \gtrsim 0.45$), because mixed crystals $Gd_{1-x}Y_xCa_4O(BO_3)_3$ had to be prepared by melting together pure $YCa_4O(BO_3)_3$ and $GdCa_4O(BO_3)_3$: Only during the first heating, single melting peaks were observed for the pure Gd- and Y-compounds at 1490 or 1504 °C, respectively. During the second heating run, the Y-rich compositions showed several (typically, 5) endothermal peaks, and after the measurements, the samples were obviously inhomogeneous.

Figure 1.27 shows the interior of three DTA crucibles after such measurements, and only pure $GdCa_4O(BO_3)_3$ crystallizes homogeneously and optically clear. The sample composition in Fig. 1.27 (b) is already in the range where multiple DTA peaks occur, and within the almost clear matrix, small orange regions can be observed (some of them are red encircled). This phase separation is further enhanced for the DTA sample of pure $YCa_4O(BO_3)_3$ in Fig. 1.27 (c).

Chemical analysis of the sample revealed that the orange region in Fig. 1.27 (c) is almost pure Y_2O_3, and the bright region is a mixture of three different borate phases. We can conclude that above the compound $YCa_4O(BO_3)_3$ inside the ternary system Y_2O_3–B_2O_3–CaO, a miscibility gap exists, which leads to this phase separation. Nevertheless, Ye and Chai (1999), Hüter et al. (2012) and many other authors demonstrated that the growth of bulk crystals from such 2-phase melts is often possible.

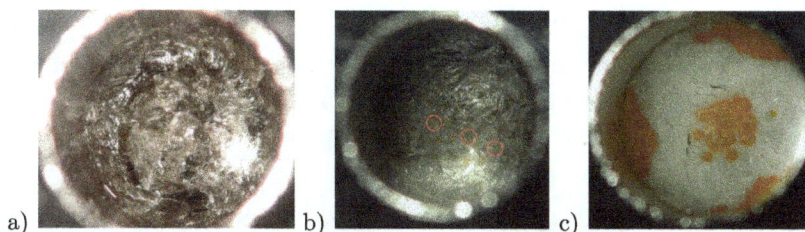

Figure 1.27: Oxyborates $RECa_4O(BO_3)_3$ (RE = Gd, Y) after DTA measurements above their melting points inside their DTA crucibles. (a) $GdCa_4O(BO_3)_3$; (b) $Gd_{1-x}Y_xCa_4O(BO_3)_3$ (x = 0.4615); (c) $YCa_4O(BO_3)_3$. The Pt crucibles have the same dimensions as the Al_2O_3 crucible in Fig. 2.2 (a).

The transformation from two liquid phases to one solid phase and vice versa, like shown with $\boxed{i}$ $YCa_4O(BO_3)_3$, is called a *syntectic* reaction, which is a particular case of a monotectic. Also, HgTe and HgI_2 show syntectic melting (Klimm, 2017).

1.5.2 Phase diagrams with intermediate compound

If the components A and B of a binary system are strongly interacting under the release of energy, then they can form one or more chemical compounds. A and B may be chemical elements; then the intermediate compound(s) are called binary. This holds, e. g., for NaCl with $x_{Na} = x_{Cl} = 0.5$. Different and multiple compositions for intermediate compounds are also possible: In the system Ni–Mg, two intermediate compounds Mg_2Ni ($x_{Ni} = \frac{1}{3}$) and $MgNi_{2\pm\delta}$ ($x_{Ni} \simeq \frac{2}{3}$, because the composition is slightly variable) are formed (Nayeb-Hashemi and Clark, 1985).

Often, chemical compounds themselves can be used as components. In the system SiO_2–Al_2O_3, one intermediate compound "mullite" with approximate composition $3\,Al_2O_3 \cdot 2\,SiO_2$ ($x_{SiO_2} \approx 0.4$, FactSage 8.0, 2020) exists, which can be called either "pseudo-binary" (based on the components $SiO_2 + Al_2O_3$) or ternary (based on the components Al, Si, O).

Some organic (especially macromolecular) and a few inorganic compounds undergo decomposition upon heating without previous melting. Such chemical reactions are often and extensively studied by DTA and TG methods, but usually the description of those reactions in terms of phase diagrams is not appropriate. Just as an example, the reader is referred to the paper by Herbstein et al. (1994) on the popular production of oxygen by thermal decomposition of potassium permanganate, where the authors showed that the simple equation

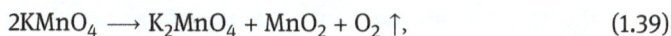

$$2KMnO_4 \longrightarrow K_2MnO_4 + MnO_2 + O_2 \uparrow, \tag{1.39}$$

which is often used to describe this reaction is incorrect, and complicated parallel reactions with intermediate products like K_2MnO_4 and $K_3(MnO_4)_2$ take place instead.

Rather than nonequilibrium decomposition reactions of type (1.39) in the solid state, which can hardly be represented in equilibrium phase diagrams, in the following sections, we will deal with the behavior of intermediate compounds in phase diagrams that can coexist with melts in thermodynamic equilibrium.

1.5.2.1 Compound with congruent melting

In the easiest case an intermediate compound, e. g., AB in the system A–B shown in Fig. 1.28 (a), becomes unstable at its melting point T_f^{AB} and forms a liquid (melt) of the same composition. It is not relevant whether T_f^{AB} is between T_f^A and T_f^B, like shown in Fig. 1.28 (a), or whether T_f^{AB} is higher or lower than the melting points of both components: GaAs melts above 1200 °C, way beyond the melting points of its components gallium (30 °C) and arsenic (817 °C under pressure). In contrast, the melting point of wollastonite[54] ($CaSiO_3$) is 1540 °C, significantly below the melting points of its components SiO_2 (1713 °C) and CaO (2580 °C).

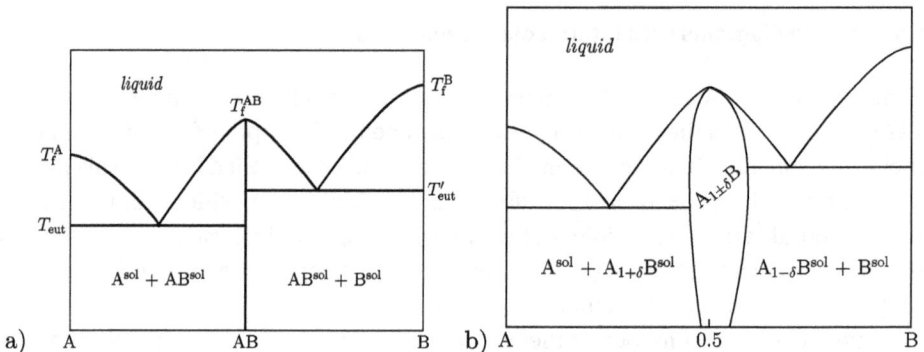

Figure 1.28: Binary phase diagram A–B with formation of an intermediate compound that melts congruently. (a) Intermediate daltonide compound AB with eutectics to both sides; (b) intermediate berthollide compound $A_{1\pm\delta}B \approx AB$ with eutectics to both sides.

The only important feature of a congruently melting compound is that a solid and a liquid phase of identical ("congruent") composition are in equilibrium at some melting (fusion) temperature T_f. It plays no role if the compound has a fixed composition ("line compound", Fig. 1.28 (a)) or if it is variable (Fig. 1.28 (b)). If we apply the lever rule (Section 1.5.1.1) to such a phase diagram left or right from a congruently melting intermediate compound, then it turns out that the "lever", and hence the liquid and solid phase(s), remain left from that compound for compositions starting left. Vice versa, the lever remains right for all compositions from the right side. Consequently,

---

54 William Hyde Wollaston (6 August 1766–22 December 1828).

both sides are independent from each other, and the phase diagram A–B can be divided into independent partial systems A–AB and AB–B.

The lever rule makes also clear that from melts with a composition between both neighboring eutectics the congruently melting compound will always crystallize. In the case of Fig. 1.28 (a) with negligible phase width, stoichiometric AB with fixed composition always crystallizes. In contrast, in the case of Fig. 1.28 (b) the resulting crystal will be A-rich ($A_{1+\delta}B$) upon crystallization left from the congruent melting point and will be B-rich ($A_{1-\delta}B$) upon crystallization right from the congruent melting point.

1.5.2.2 Compound with peritectic melting

Also, in the A–B system shown in Fig. 1.29 an intermediate compound AB exists. In this case the thermal stability of AB is rather limited, and it decomposes at T_{per}. However, this decomposition temperature is so low that a homogeneous A–B melt (like in Fig. 1.28) cannot be formed, and B(sol) remains because T_f^B is too high. The result is the "peritectic reaction"

$$AB(sol) \rightleftarrows B(sol) + liquid, \tag{1.40}$$

which proceeds at the peritectic temperature T_{per} in equilibrium. The lever rule shows that from melts with $x > x_{per}$, B(sol) crystallizes at the liquidus line of that phase, and from melts $x_{eut} < x < x_{per}$, AB(sol) crystallizes. The point (x_{per}, T_{per}) is the peritectic point of the system, and at T_{per} the three phases $liquid + AB(sol) + B(sol)$ are in equilibrium. In analogy to Fig. 1.28 (b), also a compound that melts peritectically can be a berthollide with finite phase width $\pm\delta$. However, then also the left and right boundaries of the stability field must converge to one point at the peritectic line ($\delta \to 0$ for $T \to T_{per}$).

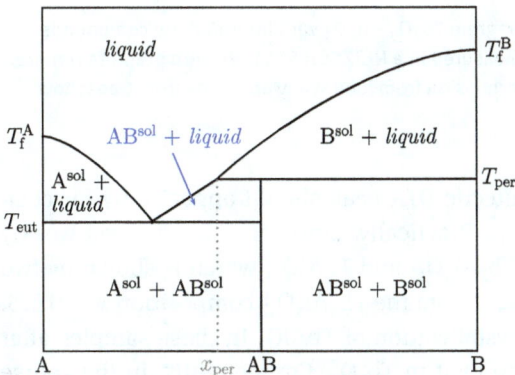

Figure 1.29: Phase diagram A–B with one intermediate compound AB. At T_{per}, AB melts peritectically to a melt with composition x_{per} under the release of solid B.

? Why the phase width of congruently and peritectically melting compounds AB must vanish at their melting temperatures under the typical isobar conditions?

For the thermal analysis of a peritectic system in the kind of Fig. 1.29, under equilibrium conditions, we would expect strong melting peaks at T_{eut} and T_{per}, because there the totally solid sample with composition left or right from AB starts to melt. However, experimental results are often not so straightforward. Ganschow et al. (1999) performed DTA measurements in the system $Tb_2O_3-Al_2O_3$, and Fig. 1.30 shows that it contains the intermediate phases $TbAlO_3$ (perovskite structure, congruent melting point 1931 °C) and $Tb_3Al_5O_{12}$ (garnet structure, peritectic melting around 1840 °C under the release of solid $TbAlO_3$). Between $Tb_3Al_5O_{12}$ and Al_2O_3, a eutectic was found at T_{eut} = 1688 °C. Another eutectic point around 1790 °C between $TbAlO_3$ and Tb_2O_3 is out of the concentration scale of Fig. 1.30.

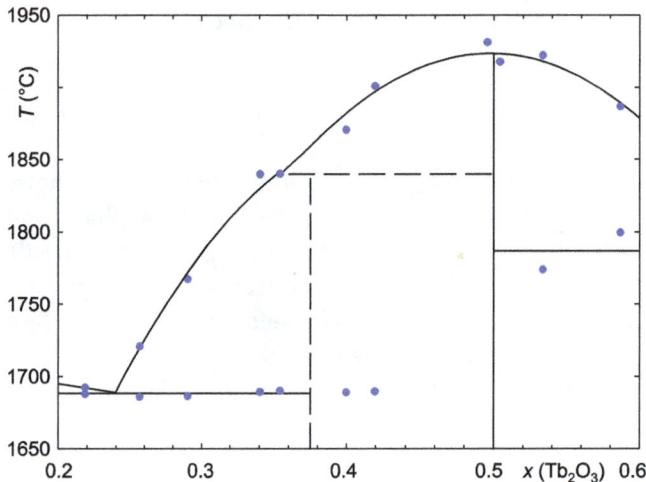

Figure 1.30: Part of the experimental phase diagram $Tb_2O_3-Al_2O_3$ with intermediate compounds $TbAlO_3$ (x = 0.5) and $Tb_3Al_5O_{12}$ (x = 0.375). Measured in a NETZSCH STA409C with graphite furnace in Ar flow, $\dot{T}$ = 10 K/min. Reproduced with permission from John Wiley and Sons from Ganschow et al. (1999).

Under equilibrium conditions, this eutectic DTA peak should appear only for compositions between Al_2O_3 and $Tb_3Al_5O_{12}$. Practically, however, it is observed weakly also for some compositions between $Tb_3Al_5O_{12}$ and $TbAlO_3$, which is shown by two experimental points around 1688 °C right from the $Tb_3Al_5O_{12}$ composition x = 0.375. These points result from the initial crystallization of $TbAlO_3$ in these samples after a first heating run, which depletes the melt in Tb_2O_3. Consequently, in the course of the crystallization process, the melt composition moves downward the $TbAlO_3$

liquidus toward the peritectic point $x_{per} \approx 0.35$, where $Tb_3Al_5O_{12}$ starts to crystallize. No complete equilibration of this Al_2O_3-rich melt with the first crystallizate $TbAlO_3$ can be reached, because the sample cools down with 10 K/min rather fast. As a result, the Al_2O_3–$Tb_3Al_5O_{12}$ eutectic at 1688 °C appears as a nonequilibrium phenomenon.

Figure 1.31 explains this typical nonequilibrium process schematically for a melt composition x_0 right from the intermediate phase AB, which melts peritectically. ("right" means in this figure the side of the higher melting phase formed in the peritectic decomposition.) B(sol) ist the phase that first crystallizes there. Consequently, the melt composition, which is for several intermediate temperatures marked by the (blue, dashed) tie lines, shifts in A direction down the liquidus line until the peritectic point is reached. Then under equilibrium conditions, the peritectic reaction (1.40) should proceed to the side of the educts. In other words, a significant part of the just formed B(sol) should react with the whole A-rich melt to AB(sol), and a completely solid mixture AB(sol)+B(sol) remains. However, chemical reactions including a solid phase are usually slow. Hence it is likely that the peritectic reaction is not complete, and B(sol) and some melt with the peritectic composition remain unreacted. From this melt the intermediate phase AB now starts to crystallize, which is indicated by the magenta dashed arrow from the peritectic point down to the eutectic point. There the small rest of the melt solidifies under the formation of A(sol) + AB(sol). Consequently, the sample with initial composition x_0 can contain, after complete solidification, B(sol), AB(sol), and A(sol) in a nonequilibrium state.

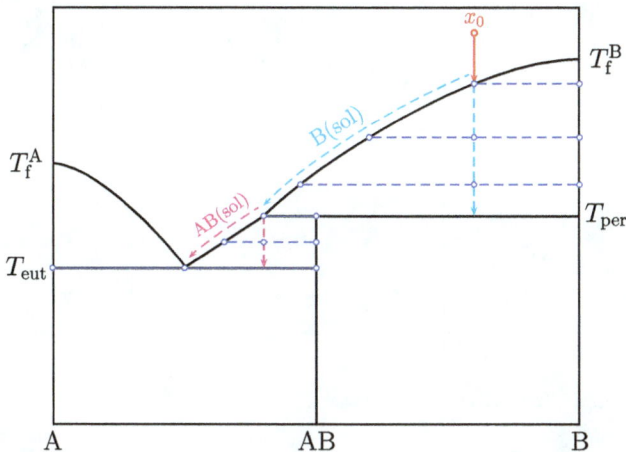

Figure 1.31: During cooling of the B-rich melt, B(sol) crystallizes first, and the composition of the remaining melt "moves" downwards the liquidus to the peritectic point. In a fast DTA/DSC experiment, it is unlikely that complete equilibrium is obtained; instead, AB(sol) can now crystallize with further shifting the melt composition toward the eutectic point.

ⓘ It is likely that during crystallization of melts in peritectic systems, nonequilibrium states are formed. Then DTA/DSC heating curves of such samples can show features like eutectic and peritectic melting that should not occur simultaneously under equilibrium conditions.

It should be noted that the crystallization of the $Tb_3Al_5O_{12}$ phase (which is indicated in Fig. 1.30 by dashed lines at its composition and its peritectic melting temperature only) was proven by Ganschow et al. (1999) with some Tammann plots (see Section 1.5.1.1). These plots showed characteristic kinks around $x = 0.375$.

1.5.3 Phase diagrams with three and more components

1.5.3.1 Concentration triangles

A graphical presentation of systems with three (or more) components is not straight-forward, because then we have two (or more) independent concentration data. If in a ternary system A–B–C, all components are equivalent, then often concentration triangles of the kind shown in Fig. 1.32 are used to show the composition. Such triangles rely on Viviani's[55] theorem, which says that for every point inside a regular triangle, the sum of the three heights is equal to the total height of the triangle:

$$h_A + h_B + h_C = h. \tag{1.41}$$

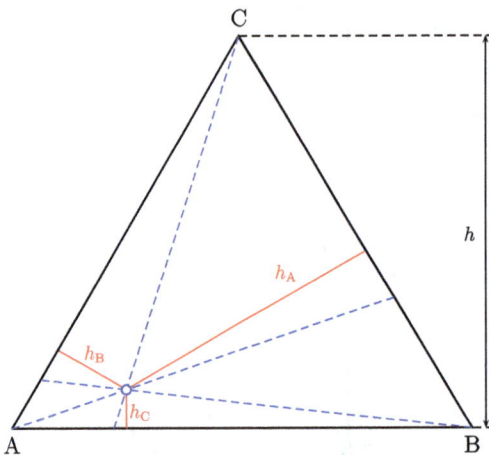

Figure 1.32: Viviani's theorem: For every point in a regular triangle, the sum of the heights over all sides equals the total height (1.41). The point marks the composition $x_A = h_A = 0.7, x_B = h_B = 0.2, x_C = h_C = 0.1$.

---

55 Vincenzo Viviani (5 April 1622–22 September 1703).

If we define $h = 1 = 100\%$, then each height h_i corresponds to the concentration of component i (i = A, B, C). This correspondence is bijective and has some implications:

---

(1) Lines parallel to one edge of the concentration triangle mark compositions where the concentration of the component opposite to this line is constant.
(2) Lines from one corner of the concentration triangle to the opposite edge mark compositions where the ratio of the components defining this edge is constant.

[i]

This means for Fig. 1.32 that on every line parallel to the edge A–B, x_C is constant, and on every line parallel to the edge B–C, x_A is constant. On the blue dashed line starting at the corner C, the ratio of A and B is constant: $x_A : x_B = 0.7 : 0.2 = 3.5$. For all compositions on the line starting at the corner A, we have $x_B : x_C = 0.2 : 0.1 = 2$.

If the temperature T has to be introduced as a third independent variable for ternary systems A–B–C (only two x_i are independent, because $x_A + x_B + x_C = 1$), then the following options are available:
1. We can try to draw perspective presentations as in Fig. 1.36. Such drawings have the benefit to be very instructive, but it is difficult to extract exact data for specific systems from them. Consequently, they are seldom used.
2. Isothermal sections T = const. show coexisting phases for the given temperature in one or more phase fields. According to the Gibbs phase rule, in every phase field, either one, or two, or three phases may coexist. Examples are given in Fig. 1.34.
3. Polythermal projections are mainly produced onto the liquidus surface. This means that the triangle in Fig. 1.32 is used as a base of a trigonal prism and has T as height coordinate, like in Fig. 1.36. Then we "look" downward from a point at high T, where all compositions of the system are molten. For every possible composition x_A, x_B, x_C inside the triangle, one specific phase Φ will first crystallize upon cooling to some temperature $T^{liq}(x_A, x_B, x_C)$. This way "primary crystallization fields" for different phases Φ_i are formed, which are separated by phase boundaries. Often, contour lines are added for several temperatures similarly to isoaltitude lines on a geographic map. An example is given in Fig. 1.33.

Draw isothermal sections through the LiF–NaF–CsF system at 450 and 800 °C.

[?]

Polythermal projections have the benefit to combine visual evidence with the possibility to extract quantitative data on the solidification behavior of the system, and here we give some basic rules:
1. Every solid phase that can exist in equilibrium with the melt has a primary crystallization field, and from melts inside this field the corresponding phase crystallizes first.

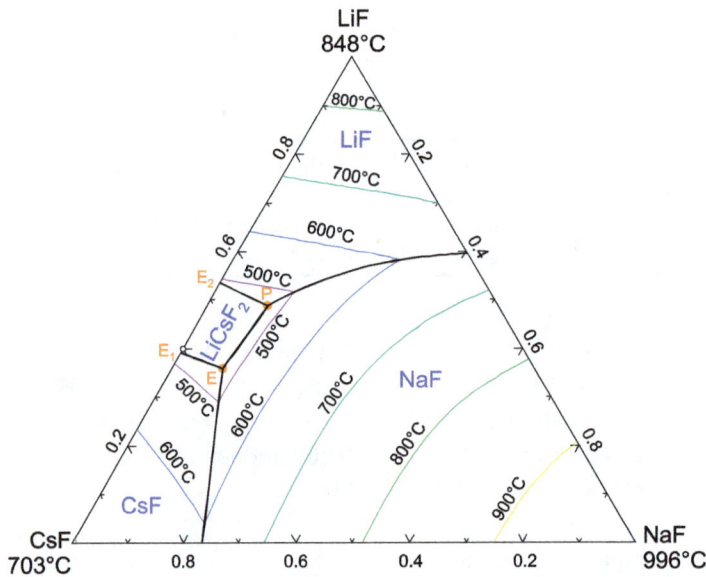

Figure 1.33: Liquidus projection of the ternary phase diagram LiF–NaF–CsF with 100 K isotherms and (blue) labels for primary crystallization fields. E_1 = 479 °C and E_2 = 490 °C are binary eutectics. The eutectic E = 455 °C and peritectic P = 462 °C are ternary invariant points. For details of the rim system LiF–CsF, see Fig. 1.35.

2. If the composition of a phase is located inside its own primary crystallization field, then the phase melts congruently. If the phase composition is outside its own primary crystallization field, then it melts peritectically.

3. During the crystallization of some phase $\Phi = A_m B_n C_p$, the residual melt depletes by the composition A:B:C = $m : n : p$. Graphically, this can be shown by a straight line starting at the initial melt composition toward the opposite of the direction to $A_m B_n C_p$ until the first phase boundary is met. This line is the beginning of the "crystallization path".

4. At the following phase boundary the primary crystallization field of another phase Φ' begins, and now Φ and Φ' start to crystallize together. The crystallization path follows the phase boundary $\Phi - \Phi'$ toward lower temperatures T to an "invariant point", where a third primary crystallization field is touched.

5. This invariant point is a ternary eutectic point if it represents a local minimum of the liquidus surface. Then the crystallization path terminates here, and the rest of the melt, with the composition of the ternary eutectic, crystallizes at the corresponding eutectic temperature.

6. Alternatively, from the invariant point another boundary between primary crystallization fields of a different phase pair (either $\Phi - \Phi''$ or $\Phi' - \Phi''$) can lead to lower T until it reaches the next invariant point.

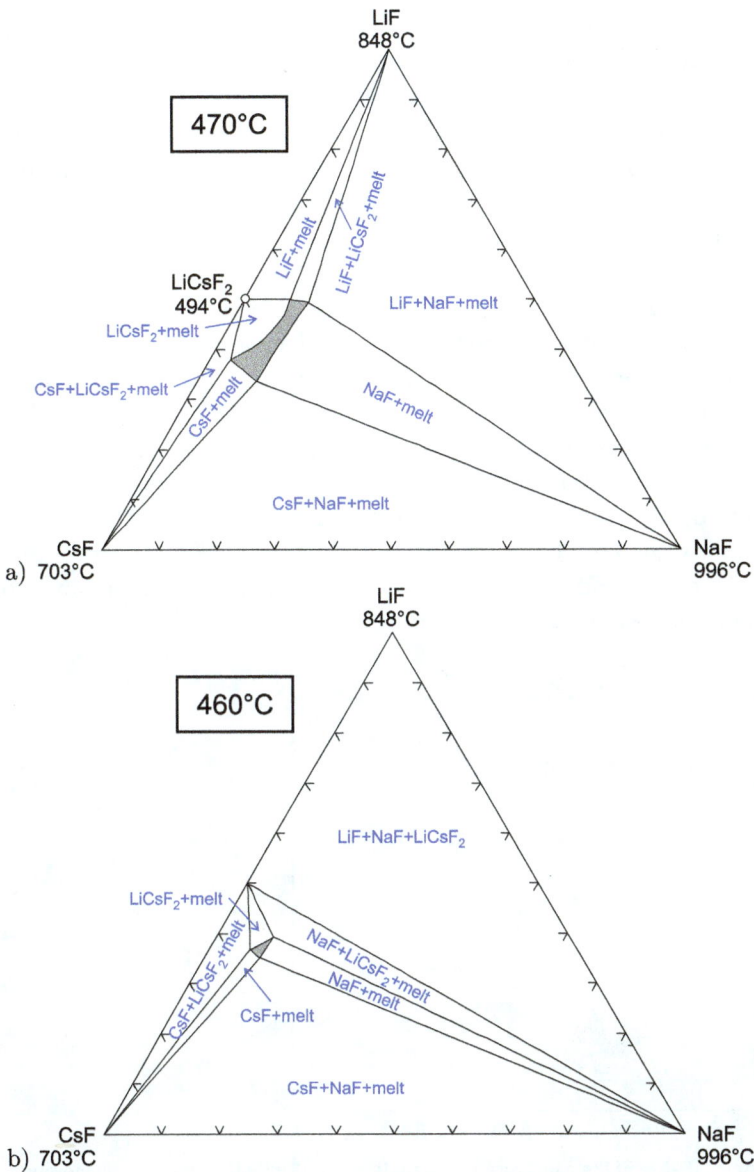

Figure 1.34: Isothermal sections through the concentration triangle LiF–NaF–CsF at 470 and 460 °C. Fields where only the liquid (melt) phase is stable are gray. (a) Above the ternary peritectic temperature (point P in Fig. 1.33, T^{per} = 462 °C), the 1-phase field "melt" extends wider, and solid LiF can coexist with melt. (b) Below T^{per} the partial triangle LiF–NaF–LiCsF$_2$ is completely solid, and these three solids coexist.

7.   The crystallization path always terminates at a eutectic point if such a point exists in the system (which is typically the case).

The depicted crystallization behavior of systems with three or even more components, which holds vice versa also for melting, often results in complicated thermoanalytic curves for arbitrary compositions. Whenever possible, partial systems should be defined. It was shown in Section 1.5.2.1 that in binary systems, partial (sub)systems are defined very straightforwardly by intermediate compounds with congruent melting point. Compositions left and right from congruently melting compounds are independent from each other.

Figure 1.36 shows that, unfortunately, in ternary systems a congruently melting compound does not always define independent partial systems. There the rim A–B contains the phase AB with congruent melting point, and indeed this rim system can be divided into partial systems A–AB and AB–B. In Fig. 1.36 (a), from the binary eutectic points of both partial systems a "eutectic valley" extends into the ternary region and meets there the eutectic valleys (hence the phase boundaries) toward the primary crystallization field of C(sol). Only in Fig. 1.36 (a) the valley between the ternary eutectic points E_1 and E_2 has a local maximum (which is a saddle point of the liquidus surface), and this maximum is situated on the direct connection (blue dashed line) between AB and C. Consequently, all compositions left from this line run finally into the eutectic E_1, and compositions right from the line run into E_2.

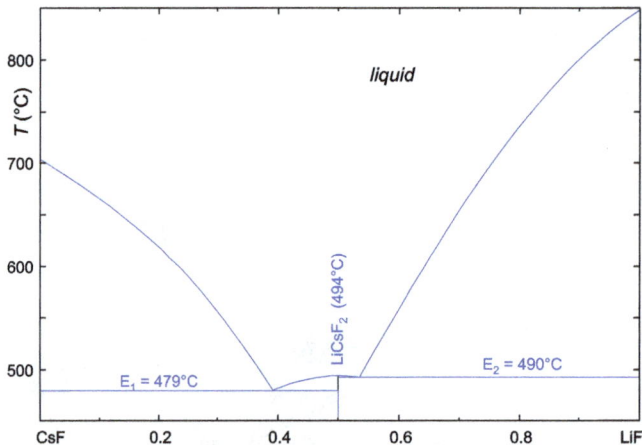

Figure 1.35: Rim system LiF–CsF of the concentration triangle Fig. 1.33 with the congruently melting intermediate compound LiCsF$_2$ and eutectics on both sides. LiCsF$_2$ separates left and right independent partial systems.

The situation in Fig. 1.36 (b) is different, because there the eutectic valley starting between AB and A runs into an invariant point P, which is not a local minimum and hence no eutectic. Instead, this is a ternary peritectic point.

The system LiF–NaF–CsF shown above gives a practical example where a ternary system cannot be divided into independent partial systems even at a congruently melting intermediate compound. The congruent melting behavior of $LiCsF_2$ is obvious from the (FactSage 8.0, 2020) calculation, as shown in Fig. 1.35, which is in agreement with data by Sangster and Pelton (1987) and ACerS-NIST (2014), entries 7608 and 7463.

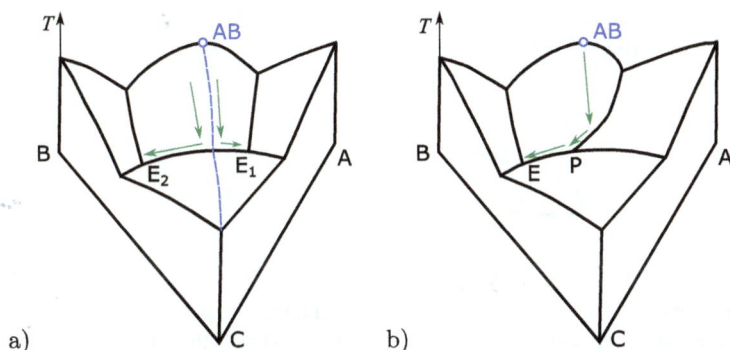

Figure 1.36: Perspective view on a ternary phase diagram A–B–C with a congruently melting compound AB. (a) The intersection line of the AB and the C liquidus surfaces has a saddle point at the straight connection AB–C, and there the system can be divided into two partial systems. (b) The connection AB–C presents no saddle point, and partial systems are not formed. Adapted from Paufler (1986).

Describe the crystallization path of a melt containing 80 % LiF and equal concentrations of NaF and CsF on the basis of the concentration triangle in Fig. 1.33. (What happens at which temperature?)

The experimental determination of phase diagrams with three or even more components is significantly more complicated than the investigation of binary systems, because the huge range of possible compositions cannot be easily handled. Whenever possible and useful, "pseudo-binary" sections should be measured, which may exist between the corners of the concentration triangle and/or congruently melting compounds on the rim systems or in the interior of the concentration triangle. However, it must be always checked that it is really possible to divide the whole system along the intended pseudo-binary sections into partial systems, because this is a precondition.

Binary systems can always be divided at congruently melting intermediate compounds into independent partial systems. For ternary systems, such a division is often possible too, but there exist exceptions where such a division is not possible; see Figs. 1.33 and 1.36 (b).

1.5.3.2 Ternary compounds
If a phase consists of all three components, then it is called a ternary phase. This is usually the case for the melt, but solid phases can also contain three (or even more) components. The systems $LiF–Me^{II}F_2–Me^{III}F_3$ in Fig. 1.37 give examples where the

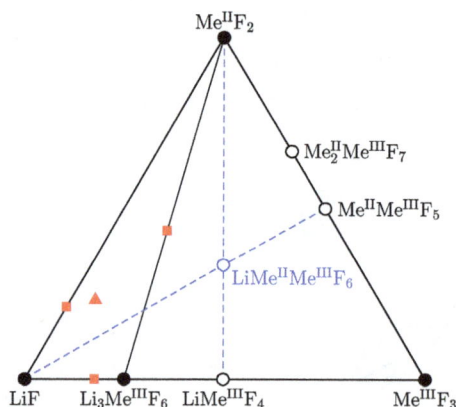

Figure 1.37: The concentration triangle $LiF-Me^{II}F_2-Me^{III}F_3$ (Me^{II} typically Ca, Sr; Me^{III} = Al, Ga, Cr) with congruently (full dots) and incongruently (empty dots) melting compounds on the rims. $LiMe^{II}Me^{III}F_6$ is the only ternary compound. Binary and the ternary eutectics in the partial system $Li_3AlF_6-LiF-CaF_2$, as reported by Vrbenská and Malinovský (1967), are marked by red squares or by a triangle, respectively.

melting behavior of intermediate compounds depends on the nature of the 2- and 3-valent metals. Klimm et al. (2005) pointed out that $Li_3Me^{III}F_6$ is the only intermediate compound that melts congruently for Me^{II} = Ca. Malinovský and Vrbenská (1967), Vrbenská and Malinovský (1967) have shown that the tie line $Li_3AlF_6-CaF_2$ represents a pseudo-binary eutectic subsystem with a eutectic point at 56.5 % Li_3AlF_6 and 43.5 % CaF_2 and that the left corner $Li_3AlF_6-LiF-CaF_2$ of the triangle in Fig. 1.37 is an independent ternary subsystem with a ternary eutectic point at 25 % Li_3AlF_6, 52 % LiF, and 23 % CaF_2.

From the positions of the eutectics in the $Li_3AlF_6-LiF-CaF_2$ subsystem (for Me^{II} = Ca; Me^{III} = Al) it follows that the dashed line from LiF to $CaAlF_5$ cannot be considered as a tie line, and hence this is not another pseudo-binary subsystem with $LiCaAlF_6$ as intermediate compound: otherwise, the crystallization path of every composition on this line should remain on the line. This is obviously not possible, because this line crosses the binary and ternary subsystems mentioned above. For example, on the $Li_3AlF_6-CaF_2$ line the crystallization path of a melt with initial composition at the intersection with the blue dashed line would go upward and terminate at the binary eutectic (red square); for compositions inside the $Li_3AlF_6-LiF-CaF_2$ subsystem, the crystallization terminates in the ternary eutectic (red triangle). Additionally, Craig and Brown (1977) reported that the compound $CaAlF_5$ melts peritectically which impedes to use it as the end member (component) of a binary subsystem (cf. Klimm and Reiche, 2002).[56] However, Meehan and Wilson (1972) reported that, in contrast, the

---

56 Peter Reiche (15 September 1942–7 December 2021).

strontium compound $SrAlF_5$ melts congruently. Nevertheless, also $LiF–SrAlF_5$ cannot be considered as a binary subsystem, because the eutectic of the pseudo-binary section $Li_3AlF_6–SrF_2$ was found by Koštenská (1976) at even higher alkaline earth fluoride concentrations than those shown in Fig. 1.37 for the system with CaF_2. Consequently, also there the crystallization of a melt located on the blue dashed $LiF–SrAlF_5$ line at its intersection with the $Li_3AlF_6–SrF_2$ tie line would move upward to the corresponding eutectic above the red square.

The only ternary compounds $LiMe^{II}Me^{III}F_6$ are considered as interesting laser host crystals. They crystallize in the trigonal colquiriite structure, which received its name from a mineral with approximate composition $LiCaAlF_6$ (with small additions of Na and Mg) reported by Walenta et al. (1980). Yin and Keszler (1992) described the crystal structures of several other colquiriite-type compounds and pointed out that the isomorphous replacement of the Me^{3+} ion by Cr^{3+} is possible with distribution coefficients close to unity, which is beneficial for the growth of laser crystals with homogeneous composition.

The growth of Cr^{3+}-doped colquiriite crystals with technically relevant size is possible only if the corresponding material melts congruently or at least "almost congruently"; but the melting behavior of colquiriites depends on their constituents (Klimm et al., 2005), which can be revealed by DTA (or better DSC) measurements. Figure 1.38 shows the first and second heating curves of single crystalline undoped $LiCaGaF_6$. The melting peak in the first measurement appears sharp, but a careful inspection shows that the DSC signal returns not immediately to the basis line after passing the peak; it rather stays for ca. 30 K slightly more endothermal before it returns completely. Obviously, the sample melts peritectically at the peak onset, and a 2-phase-field "melt+solid" is reached, where "solid" is some phase with higher melting point, probably CaF_2 in this case.

If the DTA basis line is after passing the melting peak on another (but different) level than before, then this is not necessarily a sign of incongruent melting. Rather, the specific heat capacity $c_p^{sol}(T)$ may be different from $c_p^{liq}(T)$. It will be shown in Section 2.3.2.2 that the position of the basis line shifts endothermally for larger c_p. See as examples the melting peaks of zinc in Fig. 2.28 or, even more pronounced, the LiCl curve in Fig. 3.21, where the basis line is different before and after melting.

The peritectic melting behavior of $LiCaGaF_6$ is confirmed by the second heating curve in Fig. 1.38, which shows a small additional endothermal peak near 675 °C. It is very common that the peritectic reaction (1.40) cannot return completely to the educt (here $LiCaGaF_6$): in this case the primary crystallizing phase with higher melting point (e. g., CaF_2) had to react completely at the peritectic temperature with the melt. However, this is unrealistic under the conditions of thermal analysis with cooling rates of the order 10 K/min, because the range near T_{per} (where high reaction rates between solid and melt can be expected) is passed within a few minutes.

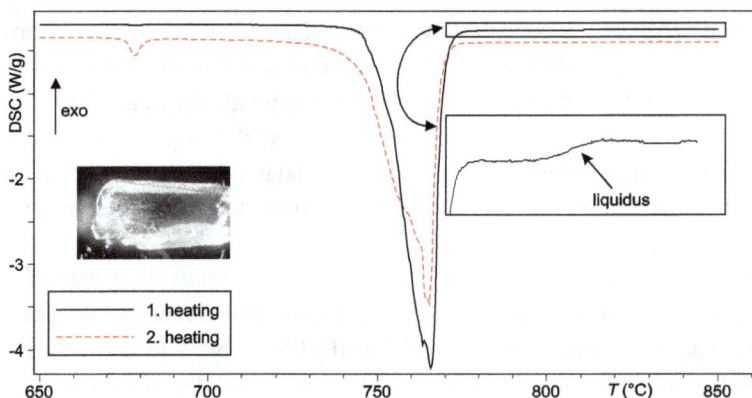

Figure 1.38: First and second DSC heating curves (NETZSCH STA 449C "Jupiter", type S thermocouples, $\dot{T}$ = 10 K/min, atmosphere flowing Ar with 99.999% purity) of an optically clear LiCaGaF$_6$ crystal fragment (32.21 mg, crystal shown in the insert, photograph taken by M. Rabe). Figure adopted with permission from Sani et al. (2005).

This case is similar to the peritectic melting of Tb$_3$Al$_5$O$_{12}$ in the pseudo-binary system Tb$_2$O$_3$–Al$_2$O$_3$ (Fig. 1.30). Two experimental points related to the Al$_2$O$_3$–Tb$_3$Al$_5$O$_{12}$ eutectic are observed on the right side of Tb$_3$Al$_5$O$_{12}$, because in a previous DTA heating/cooling cycle the higher melting phase had crystallized first. This process removes Al$_2$O$_3$ and Tb$_2$O$_3$ in the molar ratio 1:1 from the melt, and the composition of the remaining melt shifts "down the liquidus" toward x_{eut}, which is responsible for the observation of eutectic melting as a nonequilibrium process.

1.5.3.3 Reciprocal salt pairs

Around the middle of the 20th century, mixtures of molten salts were investigated especially as prospective materials for the transport of heat energy initially in nuclear reactors, but nowadays also for thermal energy storage (TES) and concentrating solar power plants (CSP); see Reddy (2011), Serrano-López et al. (2013), Bauer et al. (2021). Mixing salts with different anions and cations can sometimes lead to unexpected results. Haendler et al. (1959)[57] described mixtures of LiF with the chlorides of Li, Na, and K. Their results showed unexpected features for the latter two alkaline chlorides: Both systems, which were denominated as binary by the authors, showed liquidus curves that are bend concave over a wide concentration range (Fig. 1.39). This is very contrary to the typical convex curvature, which is shown, e. g., in Fig. 1.24. Besides, slightly lower T_{eut} values are measured close to the NaCl side.

The circumstance that the end members NaCl and LiF are no valid components of the system is the reason for the unexpected curvature of the liquidus lines in Fig. 1.39.

---

57 Helmut Max Haendler (10 June 1913–16 April 2001).

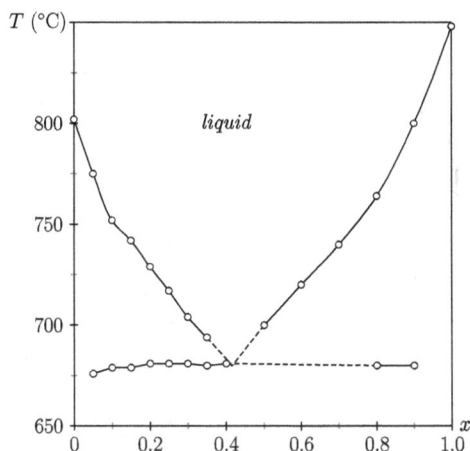

Figure 1.39: Experimental points for the system NaCl ($x = 0$)–LiF ($x = 1$) as reported by Haendler et al. (1959). Note the concave bent liquidus curves and the "suspicious" downward bend of the eutectic line from its value $T_{eut} = (680 \pm 1)\,°C$ for all points $x \geq 0.1$ to 676 °C for $x = 0.05$.

This is because NaCl and LiF contain different cations and different anions ("reciprocal"). Both salts will dissociate in the melt to Li^+, Na^+, F^-, and Cl^-. Upon crystallization, besides NaCl and LiF, also NaF and LiCl can be formed. This basically corresponds to a quaternary system. However, the condition

$$[Li^+] + [Na^+] = [F^-] + [Cl^-] \tag{1.42}$$

restricts the number F of degrees of freedom in the phase rule

$$P + F = C + 2, \tag{1.43}$$

which was developed by Gibbs (1874–1878). (The square brackets in equation (1.42) denote the amounts of the corresponding ions.) In equation (1.43), P is the number of coexisting phases, and C is the number of components. The "2" in (1.43) was introduced by Gibbs as the number of intensive physical quantities describing the system, which can be changed independently. In thermodynamics, these are usually the temperature T and the pressure p; it can be greater if other intensive quantities become significant, like electric fields E, magnetic fields H, or elastic strain ε (Nye, 1957; Schlom et al., 2014). In contrast, every constraint of the system reduces this number, and without external fields and under isobar conditions, (1.43) reduces to

$$P + F = C + 1 \quad \text{(for } p = \text{const.)} \tag{1.44}$$

For the reciprocal salt pair NaCl–LiF ($= Li^+, Na^+, F^-, Cl^-$), constraint (1.42) further simplifies (1.44) to $P + F = C = 4$. In other words, at an invariant point like a eutectic, with consequently $F = 0$, we have besides the melt (one phase) three solid phases in equilibrium. This is the same situation like, e. g., at point E in the concentration triangle in Fig. 1.33, where the melt is in equilibrium with NaF, CsF, and $LiCsF_2$. The system with four components behaves ternarily!

Often, reciprocal salt pairs are represented as square diagrams, like the central part of Fig. 1.40, which was drawn with FactSage 8.0 (2020). There each corner corresponds to one of the four end members, which can be formed from the given ions. In the whole diagram the molar fraction of Na grows from left to right, and the molar fraction of F from top to bottom. Like in concentration triangles (see Section 1.5.3.1), eutectic valleys or other boundaries separate the primary crystallization fields of different phases.

Most alkali halides, and especially all four salts from Fig. 1.40, crystallize in the rocksalt structure, and if the ionic radii are not too different, then more or less mutual solubility exists. Sangster and Pelton (1987) report in their profound discussion of binary halide systems that a small miscibility gap in the solid (Li,Na)Cl phase cannot be ruled out completely but seems not probable. Already Żemčżużny and Rambach (1909) reported in their trustworthy paper on alkali chloride phase diagrams (obtained from cooling curves of 30...40 g samples) an azeotrope point at 27 mol-% NaCl and 552 °C. Separation into Li-rich and Na-rich rocksalt solid solutions was measured at lower temperatures; the apex of this demixing ("solvus") phase boundary was measured at 314 °C. Accordingly, this (and only this) rim system is treated with complete miscibility in Fig. 1.40. The three other rim systems are eutectic in agreement with Sangster and Pelton (1987).

For LiF–NaF, because of the smaller F^- ions, the radius difference between Li^+ and Na^+ weights more, and only minor rim solubilities occur. On the left and right rim systems, in contrast, almost no mutual solubility of the solid phases exists, because the radius difference between F^- and Cl^- is too large.

? Why the solubility of Li in NaF is larger than the solubility of Na in LiF (Fig. 1.40, bottom)?

The experimental phase diagram in Fig. 1.39 corresponds to the diagonal from bottom left to top right in Fig. 1.40, which crosses only two phase fields and one phase boundary: This is the eutectic valley slightly top right from the middle of the square, almost in the middle between the 700 °C isotherms. This intersection of the diagonal with the eutectic valley is the eutectic point in Fig. 1.39 at $T_{eut} = 680$ °C. The eutectic valley has a saddle point there (local maximum) and drops toward the 610 °C eutectic on the NaF side and to the local minimum, which is situated near the LiCl–LiF eutectic (see Fig. 1.40).

The diagonal from LiF to NaCl is a tie line, because the liquidus surface bends from all points on this line downward to lower T. As a consequence, every composition on this line stays on the tie line during crystallization; every segregation process between

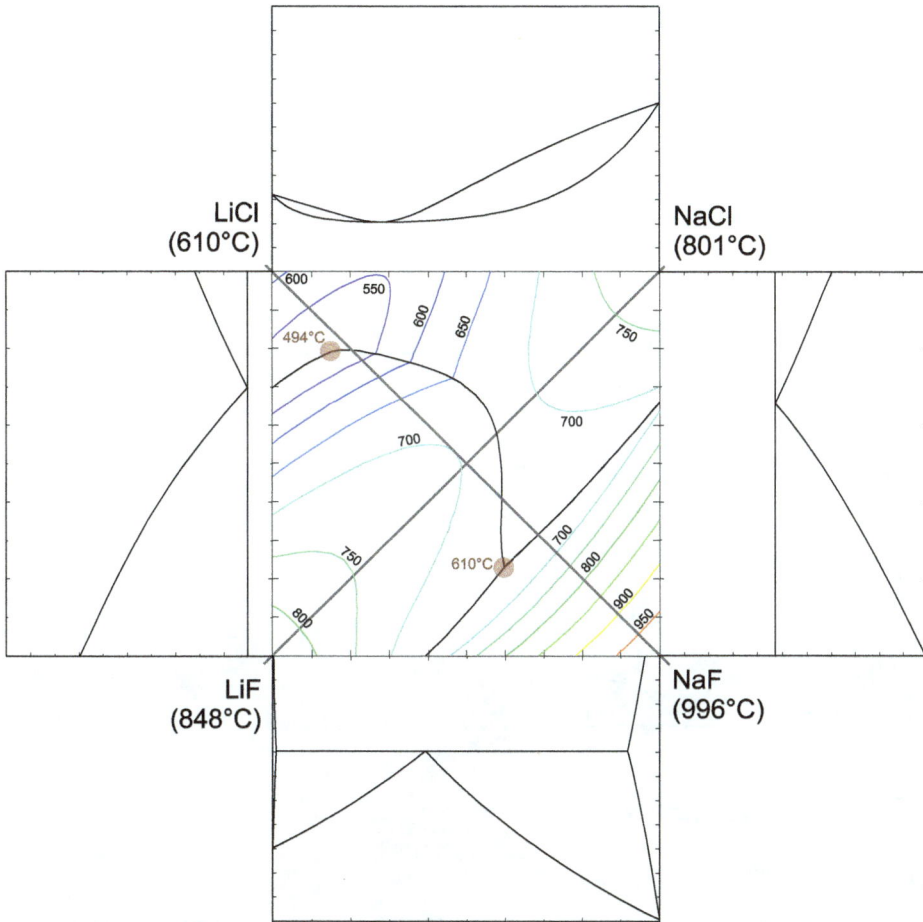

Figure 1.40: The reciprocal salt pair Li$^+$ – Na$^+$ – F$^-$ – Cl$^-$ as liquidus projection with 50 K isotherms (labels are T in °C.) Three phase fields are separated by eutectic valleys, and in all phase fields a "rocksalt" solid solution first crystallizes with different composition. On the valley that starts left and close to, the LiCl–LiF binary eutectic (which is at 500 °C) is with T = 494 °C, the coldest point where melt is stable. The only ternary eutectic point lies at the intersection of the eutectic valleys at 610 °C. The binary systems at all sides are scaled to T_{min} = 450 °C and T_{max} = 1000 °C.

melt and crystallizing phase will proceed only on the tie line. This is the reason why LiF–NaCl can be considered as an independent subsystem of the reciprocal salt pair.

Figure 1.41 shows both diagonals of the reciprocal salt pair from Fig. 1.40. The upper phase diagram, which is the diagonal LiF–NaCl, behaves almost like a normal binary eutectic system. This line separates two independent ternary partial systems shown in Fig. 1.42. Every crystallization process starting in the triangle LiCl–NaCl–LiF will remain in this partial system. Vice versa, every melt from the independent ternary partial system NaF–LiF–NaCl will remain there.

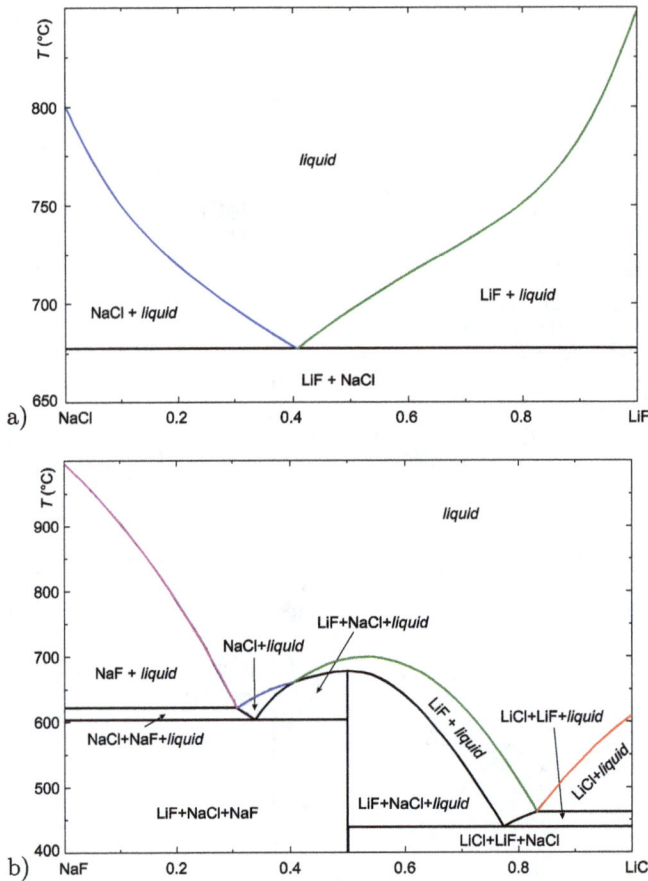

Figure 1.41: The two diagonal sections through Fig. 1.40. Only LiF–NaCl (a), compare to the experimental diagram in Fig. 1.39, behaves like a pseudo-binary system. The other diagonal (b) is not pseudo-binary but rather an isopleth section. The regions where different components prevail in the primary crystallizing rocksalt phase are marked by different colors of the liquidus: NaCl (blue), LiF (green), NaF (magenta), LiCl (red).

The other diagonal LiCl–NaF of the reciprocal salt pair is shown in Fig. 1.41 (b). It is obvious that this is not a (pseudo-)binary system. In Fig. 1.40, this diagonal crosses, starting from LiCl, three times a eutectic valley, which results in four fields of primary crystallization. These regions are also found in this isopleth section. Fig. 1.41 (b) is not a true phase diagram, but only a section; hence the lever rule cannot be used on it. The composition of a melt that starts on this isopleth will move upon cooling away from the primary crystallizing phase. This phase will be only top left or bottom right (almost, because of solid solution formation), pure LiCl or NaF, respectively; and then the segregation changes the composition of the rest melt (almost) along this section. However, in the central parts of the isopleth, a solid solution, which is rich in LiF or

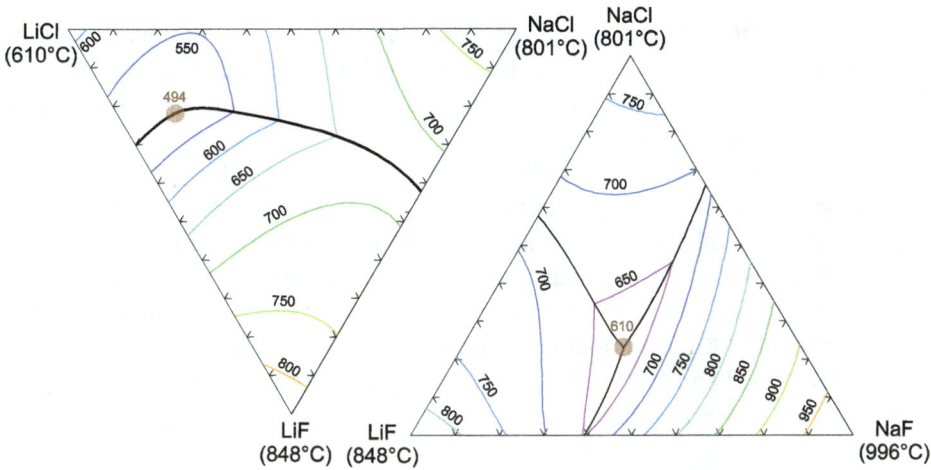

Figure 1.42: The "top left" and "bottom right" partial systems of Fig. 1.40. In the left triangle the eutectics on the LiCl–LiF and LiF–NaCl rims are connected by a eutectic valley, which has a flat minimum near the left rim; the azeotrope of the NaCl–LiCl system is near the top point of the 550 °C-isotherm. In the right triangle, all rim systems are eutectic with a ternary eutectic at 610 °C.

NaCl, will crystallize first. Then segregation shifts the remaining melt composition almost perpendicularly away from the NaF–LiCl isopleth.

An interesting aspect of reciprocal salt pairs was discussed by Acosta et al. (2017), who observed by DTA a small "premelting" peak in LiF–NaCl–CaF$_2$ mixtures. In agreement with previous reports, this system has a ternary eutectic point at ca. 8 mol-% CaF$_2$ and almost identical concentrations of LiF and NaCl with T_{eut} ≈ 665 °C. Very surprisingly, another small endothermal peak was observed ≈ 80 K below T_{eut}. Acosta et al. (2017) explained this "premelting" event by the (limited) solubility of Li$^+$ in NaCl. (The formation of this rocksalt-type solid solution containing ≪ 1% Li$^+$ was disregarded in the preceding discussion, e. g., in Fig. 1.41 (a).)

Considering the four components from equation (1.42), Li$^+$, Na$^+$, F$^-$, Cl$^-$, and two constraints of (1) constant pressure and (2) sum of cations = sum of anions, as expressed in equation (1.42), we arrive at the phase rule equation

$$P + F = C \equiv 4, \tag{1.45}$$

which means that at an invariant point with $F = 0$, as much as four phases are in equilibrium. This is in contrast to $P = 3$, which holds for the eutectic point of a normal binary system. On the other hand, if only three phases (1 melt + 2 solids, $P = 3$) are in equilibrium, then one degree of freedom ($F = 1$) remains.

Figure 9 of Acosta et al. (2017) presents an overview of the NaCl (rocksalt⟨ss⟩)–LiF phase diagram and details of the NaCl rich side. The overview is similar to Fig. 1.41 (a) with the difference that NaCl is treated as a solid solution. As a result, the eutectic line (calculated at $\approx 680\,°C$) terminates at the homogeneity limit of the rocksalt$\langle ss \rangle$ phase, which is basically Li:NaCl with $\leq 0.25\,\%$ LiCl doping. This leads to a drop of the eutectic line in the vicinity of NaCl, which is in agreement with the experimental results of Haendler et al. (1959) shown in Fig. 1.39.

We have $P = 3$ phases (melt, LiF(s), and rocksalt$\langle ss \rangle$) in equilibrium, because the rocksalt$\langle ss \rangle$ phase has a variable composition ($F = 1$). At $\approx 600\,°C$, 80 K below T_{eut}, another isothermal phase boundary appears, which is responsible for the small "premelting" peak. Below this boundary, only the solid phases rocksalt$\langle ss \rangle$, LiF(s), and NaF(s) exist. Above the boundary, minor amounts of the "melt" phase occur. According to Acosta et al. (2017), this melt contains 63 % Na^+, 37 % Li^+, 73 % F^-, and 27 % Cl^-. For ca. 99.5 % NaCl and 600 °C, these four phases are in equilibrium, with no degree of freedom left ($P = 4, F = 0$).

1.5.3.4 Systems with four and more components

Rare-earth scandates $REScO_3$ with orthorhombically distorted perovskite structure are key materials in the quest for oxide crystals with applications as substrate materials for the epitaxial deposition of functional oxides, e. g., with multiferroic properties (Schlom et al., 2014). With rare-earth elements ranging from RE = Dy to Pr, the "pseudocubic" lattice parameter of the substrates can be adjusted from $3.95\,Å \leq a_{pc} \leq 4.02\,Å$. The melting points of these scandates become higher in this order (hence with larger radii of the RE^{3+} ions) and reach for $PrScO_3$ already $T_f \approx 2200\,°C$ (Gesing et al., 2009). For the subsequent elements cerium and lanthanum, the corresponding $REScO_3$ melt so high that single crystals as basis for substrate production cannot be grown in sufficient size and quality: for $LaScO_3$, e. g., Badie (1978)[58] measured already $T_f = (2390 \pm 20)\,°C$.

Uecker et al. (2013, 2017) have demonstrated that the isostructural $RE'\,RE''O_3$ compounds (RE' = rare-earth element at least from La to Dy; RE'' = Sc or Lu) form solid solutions. For very high structural similarity (i. e., similar lattice parameters as shown for neighboring RE), the 2-phase region between liquidus and solidus (see Section 1.5.1.2) is narrow. This is very beneficial for crystal growth because then segregation is small. The possibility of growing mixed crystals allows us to adjust the lattice parameter a_{pc} to almost every desired value in the range of accessible end members just by a proper choice of RE elements. Besides, mixed crystals with lanthanum lutetate ($LaLuO_3$, $T_f = 2210\,°C$, $a_{pc} = 4.17\,Å$) expand the range of accessible lattice parameters to high values, significantly above 4 Å.

In this framework, Guguschev et al. (2020) investigated the quaternary system $La_2O_3 - Nd_2O_3 - Lu_2O_3 - Sc_2O_3$ and succeeded in the growth of homogeneous quaternary mixed crystals from the melt by the Czochralski[59] method. In analogy to ternary sys-

---

58 Jean-Marie Badie (born 1943).

59 Jan Czochralski (23 October 1885 – 22 April 1953).

tems, where the concentrations of three components can be displayed in a concentration triangle (see Section 1.5.3.1), four concentrations can be displayed in a regular tetrahedron. Analogous rules to Viviani's theorem (1.41) apply:

- In the tetrahedron ABCD the compositions with identical concentrations of A are situated on a plane that is parallel to the face opposite A; and so on for B, C, D.
- Along a line from A to the opposite face, the ratio of the concentrations [B], [C], [D] is constant.
- The lever rule, which gives the quantities of equilibrium phases, applies similarly.

Figure 1.43 shows such a tetrahedron, which is produced with software from Keesmann and Schmitz (2021).[60] There the melt composition used by Guguschev et al. (2020) is shown as a green circle (0). Compositions of three positions along the crystal growth axis are shown as dots (1)–(3). (The gray insert is an enlarged reproduction of these four concentrations.) Like usual for crystal growth of mixed crystals, segregation occurs, and the composition of the crystal is different from the composition of the melt where it is grown from. However, the concentrations of three subsequent points along the growth axis of the crystal are not very different: (1) beginning of growth, close to the seed; (2) afterward, at the "shoulder", where the cylindrical part of the crystal begins; (3) the central part of the cylinder.

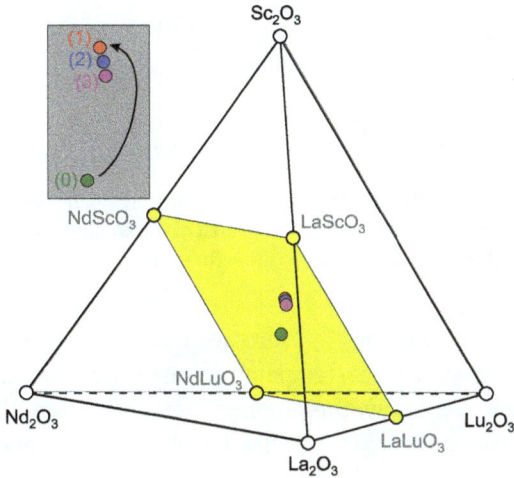

Figure 1.43: Tetrahedral presentation of segregation during crystallization in a quaternary system: From a melt (0, green circle) with composition $La_{0.500}Nd_{0.550}Sc_{0.400}Lu_{0.585}O_3$ a crystal with initial composition (1) $La_{0.557}Nd_{0.394}Sc_{0.589}Lu_{0.459}O_3$ is crystallized (arrow). With progress of crystallization, the composition is shifted over (2) $La_{0.551}Nd_{0.400}Sc_{0.575}Lu_{0.473}O_3$ to (3) $La_{0.545}Nd_{0.406}Sc_{0.561}Lu_{0.488}O_3$. See also Fig. 1.45.

---

60 Karl-Ingo Ortwin Keesmann (1937–2010).

The melt composition $La_{0.500}Nd_{0.550}Sc_{0.400}Lu_{0.585}O_3$ represented by the green circle in Fig. 1.43 proved to be well suited for the growth of mixed crystals with satisfactory size and quality, and with a lattice parameter a_{pc} = 4.086 Å this study was looking for. Already in preliminary investigations, it turned out that the primary crystallization field of such perovskite-type mixed crystals is obviously considerably large. Figure 1.44 shows the first and second DTA heating curves of a crystal grown from an equimolar mixture of the component oxides (top), together with heating curves obtained from a mixture of the starting powders (metal oxides). It turns out that all four curves are similar, which is an indication that segregation is not very strong and that the mixed crystal (= solid solution) phase is formed always, with a narrow melting range in the order of 50 K (slightly more for the powder).

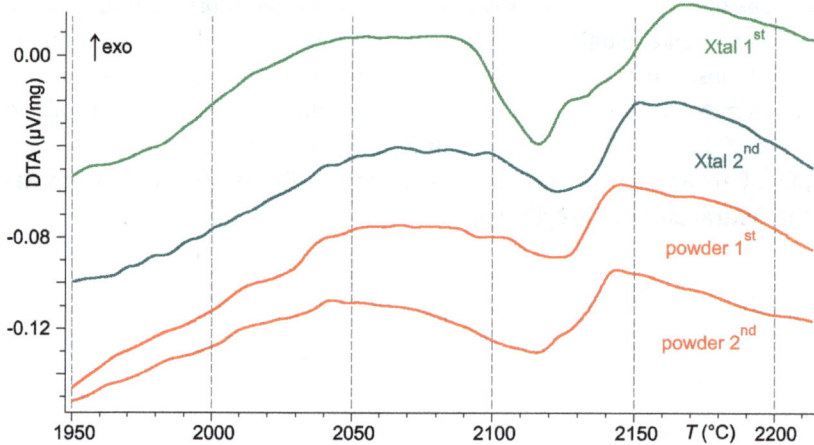

Figure 1.44: 1st and 2nd DTA heating curves (15 K/min, NETZSCH STA 429, static He atmosphere) with melting peaks of a $(La,Nd,Sc,Lu)_2O_3$ crystal (63.31 mg) grown from a 1:1:1:1 melt of the component oxides (top), and of a powder mixture (67.13 mg) with the 1:1:1:1 composition (bottom).

The 3D representation of four concentration values in Fig. 1.43 is accurate but does not allow to easily read quantitative data from it. Besides, temperatures are not included there at all, which is another drawback. Generally, the graphical representation of multicomponent systems becomes more complicated (and often less vivid) with every added component.

Whenever possible, graphical representations with reduced number of free parameters should be preferred for clarity. In the case of perovskite-type solid solutions from the La_2O_3–Nd_2O_3–Lu_2O_3–Sc_2O_3 system Guguschev et al. (2020) revealed by X-ray crystal structure analysis that in the ABO_3 perovskites the A–O distance is larger (235 pm…292 pm) than the B–O distance (223 pm…225 pm). As a result, the A site, which corresponds to Wyckoff position 4c, is preferably occupied by the larger La^{3+} and Nd^{3+} ions (to ca. 87 %). In contrast, the B site (Wyckoff position 4a) is to 90 %

occupied by the smaller Lu^{3+} and Sc^{3+} ions, which are octahedrally coordinated to oxygen. The closer coordination of the latter ions allows us to call this mixed crystal a "lanthanum-neodymium scandate-lutetate", rather than a "lanthanum-neodymium-scandium-lutetium oxide".

This situation has some analogy to the reciprocal salt pairs introduced in Section 1.5.3.3, where the sum of the cations is in a fixed ratio to the sum of the anions. Similarly, all compositions that are relevant for the crystallization of $(La,Nd)(Lu,Sc)O_3$ mixed crystals can be displayed in the quadratic phase diagram in Fig. 1.45, which corresponds nearly to the yellow square plane in Fig. 1.43. The four compositions shown in this tetrahedron can be found here (almost) in the paper plane. Just the yellow plane from Fig. 1.43 is slightly shifted toward the Lu_2O_3–Sc_2O_3 edge, where the congruently melting pseudo-binary compounds are situated, as mentioned in the figure caption.

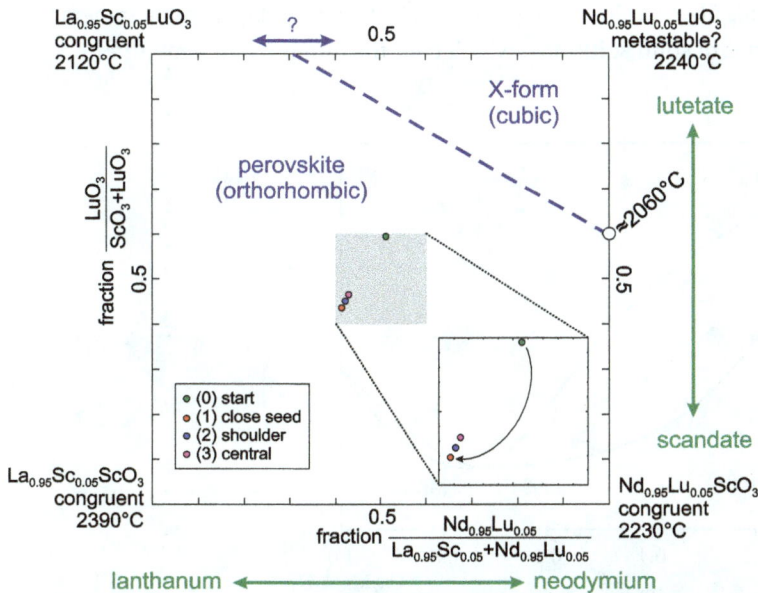

Figure 1.45: This phase diagram represents (almost) the yellow square $LaLuO_3$–$NdLuO_3$–$NdScO_3$–$LaScO_3$ from Fig. 1.43. The four components are handled as end members of a reciprocal salt pair; three of them crystallize as orthorhombically distorted perovskites (space group $Pbnm$); only $NdLuO_3$ crystallizes in the "X-type" structure with space group $Im\bar{3}m$ (Aldebert and Traverse, 1979). "Almost" relates to the circumstance that in Fig. 1.43 the small excess ($\approx$ 5%, Veličkov et al., 2008; Uecker et al., 2008) of the "anions" $(Sc,Lu)O_3^{3-}$ over the "cations" $(La,Nd)^{3+}$ was disregarded but is taken into account here. Slightly amended reproduced with permission from Guguschev et al. (2020).

For the starting composition (green circle), we calculate the "anion/cation ratio" $z =$ $([Lu] + [Sc])/([La] + [Nd]) = 0.97$. This means that the starting melt was slightly depleted in the "anions" $(Sc,Lu)O_3^{3-}$. In contrast, for all three compositions at different positions of the grown crystal, $z = 1.10$ was observed. A similar excess of Sc was reported by Veličkov et al. (2008), Uecker et al. (2008) for pure rare-earth scandates $REScO_3$ (RE = Nd, Sm, Gd, Tb, Dy). We can assume that also in the quaternary system, from the slightly off-congruent melt with $z = 0.97$ a congruently melting mixed crystal with $z = 1.10$ crystallizes, in analogy to the idealized case of an AB compound with congruent melting discussed in Section 1.5.2.1.

Sometimes, it is possible to restrict the presentation on simple ternary or even binary isopleth sections of multicomponent systems, which can reduce the number of components to be taken into account. Schairer (1944)[61] reports some examples from important rock-forming systems based on oxides like CaO, MgO, FeO, Fe_2O_3, Al_2O_3, SiO_2. One pseudo-binary section of the quaternary system $MgO–CaO–Al_2O_3–SiO_2$ is shown in Fig. 1.46. This phase diagram describes the melting behavior of mixtures between the minerals åkermanite[62] and gehlenite.[63] Obviously, both end members form an unlimited series of solid solutions, which bear the mineral name melilite.

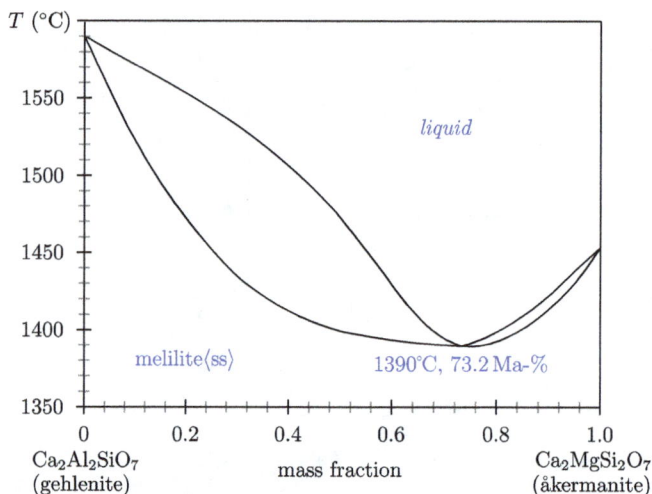

Figure 1.46: Pseudo-binary phase diagram åkermanite–gehlenite as reported by Schairer (1944). Drawn and adopted with data from ACerS-NIST (2014), entry 918. The melilite group is solid solutions of the type $A_2B(T_2O_7)$, where A can be Ca, Na, Ba; B = Mg, Al, Fe, Be, B, Zn; T = Si, Al, B. Åkermanite and gehlenite are the main constituents.

---

61 John Frank Schairer (13 April 1904–19 September 1970).

62 Anders Richard Åkerman (10 April 1837–23 February 1922).

63 Adolph Ferdinand Gehlen (5 September 1775–16 July 1815).

Calculate the azeotrope point in Fig 1.46 for a scaling in mol-%! | **?**

The pseudo-binary phase diagram in Fig. 1.46 agrees well with more recent measurements by Mendybaev et al. (2006). It is interesting to note that the latter authors did not investigate this system by DTA measurements, because this silicate system tends to the formation of glasses. Rather, Mendybaev et al. (2006) annealed samples of different composition to ≈1230 °C to create the crystalline melilite⟨ss⟩ phase. Subsequently, they were brought to a temperature of interest for 10 to 20 hours and then were quenched into water. Quenched products, in the form of pure glass from the experiments conducted above the liquidus, pure crystalline melilite⟨ss⟩ from experiments below the solidus, or coexisting glass and melilite⟨ss⟩ at intermediate temperatures, were examined by a scanning electron microscope (SEM) with X-ray microanalysis system.

The unit cells of the tetragonal gehlenite (a = 7.6850 Å, c = 5.0636 Å) and åkermanite (a = 7.8288 Å, c = 5.0052 Å) are very similar. In the latter, Mg^{2+} occupies almost completely (96 %) tetrahedral sites at the unit cell corners and the centers of the basis plane, and Si^{4+} occupies tetrahedral sites inside the cell. In gehlenite a coupled exchange

$$Mg^{2+}, Si^{4+} \rightarrow 2\,Al^{3+} \tag{1.46}$$

takes place, which concerns all positions of Mg^{2+} and 50 % of the Si^{4+} positions, resulting in mixed occupancy there. This exchange is shown in Fig. 1.47; it can proceed also partially. These are the melilite⟨ss⟩ solid solutions in the phase diagram Fig 1.46.

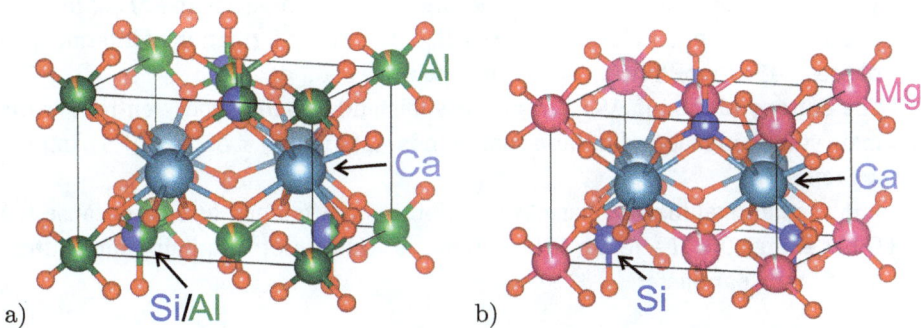

Figure 1.47: One unit cell of (a) gehlenite $Ca_2Al_2SiO_7$ (X-ray data by Louisnathan, 1971) and (b) åkermanite $Ca_2MgSi_2O_7$ (neutron powder diffraction data by Swainson et al., 1992). Space group $P\bar{4}2_1m$, the c axis is directed to the top. Plotted with VESTA by Momma and Izumi (2011).

1.5.3.5 High-entropy alloys (HEAs)

In the introduction of their highly cited paper "Microstructural development in equiatomic multicomponent alloys" (>2700 times), Cantor et al. (2004)[64] wrote: "For quaternary, quinary and higher order systems, information about alloys in the centre of the phase diagram is virtually non-existent. ... This paper describes an initial attempt to investigate the unexplored central region of multicomponent alloy phase space, concentrating particularly on multicomponent transition metal alloys which are found to exhibit a surprising degree of intersolubility in a single fcc phase." The authors prepared in this paper alloys with up to 20 components by melting them together and ejecting them onto a rotating Cu drum for quick cooling. Alloys with as much as 20 (5 at.-% each of Mn, Cr, Fe, Co, Ni, Cu, Ag, W, Mo, Nb, Al, Cd, Sn, Pb, Bi, Zn, Ge, Si, Sb, Mg) or 16 elements were multiphase, but the number of phases was always much lower than the phase rule would allow. Predominantly, the alloys consisted of a face-centered cubic (f. c. c.) phase rich in transition metals, notably Cr, Mn, Fe, Co, and Ni. Chemical analysis revealed in the multicomponent phase space a wide f. c. c. phase field, which includes the composition $Fe_{20}Cr_{20}Mn_{20}Ni_{20}Co_{20}$.

In the same year, Yeh et al. (2004) described for a six-component $CuCoNiCrAl_xFe$ alloy the gradual transition from f. c. c. (exclusively for $x \leq 0.5$) to b. c. c. (for $x > 2.8$), with coexistence of both phases for intermediate concentrations of Al. Figure 1.48 shows the atomic arrangement for another six-component crystalline phase with b. c. c. structure. Yeh et al. (2004) pointed out that many phases with very high numbers of components exhibit high hardness and excellent thermal and chemical stability and wrote "...long-range diffusion for phase separation was sluggish in solid HE (high entropy) alloys that are devoid of a single principal matrix element." Tsai and Yeh (2014) report several parameters relevant for the formation of phases with $C = 5$ or more components:

- The enthalpy of mixing may not be too large, because large positive ΔH_{mix} leads to phase separation, and large negative ΔH_{mix} favors the formation of intermediate phases. $-15\,kJ/mol \leq \Delta H_{mix} \leq 5\,kJ/mol$ are preferred.
- The entropy of mixing ΔS_{mix} has to be sufficiently high, because entropy is the main stabilizing factor for these compounds. $12\,J/(mol\,K) \leq \Delta S_{mix} \leq 17.5\,J/(mol\,K)$ are reported.
- Entropy can stabilize a phase at sufficiently high temperature T, because the Gibbs energy $G = H - TS$ has to be minimized. This entropy-enthalpy competition leads to the parameter

$$\varepsilon = \left| \frac{T\Delta S_{mix}}{\Delta H_{mix}} \right|,$$

which stabilizes a "high-entropy alloy" (HEA) if ε is large.

---

64 Brian Cantor (born 11 January 1948).

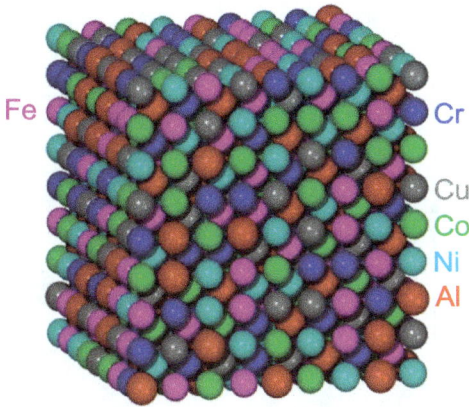

Figure 1.48: Simulated atomic arrangement of the alloy $Al_{\frac{1}{6}}Co_{\frac{1}{6}}Cr_{\frac{1}{6}}Cu_{\frac{1}{6}}Fe_{\frac{1}{6}}Ni_{\frac{1}{6}}$ in the body-centered cubic (b. c. c.) structure. Reprinted with permission from Kleber et al. (2021), produced with data by Wang (2013).

- To allow miscibility, atomic sizes differences of the components may not be too large; typically, $\delta \leq 8.5\,\%$ are required.
- The valence electron concentration VEC, which is calculated from the weighed average of the VEC of the components, determines the structure of the HEA: For VEC > 8, the f. c. c. structure is stabilized. VEC < 6.87 stabilizes b. c. c., and for intermediate values, both phases coexist.

In the ideal case, $\Delta H_{mix} = 0$, and the Gibbs energy of a mixture phase is only determined by the entropy. This entropy has its maximum at a composition where the concentrations of all C components are identical: $x_i = 1/C$ ($i = 1, 2, \ldots, C$). Figure 1.49 shows that the contribution of ΔS_{mix} to G can be large for high component numbers C, especially, at high T. Now the names "high-entropy alloys" (HEAs), "high-entropy oxides" (HEOs), or other "high-entropy compounds", respectively, are given to multi-component phases with ≥ 5 components, which are stabilized by high configuration entropies. Schneider (2021)[65] pointed out that the very high entropy gain shown in Fig. 1.49 does not fully apply for compounds like oxides, carbides, nitrides, etc., because there a non-metal sublattice decreases the entropy gain by mixing significantly. Nevertheless, somewhat incorrectly, the term "high-entropy alloy" is also used for multicomponent mixtures of chemical compounds.

Rost et al. (2015) extended the concept of entropic phase stabilization to an oxide phase with composition $Mg_{0.2}Co_{0.2}Ni_{0.2}Cu_{0.2}Zn_{0.2}O$. Figure 1.50 is copied from their paper and shows in panel (a) the development of X-ray diffraction patterns with tem-

---

65 Jochen M. Schneider (born 25 September 1969).

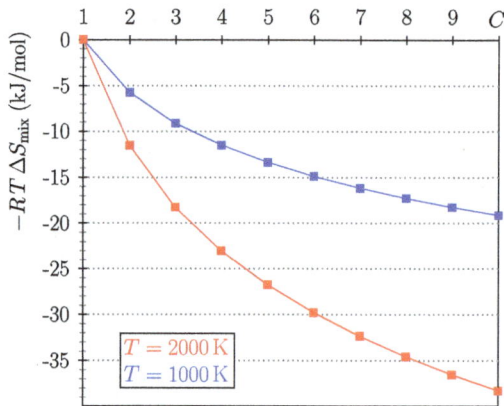

Figure 1.49: The Gibbs energy of an ideal mixture phase as a function of the number of equally concentrated components C, calculated for two temperatures by eq. (1.28).

perature. The labels (letter + 3 numbers) stand for R = rocksalt structure (initial MgO, NiO, CoO; and the final high entropy oxide "HEO" phase), W = wurtzite (initial ZnO), S = spinel, an intermediate product which is responsible for the TG effect close to its disappearance, and the final formation of the HEO phase ≤900 °C. The three numbers are the Miller[66] indices hkl of the X-ray reflection. Figure 1.50 (b) shows DSC/TG results of an identical sample. Here a pronounced endothermal effect occurs between 825 and 875 °C, exactly there where the strong R111 reflex in the X-ray pattern indicates the formation of a rocksalt-type $Mg_{0.2}Co_{0.2}Ni_{0.2}Cu_{0.2}Zn_{0.2}O$ solid solution. McCormack and Navrotsky (2021) attributed this endothermic effect mainly to the necessary transformation of CuO from tenorite[67] and of ZnO from wurtzite to the rocksalt structure of the five-component solid solution phase.

It is very difficult to present a vivid graphical presentation of phase diagrams with five or more components; fortunately, even the multidimensional phase space of systems with very high numbers of components can be handled by thermodynamic simulation software. We cannot hope that in the near future, complete thermodynamic datasets are available for most multicomponent systems, which would allow their accurate calculation. However, interactions in the systems occur basically between two or possibly three species. Correspondingly, the knowledge of binary and perhaps ternary interactions of system components is often sufficient, at least for an estimation of the thermodynamic properties of multicomponent systems. Senkov et al. (2015) write that "calculations for a C-component alloy are considered to be fully credible if the thermodynamic data are available for all the binary and ternary systems embed-

---

66 William Hallowes Miller (6 April 1801–20 May 1880).

67 Michele Tenore (5 May 1780–19 July 1861).

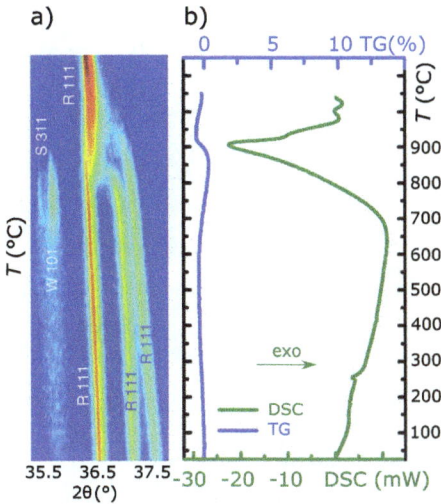

Figure 1.50: (a) In situ X-ray diffraction intensity map of a sample with composition $Mg_{0.2}Co_{0.2}Ni_{0.2}Cu_{0.2}Zn_{0.2}O$. (b) Simultaneous DSC/TG curves for the same composition (Netzsch STA 449 F1 "Jupiter", Pt crucible, air flow). For (a) and (b), the heating rate was 5 K/min. Copied and slightly modified from Rost et al. (2015).

ded in this alloy". More recently, Ushakov et al. (2020) showed that even the knowledge of half the binary interactions results often in useful approximations.

Figure 1.51 gives an explanation of the experimental results obtained by Rost et al. (2015) on the basis of FactSage 8.0 (2020) calculations. This software package contains data for the pure solid and liquid phases of all five component oxides, besides data on a rocksalt-type solid solution of all five components. Data on excess energies of a melt containing all components are not available; certainly, this is difficult to handle because copper(II) oxide is partially reduced even under oxygen rich conditions above ca. 1000 °C (see Fig. C.30). This is beyond the experimental limits of Rost et al. (2015) but should be included in the present calculation. The melt was handled as an ideal solution, because exact data are not available.

Three of the five component oxides crystallize as pure substances in the rocksalt structure: MgO, CoO, and NiO. CuO crystallizes in the monoclinic tenorite structure, and ZnO in the hexagonal wurtzite structure. Isotype MgO–NiO (ACerS-NIST, 2014, entry 10365), MgO–CoO (ACerS-NIST, 2014, entries 51 and 52), and NiO–CoO (ACerS-NIST, 2014, entries 5149, 9223, and 53) form solid solution series at least at elevated $T \gtrsim 750\,°C$. The situation is worse for the nonisostructural components CuO and ZnO, because then (under ideal conditions, i. e., neglecting excess contributions) the solubility limit

$$\tilde{x}_A^\beta = \exp\left[-\frac{\Delta G_A^{\alpha \to \beta}}{RT}\right] \tag{1.47}$$

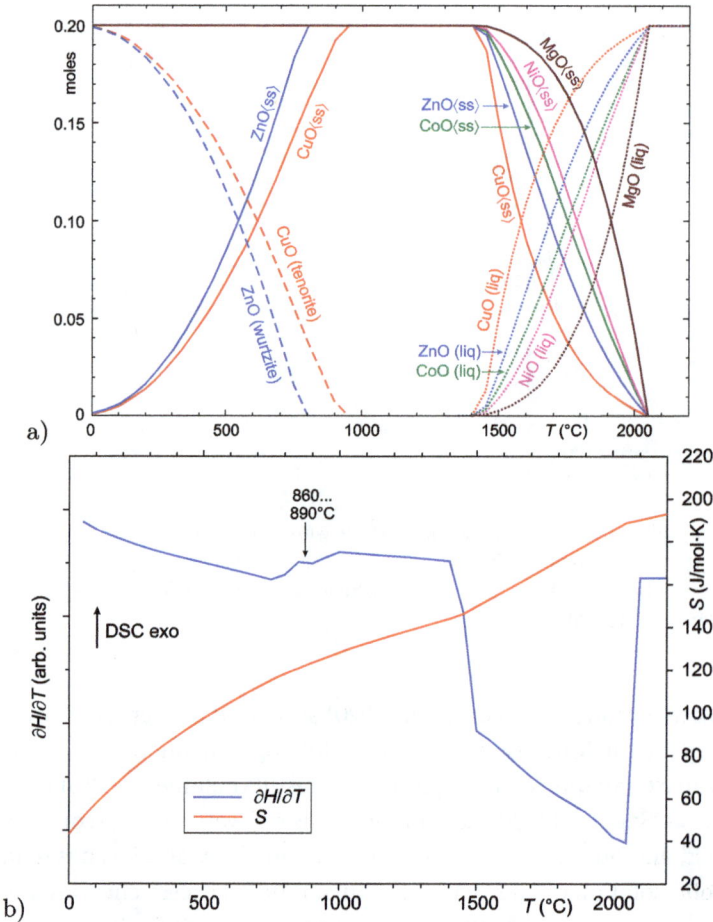

Figure 1.51: Equilibrium calculation for the $Mg_{0.2}Co_{0.2}Ni_{0.2}Cu_{0.2}Zn_{0.2}O$ phase described by Rost et al. (2015). FactSage data for the pure MeO and for the MeO⟨ss⟩ (rocksalt type) were used. For the liquid phase, formation of an ideal solution was assumed. (a) Moles of the phase constituents: at low T, only ZnO (wurtzite) and CuO (tenorite) exist as pure phases. (b) First derivative of H (proportional to the DSC signal) and entropy of the system.

must be expected (McCormack and Navrotsky, 2021), where $\bar{x}_A^\beta$ is the ideal limiting solubility of component A with structure α in a solution with different structure β, and $\Delta G_A^{\alpha \to \beta}$ is the free energy of transformation of A in structure α to A in structure β. Even if the MgO–ZnO phase diagram presented by Segnit and Holland (1965)[68] is incorrect on the ZnO side (Schulz et al., 2011), their solubility limit of 60 % ZnO in MgO is certainly correct. In CoO the solubility of ZnO is smaller ($\approx$ 20 %, ACerS-NIST, 2014, entry 11052)

---

68 Edgar Ralph Segnit (5 September 1923–13 July 1999).

but still high. On the same level, ACerS-NIST (2014), entry 93-220, reports the solubility limit on CuO in MgO. All these data are taken into account for the MeO⟨ss⟩ phase in Fig. 1.51.

Figure 1.51 (a) demonstrates that initially only the rocksalt-type components dissolve into the MeO⟨ss⟩ phase; however, with higher T the entropic contribution allows us to lower the Gibbs energy $G = H - TS$ of this phase further by the dissolution of ZnO and slightly later also of CuO. From $950 \leq T \leq 1400\,°C$ the HEO MeO⟨ss⟩ phase is stable. The subsequent melting to the ideal "liq" phase occurs, like typical, with segregation: components with lower melting point enter the molten phase first. It should be noted that the calculated data about melting are not yet supported by experiments.

Figure 1.51 (b) shows the first derivative of the function $H(T)$ for this equilibrium reaction. It is obvious, and will be described in Section 2.8.2.1 in more detail, that this function describes the heat flow between the material and its environment, and hence the DSC signal. $\partial H / \partial T$ shows a continuous downward (endothermal) trend, resulting from the simultaneous dissolution of ZnO and CuO in MeO⟨ss⟩. Like predicted by equation (1.47), their solubility rises for higher T. In the experimental DSC curve in Fig. 1.50 (b), the strong endothermal effect starts not before ca. 700 °C, because at lower T, all component oxides are present as separate phases, as shown by the three separate R111 peaks shown in Fig. 1.50 (a). This, however, is a nonequilibrium situation due to the slow interdiffusion of the solid phases. The endothermal effect occurs only in the more confined range $700 \leq T_{\text{formation}} \leq 950\,°C$ but is much stronger, because the MeO⟨ss⟩ phase is formed there. The calculated DSC curve in Fig. 1.51 (b) shows a small but significant kink at 860...890 °C, exactly where ZnO is already completely dissolved in MeO⟨ss⟩, and only some remaining CuO continues to enter this rocksalt phase. It is remarkable that the experimental DSC curve in Fig 1.50 (b) shows a small kink after passing the maximum of the endothermal peak, almost at the same temperature. This kink was not discussed by Rost et al. (2015).

Ushakov et al. (2020) performed DTA measurements with high entropy rare-earth oxide mixtures $(La_{0.2}Sm_{0.2}Dy_{0.2}Er_{0.2}RE_{0.2})_2O_3$ (RE = Nd, Gd, Y) in a modified Setaram Setsys 2400 instrument up to 2500 °C, and the results were compared with thermodynamic equilibrium calculations. Temperature and sensitivity calibrations (see also Sections 2.5.1 and 2.5.2) of the DTA device were performed using melting and phase transition temperatures and enthalpies of Au (1064 °C), Al_2O_3 (2054 °C), Nd_2O_3 (A-H, H-X, and X-Liquid at 2077, 2201 and 2308 °C, respectively), and Y_2O_3 (C-H and H-Liquid at 2348 and 2439 °C, respectively). Figure 1.52 shows on a time scale such DTA curve of a sample with composition $La_{0.20}Sm_{0.20}Dy_{0.21}Er_{0.20}Nd_{0.19})_2O_3$. Calibration in the "hot" range $T > 2400\,°C$ was done in situ at the melting point of Y_2O_3. A sample and calibration substance were placed simultaneously into reference or sample crucible, respectively.

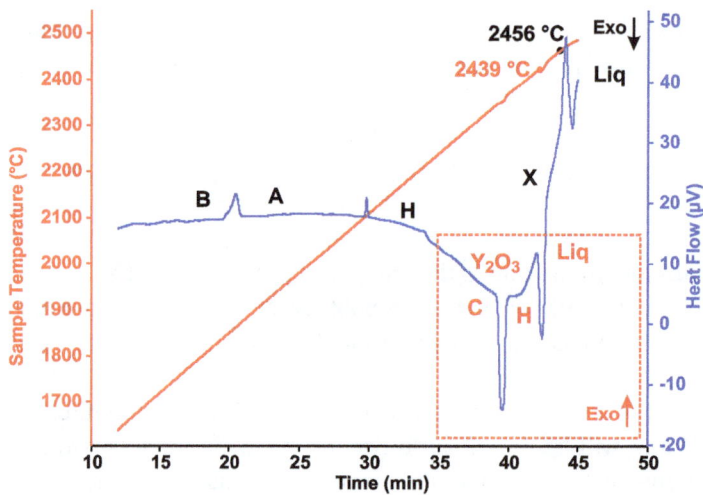

Figure 1.52: Heat flow curve by Ushakov et al. (2020): $La_{0.20}Sm_{0.20}Dy_{0.21}Er_{0.20}Nd_{0.19})_2O_3$ powder on the "reference" side, and pure Y_2O_3 (for calibration) on the "sample" side were measured simultaneously. Subsequent B → A → H → X → Liquid transitions appear endothermic; C → H → Liquid transitions of Y_2O_3 appear in opposite direction, and are red labeled. $T_f^{Y_2O_3}$ = 2439°C was used for calibration of T_f^{HEO} = 2456°C of the HEO phase. Original file supplied and reuse permitted by Ushakov and Navrotsky (2021).

? If sample and reference are placed in different crucibles, then they are not necessarily at the same temperature. Estimate this T difference for the measurement shown in Fig. 1.52 under the assumption that "type D" thermocouples were used. Use the thermovoltage data from Table 1.4.

Although the phase rule allows for systems with C = 5 components up to P = 6 phases in equilibrium at constant pressure and temperature, Ushakov et al. (2020) observed no more than two coexisting phases. Also, the thermodynamic calculations showed only a few very narrow three-phase fields and only narrow two-phase fields, in agreement with the experiments. Another unexpected and not yet fully understood result is the very high T_f = (2456 ± 12) °C, which was measured, e. g., for the $(La_{0.2}Sm_{0.2}Dy_{0.2}Er_{0.2}Nd_{0.2})_2O_3$ phase. If the system behaves ideally, then the melting point is expected near the average $\overline{T_f}$ of the component melting points; cf. Fig. 1.22 for the binary case. With data reported by Coutures and Rand (1989), Ushakov et al. (2020) calculated $\overline{T_f}$ = (2357 ± 16) °C, almost 100 K lower! If deviations from ideality occur, then they originate usually from a positive interaction energy in the solid state. This might be, e. g., repulsive interactions resulting from different ionic radii, but then the solid is destabilized with respect to the liquid, and a downward bend of the phase boundaries occurs. This can lead to an azeotrope point, like shown by the LiCl–NaCl subsystem in Fig. 2.49 (b). In any case, such behavior leads to lower T_f, which is in contrast to the observation with $(La_{0.2}Sm_{0.2}Dy_{0.2}Er_{0.2}Nd_{0.2})_2O_3$.

The observed increase of the melting temperature T_f over the average melting temperature $\overline{T_f}$ of the components is rather untypical, because positive deviations of the Gibbs energy are expected to occur preferably in the solid state. However, a resulting destabilization of the solid phase with respect to the liquid should decrease T_f. Indeed, from the DTA curves for the $(La,Nd,Sc,Lu)_2O_3$ phase in Fig. 1.44, T_f is measured and is 145 K lower than $\overline{T_f}$, which can be calculated from the four endmembers in Fig. 1.45. However, this system with $C = 4 < 5$ is no HEO by definition. The system $Y_2O_3-ZrO_2$ is mentioned by Ushakov et al. (2020) as one of the few examples where a mixed crystal phase (fluorite type) melts higher than the pure components (cf. also the thermodynamic assessment of this system by Fabrichnaya and Aldinger (2004)[69]).

---

69 Fritz Aldinger (born 30 April 1941).

2 Techniques

A wide range of thermal analysis techniques is available nowadays. Some of them measure any physical or chemical property of the sample as a function of the temperature T. This can be the sample mass (thermogravimetry; see Section 2.2), the chemical composition of gas species evaporating from a heated sample (evolved gas analysis; see Section 2.4), thermal conductivity, thermal expansion, and others.

However, in a closer sense, "thermal analysis" means the measurement of thermal properties (temperatures, heat fluxes, specific heat capacity) as a function of T. Already the historical setup by Le Chatelier (1887) shown in Fig. 1.1 was a thermal analyzer, and the principle to gather thermal data from the $T(t)$ (t is time) function that is delivered by just one thermal sensor is sometimes still used (Schindler et al., 2017). However, the accuracy of such measurements is limited because the physical effect (e. g., a slightly slower heating rate $\partial T/\partial t = \dot{T}$ of the sample during a constant heat input $\partial Q/\partial t = \dot{Q}$) is rather small compared to the heating rate itself, $\Delta \dot{T} \ll \dot{T}$.

A significant improvement was obtained by Roberts-Austen,[1] who introduced a "neutral body" of platinum in the vicinity of the sample (Schultze, 1969; Kayser and Patterson, 1998). Then the temperature of the sample T_S and the temperature difference ΔT between sample and "neutral body" can be measured independently. The small signal ΔT can be determined with significantly better accuracy than fluctuations of the large signal T_S. Contemporary DTA ("Differential Thermal Analysis") instruments still rely on a "neutral body" or reference sample.

Figure 2.1 (a) shows the principle of DTA. A sample and a reference are placed close together in a furnace and exposed to a desired temperature program $T(t)$. Often, this is linear heating or cooling (e. g., with 10 K/min), but other programs such as isothermal measurements or periodic fluctuations are possible too.

! In modern DTA (Fig. 2.1 (a)) and heat flow DSC instruments (Fig. 2.1 (b)), the temperature T is measured at the reference **R**, and the DTA (or DSC) signal is the temperature difference between **R** and sample **S**.

For DTA and DSC measurements, a sample **S** is compared to a reference **R**. **S** and **R** are situated inside identical crucibles placed on a sample holder. The sample and reference thermocouples are situated nearby, with good thermal contact. The thermovoltage between poles (1)–(2) is a function of T at the point **R** (see Table 1.4). Voltage (2)–(3) depends on T at point **S** but is usually not measured. Instead, voltage (1)–(3) gives directly the temperature difference $\Delta T^{(\text{DTA})} = T(\mathbf{R}) - T(\mathbf{S})$. This principle drastically increases the accuracy compared to the measurement of $T(\mathbf{R})$ and $T(\mathbf{S})$ with subsequent calculation of the difference.

---

1 Sir William Chandler Roberts-Austen (3 March 1843–22 November 1902).

https://doi.org/10.1515/9783110743784-002

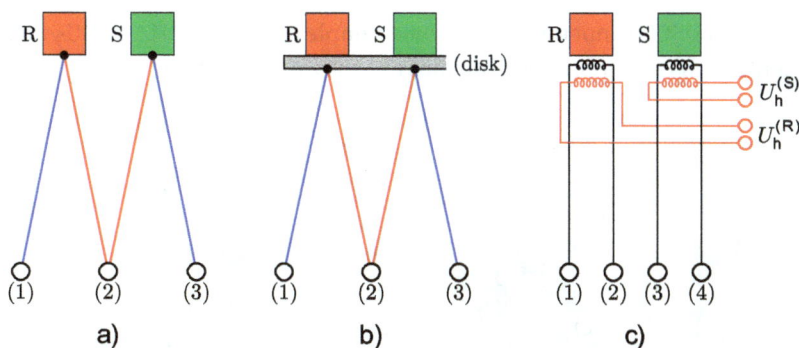

Figure 2.1: Principles of techniques for the analysis of thermal signals (R = reference, S = sample). (a) Differential Thermal Analysis: (1)–(3) = DTA signal, (1)–(2) = T, (b) Differential Scanning Calorimetry (heat-flux): (1)–(3) = DSC signal, (1)–(2) = T, (c) Differential Scanning Calorimetry (power compensation): (1)–(2) T_R, (3)–(4) T_S, $U_h^{(S)}$ = sample heater, $U_h^{(R)}$ = reference heater.

It is somewhat arbitrary if the DTA signal should be defined as $T(\mathbf{S}) - T(\mathbf{R})$ or $T(\mathbf{R}) - T(\mathbf{S})$. In the former definition an exothermal effect in the sample results in a positive (upward) DTA signal, which complies with norms ASTM E793 and E794. In the latter definition an exothermal effect results in a negative (downward) DTA signal, which complies with norms DIN 51007 and ISO 11357-1. To avoid confusion, it is recommended to place an arrow with the word exo in the DTA diagram, e. g., "↑ exo".

DTA allows us to measure the temperature of thermal events such as melting or other phase transformations with high accuracy. The quantization of enthalpies (e. g., heat of fusion); however, it is less accurate: A badly reproducible share of heat is exchanged by radiation or with the gas phase surrounding sample and reference. Two options were developed for improvement, where the uncontrolled exchange of heat with the surrounding is restricted.

For heat flow DSC, reference and sample are connected by a thermal bridge ("disk", or "plate"), which carries a significant part of the exchanged heat energy. Temperature measurement is then performed not directly at the crucibles, but at two points of the disk that are close to sample or reference (see Fig. 2.1 (b)). Under such conditions, the heat exchange can be quantitatively measured. The external electrical contacts for DTA and heat flow DSC are identical, and sample carriers can usually be quickly exchanged.

An even better alternative is shown in Fig. 2.1 (c), where separate heaters close to sample (heater voltage $U_h^{(S)}$) and reference (heater voltage $U_h^{(R)}$) are used. $T(\mathbf{S})$ and $T(\mathbf{R})$ can be measured independently, e. g., with resistivity thermometers, and a bridge circuit controls $U_h^{(R)}$ and $U_h^{(S)}$ during the measurement so that the temperature difference remains negligible. As the required heat energy can be measured easily by the requested electric heating power, it gives a direct measure of the amount of heat produced or consumed by the sample. Hence sensitivity calibration (see Section 2.5.2) for power compensation DSC's is not necessary. Some sample carriers for DTA-, DSC-, and TG-measurements with corresponding crucibles are shown in Fig. 2.2.

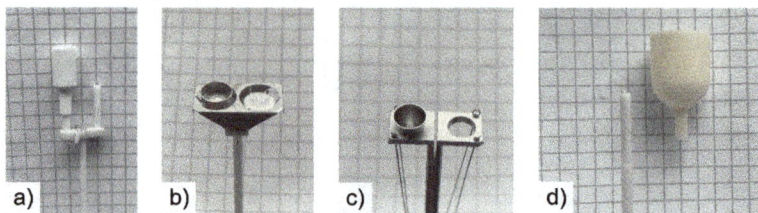

Figure 2.2: Different types of sample carriers for thermal analysis by NETZSCH (2020). (a) DTA/TG carrier with type S thermocouples (see Table 1.4) and one Al_2O_3 crucible 0.3 ml; (b) DSC/TG carrier type S with one Pt crucible 85 µl; (c) DTA/TG carrier for highest $T \leq 2400\,°C$ with W/Re thermocouples and one W crucible 80 µl (thermocouple wires visible on the sides); (d) TG carrier with one type S thermocouple and Al_2O_3 crucible 3.4 ml for large samples (5-mm grid for all pictures).

! For all DSC measurements, covering the crucibles with lids is highly recommended to reduce the exchange of heat by radiation.

2.1 Differential thermal analysis (DTA)

2.1.1 DTA: technical details

The principle of DTA is shown in Figs. 2.1 (a) and relies on imposing a $T(t)$ program simultaneously on a sample and a reference. Different commercial suppliers offer several technical realizations of this idea, e. g.:

1. As shown in Figs. 2.1 (a) and 2.2 (a), with sample and reference thermocouples on the top of a sample holder (or "sample carrier") and electrical contacts as a plug at its bottom. Identical crucibles (see Section 2.6.1) containing the sample (and optionally a reference substance) are placed close to the thermocouples. Then a furnace to heat **S** and **R** is be shifted over the sample holder from the top.
2. Vice versa, **S** and **R** can hang at the bottom of the sample holder with electrical contacts on top of it. Then the furnace remains on its place, and the sample holder is moved downward into the furnace.
3. Thanks to technical progress, it is a general trend during the last decades that sample masses m_s became smaller. Consequently, m_s is nowadays often negligible compared to the mass of the crucible. This has (also for the realizations 1 and 2 above) the effect that often a "reference sample" in the closer sense is not used; instead, an empty crucible of the same type that is used for the sample serves as reference. Outgoing from this, as a third alternative, it is possible to perform DTA

measurements with just one crucible for the sample, and the reference thermo-couple is placed near some thermal balance body close to it.

Often, the DTA sample holder is placed on top (variants 1 and 3 above) or below (variant 2) a balance. If the mass of the sample holder itself, including reference and crucibles, can be regarded constant, then the observed mass change must result from a change of m_s. Such a combination of thermogravimetry (see Section 2.2) with DTA (or DSC; see Section 2.3) is called simultaneous thermal analysis (STA). A combination of two or more signals is often very beneficial for the interpretation of DTA curves.

Crucibles, sample holder, and furnace are the heart of a DTA device. Furnaces are usually based on resistance heaters with different heater elements: silicon carbide, platinum-rhodium alloys, graphite, kanthal,[2] or tungsten are typical. Some furnaces can be precooled with liquid nitrogen and allow the performance of measurements below room temperature (ca. −150 °C) up to a few 100 °C. Such furnaces use, e. g., a protective tube made of steel or silver between heater and sample. Silver flattens the temperature field inside the furnace very well, because it has the highest thermal conductivity among all metals (for 327 °C, The Engineering Toolbox, 2021 reports $\lambda = 405 \text{ W m}^{-1}\text{K}^{-1}$ for Ag, to be compared with $\approx 15 \text{ W m}^{-1}\text{K}^{-1}$ measured by Hidde et al., 2018 at this temperature for α-Al_2O_3 single crystals, as an upper limit for the alumina ceramic which is a standard material for protective tubes). The highest temperatures of about 2400 °C can be reached with tungsten heaters. Special furnaces allow measurements with extremely high heating rates, much beyond the typical rates in the order of 5...40 K/min, or can perform measurements in atmospheres with specific compositions, e. g., high humidity or high gas pressure.

With a few exceptions, DTA measurements are performed in a gas flow; see Section 2.6.2. The choice of this gas should be taken with care, because several limitations may apply:
- The furnace heating element, the sample carrier, or the crucibles can be made of components that are sensitive with respect to oxidation. These are, e. g., graphite, iron, or tungsten (for the latter, see the thermocouples in Table 1.4). Then an atmosphere virtually free of oxygen is mandatory. Often nitrogen (5N = 99.999 % purity) is sufficient, but graphite or tungsten require noble gases like Ar or He.
- Some samples are sensitive with respect to typical impurities of technical gases; often O_2 or humidity are critical. (Depending on the geographical location and weather, ambient air often contains 1 % H_2O or even more.)
- If a simultaneous technique is performed, then its effectivity can be influenced by the atmosphere: TG relies on an accurate mass measurement, which is influenced by buoyancy (see Section 2.2), and this effect becomes smaller in a gas with lower

---

[2] Hans von Kantzow (9 June 1887–12 April 1979).

density, such as He. EGA (see Section 2.4) analyzes the gas flow through the thermal analyzer; then also interference of the rinsing gas with the species that have to be found should be taken into account.

– If the behavior of the sample upon heating in a specific atmosphere is under investigation, then there is no freedom of choice. In this case the DTA equipment itself (crucible, sample holder, possibly a protective tube that separates the heater element from the sample) has to be adjusted. For example, for not too high temperatures, graphite heaters can be separated from the sample by Al_2O_3-based ceramic tubes, which allows measurements in air flow.

⚡ At least for $T \gtrsim 400 \ldots 500\,^{\circ}C$ the flowing gas tends to approach thermodynamic equilibrium with the components of the thermal analyzer. To avoid damage, it is a good idea to calculate the relevant equilibrium (e. g., as Ellingham-type predominance diagram; see Section 2.8.3) to see whether all relevant parts of the thermoanalytic setup (crucible, thermocouples, and other parts of the sample holder, heater, or protective tube) are in equilibrium with the sample under the given atmosphere. For many chemical elements, Ellingham-type predominance diagrams are given in Appendix C.

2.1.2 DTA: examples

2.1.2.1 Zinc

Typically, DTA curves are plotted versus temperature T, but Fig. 2.3 shows all curves as functions of time t instead. Such plots give some insight into the technical quality of the whole device, because a reliably constant heating rate $\dot{T}$ is the precondition for the quantitative interpretation of DTA (and even more so DSC) curves. Especially, in the beginning of a new temperature program segment, T begins to change with some delay. This results from the construction of the devices where the temperature T (which is plotted in Fig. 2.3) is measured near the reference crucible, but heating energy is produced in a distance of several centimeters by the heater element, and thermal conduction requires time. The T controller "sees" the temperature lag and compensates it by a temporarily higher heating rate. For the measurement in Fig. 2.3, $\dot{T} = 10\,K/min$ was programmed, and from the dashed curve we can see that it was raised quickly within $\approx 5\,min$ to $\dot{T} = 14\,K/min$ and with some fluctuation arrived at the desired value ca. 15 min after the start of the measurement.

❓ Why is T measured with the reference thermocouple and not with the sample thermocouple?

The DTA signal between contacts (1) and (3) in Fig. 2.1 (a) can be displayed in its initial unit µV. However, for better comparability, it is often scaled by the sample mass, because a larger sample of course produces a stronger thermal effect. Hence an alternative unit for the DTA signal is µV/mg. Scaling of the DTA signal in Kelvin (see question

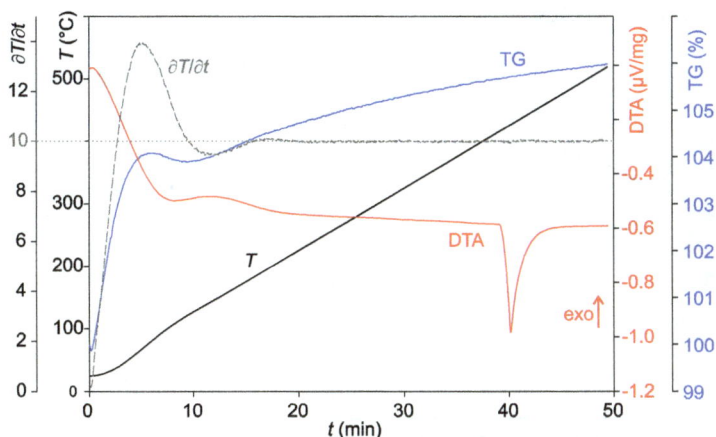

Figure 2.3: Simultaneous DTA/TG (= STA) measurement of Zn with scaling vs. time t. Besides T, also its first derivative, the heating rate $\dot{T} = \partial T/\partial t$ is plotted, which for this measurement, was set to 10 K/min.

below) is possible but nowadays unusual. However, also DTA sample carriers can be calibrated for sensitivity (see Section 2.5.2), and then scaling of the DTA signal in mW (or more commonly W/g) is possible.

$T(t)$ control is usually performed with PI or PID controllers (proportional, integral, differential) of the difference between the program temperature and the measured temperature. To ensure quick response, the time constant between heater and T measurement must be short, which can be obtained best with an additional control thermocouple in the vicinity of the heater; see Fig. 2.4. However, then the actually reached sample temperature is sometimes slightly different from the programmed T. To overcome this problem, a second control circuit can gather input from T_r for adaption.

Figure 2.4 demonstrates the huge benefit of "differential thermal analysis" over the classical "heating curves" introduced by Le Chatelier (1887), as mentioned in Section 1.1.1. During the melting process of the Zn sample peaking at ca. 99.7 min, the sample temperature T_s (which is there always slightly above the reference temperature T_r) lags only marginally, and thus it is even hard to recognize that the red curve approaches the blue one. In contrast, DTA measures the difference $T_s - T_r$ directly, and the black curves easily shows that it reduces from ca. 0.9 K outside the melting peak to ca. 0.45 K.

Figure 2.4 demonstrates also the value of the "neutral body", or reference, for thermal analysis: Only $\dot{T}_r = \partial T_r/\partial t$ through the melting process remains also fairly constant at the set value 10 K/min. This cannot be achieved for $\dot{T}_s = \partial T_s/\partial t$ because the consumed heat of fusion prevents constant heating. It is interesting to note that the furnace heating rate can be slightly different from $\dot{T}_s \approx \dot{T}_r$ because of thermal trans-

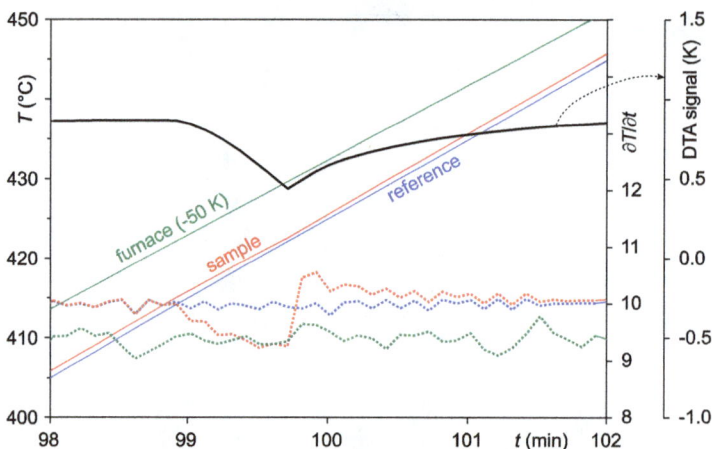

Figure 2.4: DTA measurement with 10.92 mg zinc (NETZSCH STA 409C, SiC furnace, Al_2O_3 crucibles, 10 K/min, Ar flow). Besides the DTA curve (black, exo up, scaled in Kelvin), the true temperatures at the reference and sample crucibles, and the "furnace temperature" at the heater element (lowered by 50 K) are shown. The dotted lines are the 1st derivatives of these $T(t)$ curves.

port conditions between heater element and sample change during the measurement, resulting from different thermal conductivity and thermal radiation.

> [?] Give an estimation for the DTA melting peak height in Fig. 2.3 in Kelvin! The sample mass is $\approx$ 10 mg, and data for the type S thermocouple can be found in Table 1.4.

2.1.2.2 Calcium oxalate hydrate

Calcium oxalate hydrate is a white crystalline powder showing an interesting behavior upon heating, which makes it suited as a test material for thermoanalytic devices. The material decomposes into three subsequent steps

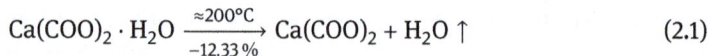

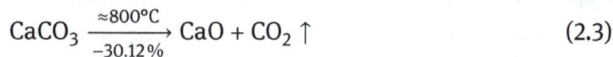

$$Ca(COO)_2 \cdot H_2O \xrightarrow[-12.33\%]{\approx 200°C} Ca(COO)_2 + H_2O \uparrow \tag{2.1}$$

$$Ca(COO)_2 \xrightarrow[-19.17\%]{\approx 500°C} CaCO_3 + CO \uparrow \tag{2.2}$$

$$CaCO_3 \xrightarrow[-30.12\%]{\approx 800°C} CaO + CO_2 \uparrow \tag{2.3}$$

under the release of volatile species finally to CaO. (The mass losses given in equations (2.1)–(2.3) refer to the initial mass of $Ca(COO)_2 \cdot H_2O$.) This series of quantitatively well-described mass losses offers a convenient way for testing not only the DTA/DSC signal (which is possible with typical calibration substances; see Setion 2.2), but also simultaneously the TG signal.

Figure 2.5 shows two sets of simultaneous DTA/TG measurements with calcium oxalate hydrate in different atmospheres. In Fig. 2.5 (a) the TG curves are similar, with slightly steeper TG steps resulting from the decomposition reactions (2.3) for the blue curves. It should be taken into account that not only the gas surrounding the sample, but also the sample mass m_s is different for both measurements. In addition to the general effect that larger m_s result in broader "smeared" thermal effects, which will be described in Section 2.6.3, also the gases evolving from the sample itself can influence the kinetics of the decomposition: The decomposition reactions (2.1)–(2.3) represent chemical equilibria. This means that a high concentration of the product H_2O, CO or CO_2, respectively, in the atmosphere will inhibit them. Transport of these reaction products with the flowing gas (50 ml/min in both cases) away from the sam-

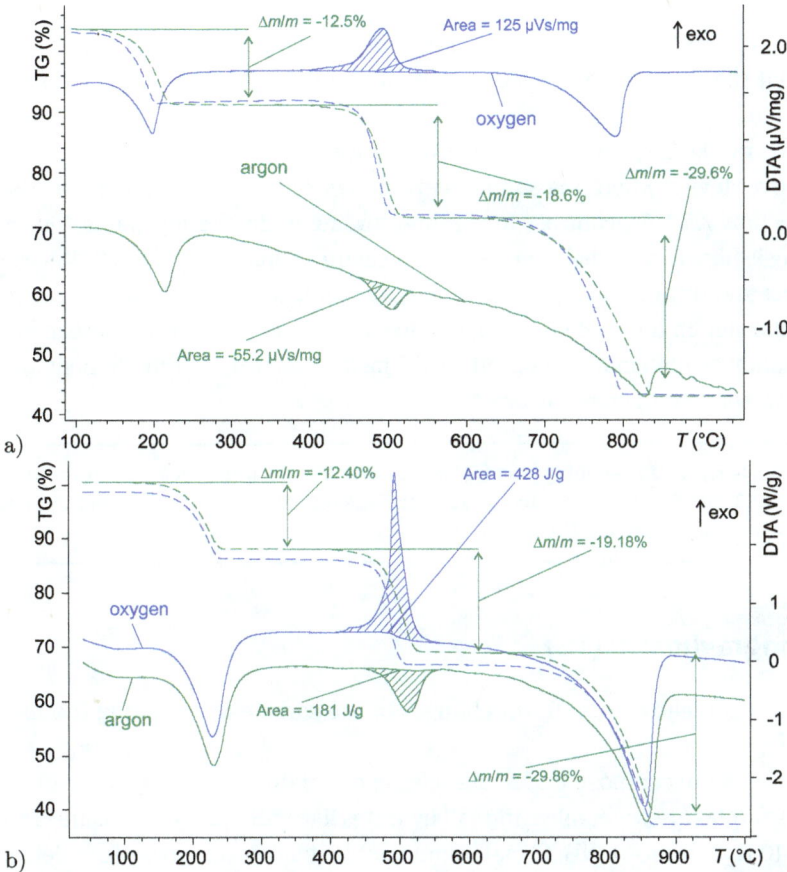

Figure 2.5: Simultaneous DTA/TG (= STA, DTA: full lines, TG: dashed lines) measurements of $Ca(COO)_2 \cdot H_2O$ in argon (green) or oxygen (blue), respectively. Pay attention to the opposite direction of the DTA peak around 500 °C. (a) NETZSCH STA409, Ar: m_s = 25.69 mg, O_2: m_s = 15.47 mg. (b) NETZSCH STA449 "F3", Ar: m_s = 64.47 mg, O_2: m_s = 62.45 mg.

ple will proceed more slowly for larger m_s, which can indeed be seen in the TG curve of the Ar measurement. The total height of the TG steps (numerically shown in Fig. 2.5 only for the measurements in Ar) is almost identical and in fairly good agreement with the expected values shown in equations (2.1)–(2.3).

An interesting difference is obvious if one compares the DTA curves. All decomposition reactions (2.1)–(2.3) are endothermal, and correspondingly endothermal DTA peaks are expected parallel to the TG steps. However, this behavior is shown only for the measurement in Ar. The measurement in O_2 shows endothermal DTA peaks only for the first and third TG steps. In contrast, the second TG step is accompanied by an exothermal peak! This behavior can be explained by the secondary oxidation

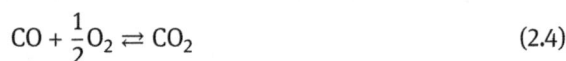

$$CO + \frac{1}{2}O_2 \rightleftarrows CO_2 \tag{2.4}$$

of the emanating carbon monoxide, which is strongly exothermal and produces so much heat that the endothermal reaction enthalpy of the oxalate decomposition is overcompensated.

The curves in Fig. 2.5 (b) were obtained under almost identical conditions with a more sensitive device where the sample carrier was sensitivity calibrated, as described in Section 2.5.2. In contrast to the measurements in the top panel, $m_s \approx 62.5 \ldots 64.5$ mg is here almost the same for both measurements, and indeed the steepness of the first and third TG steps is almost the same. It is not so for the second step, where again the step in the blue curve (O_2) is steeper. This is also a result of reaction (2.4), which removes the emanation product CO faster from the sample than simple rinsing with Ar; see also the next section 2.2.

? The second TG step in Fig. 2.5 (b) resulting from CO loss has a height of $\approx 0.192 \times m_s = 12$ mg, and the oxidation of these 12 mg CO at 500 °C produces 121 J (FactSage 8.0, 2020). Which share of this heat energy flows from the sample to the DTA thermocouple?

2.2 Thermogravimetry (TG)

Different effects can lead to significant changes of the sample mass during thermal analysis:

- The sample can decompose under the release of volatile species upon heating. Typical examples are water releasing salt hydrates like $CaSO_4 \cdot 2\,H_2O$ (gypsum) and $Na_2CO_3 \cdot 10\,H_2O$ (natron). NH_4Cl (salammoniac) instead decomposes completely to gaseous NH_3 and HCl.
- The sample can react with the atmosphere. Examples are, e. g., substances reacting with O_2 under the formation of solid oxides, which may lead to a growing sample mass (Fe $\rightarrow \frac{1}{3} Fe_3O_4$). In contrast, the formation of gaseous oxides can

lead to a partially or completely disappearing sample ($C \rightarrow CO/CO_2$). Growing and shrinking sample mass can also be combined ($Cu \rightarrow CuO \rightarrow \frac{1}{2} Cu_2O$).

– Species that are (often only weakly) bound to the surface can emanate during desorption processes; see, e. g., de Jong and Niemantsverdriet (1990), who give an overview how thermal desorption spectra can be used for the determination of activation energies, preexponential factors, and reaction orders of such processes.

Of course, the composition of atmosphere during a TG measurement has a significantly larger influence on the TG curve for the second case. However, also for decomposition reactions, such influence cannot be ruled out. Like mentioned above, the second TG step in Fig. 2.5 (b) is a good example, where anhydrous calcium oxalate decomposes to carbonate under the release of carbon monoxide; see equation (2.2). The rate of this equilibrium reaction increases if CO is quickly removed from the sample. Here O_2 is more efficient than Ar with identical flow rates, because then equation (2.4) removes CO by oxidation, in addition to just rinsing it away with the noble gas Ar.

2.2.1 TG: technical details

Pure TG devices, which are also named "Thermogravimetric Analyzers" (TGA) or "thermobalances", consist of a precise electronic balance (typical resolution 0.1 µg), which determines the mass of a sample holder with one thermocouple for the determination of T_s. The sample can be heated in a similar way like for DTA (Section 2.1). Also, the choice of furnaces and atmospheres is in analogy to DTA. Often, this method is used if only mass losses (e. g., resulting from sample decomposition) are interesting.

Some TG devices offer the possibility to calculate a so-called c-DTA signal from the difference between the real sample temperature and the programmed temperature. This has some similarity with the setup of Le Chatelier (1887) in Fig. 1.1, where an endothermal event like melting results in a lag of T_s behind the programmed temperature.

As mentioned above, TG and DTA (or DSC) are often combined, and then simply the mass of the DTA sample holder is measured by a balance. This combination of two independent signals often significantly simplifies the interpretation of results.

2.2.2 TG: examples

2.2.2.1 Calcium scandate $CaSc_2O_4$ and lanthanum oxide La_2O_3

Figure 2.6 shows the results obtained with an STA (DTA/TG) device, but only TG signals are displayed. The sample measurement (blue curve) shows that the mass changes during this measurement by > 1 mg. However, the major part of the measured mass change is an artifact that is evoked mainly by the buoyancy of the gas, superimposed by the viscosity of the flowing gas that imposes a small mechanical force on the sample carrier.

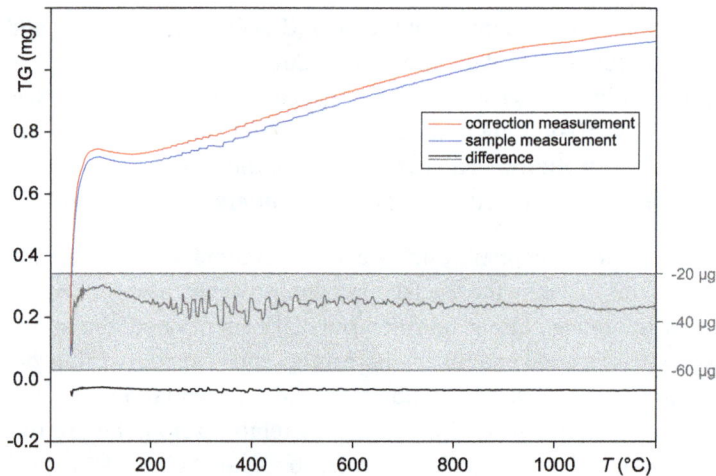

Figure 2.6: Thermogravimetry of a $CaSc_2O_4$ crystal in Ar, m_s = 65.71 mg, $\dot{T}$ = 10 K/min, compared to a "correction measurement". The difference (black) is reproduced on a finer scale in the gray insert.

? Give an estimation for the magnitude of the buoyancy of the sample carrier in Ar if its volume is assumed to be 1 cm^3. How does the buoyancy change if T rises from 300 K (room temperature) to 600 K?

The onset of temperature enforced gas convection, together with the reduced buoyancy of the sample carrier–crucible–sample system if the device is heated, results in a seemingly increased sample mass. Depending on the gas flowing through the device and on the device geometry, this effect often amounts to several 100 µg. This problem can be reduced by several means:

- The sample mass m_s can be increased, e. g., using very large crucibles of the kind shown in Fig. 2.2 (d). Then real changes of m_s can be detected better, because the volume of the experimental setup (carrier, crucible) is less significant.
- The gas flowing through the device can be changed from the typical Ar (molar mass $M \approx 40$), N$_2$ ($M \approx 28$), or air ($M \approx 29$; this is the weighed average of N$_2$ and O$_2$) to He ($M \approx 4$). This leads not only to reduced buoyancy, because this effect changes $\propto M$; additionally, the mechanical force imposed on the carrier by the flowing gas is lower than for Ar; according to Kestin et al. (1972), He has the lowest viscosity at least of all noble gases. Relevant properties of some gases can be found in Fig. 2.26 and Table 2.6.
- Another feasible and advised method is the performance of a "correction measurement". Usually this means a measurement under identical conditions (crucible(s), T program, gas flow) but without a sample. Often, the control software of the thermal analyzer (TG, DTA/TG, or DSC/TG) allows us then to reopen this correction measurement file and run it again with sample. As the result, then the difference between correction and sample measurement is shown. This procedure

has the benefit that for simultaneous DTA/TG or DSC/TG, also correction data for the thermal signal are obtained, and spurious DTA effects like, e. g., for $t < 20$ min in Fig. 2.3 are compensated.

– If an irreversible mass change is expected only for the first heating of a sample, then a T program of the following kind can be a useful alternative to a separate correction measurement: $\boxed{\text{Start}} \rightarrow \boxed{\text{5 K/min to 40 °C}} \rightarrow \boxed{\text{isothermal} \approx 30 \text{ min}} \rightarrow$ $\boxed{\text{10 K/min to } T_{max}} \rightarrow \boxed{\text{–10 K/min to 40 °C}} \rightarrow \boxed{\text{isothermal} \approx 120 \text{ min}} \rightarrow$ $\boxed{\text{10 K/min to } T_{max}} \rightarrow \boxed{\text{–10 K/min to 400 °C}} \rightarrow \boxed{\text{Stop}}$; and subsequently free cooling to room temperature. Very often, the sample remains after the first cooling from T_{max} in the state that was reached there (e. g. dehydrated), and the "corrected" TG curve can be obtained from the difference between the first and second heating. This difference can often be calculated inside the analysis software, which comes with the thermoanalytic device.

The TG signal of an STA measurement with $CaSc_2O_4$ that follows the rule of the last point is shown in Fig. 2.6; only two heating segments to $T_{max} = 1200$ °C are displayed. The first slow heating to 40 °C with subsequent 30-min equilibration there guarantees that both subsequent heating segments start under identical conditions. The intermediate equilibration requires more time (here 120 min) because cooling from the high T_{max} cannot be performed with very high rate in air-cooled furnaces.

Sc_2O_3 is known to be a hygroscopic substance (see, e. g., Jur et al., 2011 or chemistry textbooks) and the aim of this measurement was to reveal if also the scandate $CaSc_2O_4$ absorbs significant amounts of H_2O and/or CO_2 at its surface. The black curve in Fig. 2.6 shows the difference between the first heating (blue, where the mass loss should occur) and the second heating (red, where the substance was assumed to be calcined, hence to show no mass loss). On the scaling of the left ordinate, this signal seems almost flat. The somewhat statistical fluctuations, especially around 400 °C, are of course stronger than in the original curves because the experimental scatter of both measurements is added. The magnified insert in the gray field (scale on the right ordinate) shows that in addition to this scatter, a real minor mass loss around 10 µg occurs from 100–200 °C, demonstrating the high accuracy that can be obtained with such measurements. Section 2.2.2.2 shows that an even better resolution can be obtained if the measurement is performed in helium instead of argon.

However, very often, a very high accuracy is not required, especially if commercial chemicals have to be checked. Figure 2.7 shows measurements with commercial lanthanum oxide. The TG of the first heating shows two strong steps below 400 °C and above 500 °C. These steps are absent in the second heating, but also this curve shows in its beginning the apparent mass gain resulting from buoyancy and other spurious effects. The red difference curve eliminates these flaws and allows the reliable determination of different TG steps. Like for many other powdered materials, adsorbed humidity leaves the sample first between room temperature and $\approx 150...300$ °C

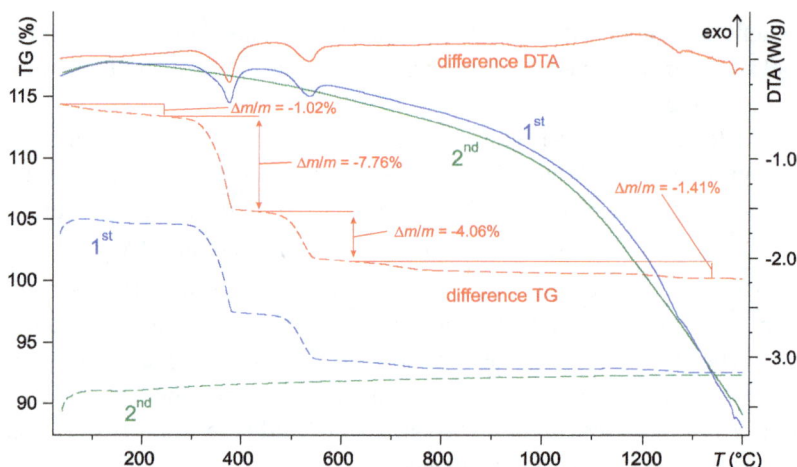

Figure 2.7: Two subsequent DTA/TG measurements (with NETZSCH STA 409C, blue and green curves) of 55.67 mg La_2O_3 powder in Ar, $\dot{T} = 10$ K/min from 40 °C to $T_{max} = 1400$ °C.

($\Delta m/m = -1.02\%$). The following steps result from the decomposition of different oxo-, hydroxo-, and carbonate phases present in the sample; see the discussion in Section 3.1.2.

Often, such measurement saves time for the experiments, because the whole T routine can be programmed at once, without intermediate opening and restarting the device. Practically, this means that the measurement can be started in the afternoon and will be ready on the next morning. However, we should work with some caution and take into account that impurity traces (here moisture; see Section 2.6.2) are present in every technical gas. This holds also for the Ar with 99.999 % purity, which was used for the measurement from Fig. 2.6. Consequently, it must be ensured that during the first cooling or the following isothermal segment at low T, contamination of the calcined sample does not occur.

2.2.2.2 Nitrogen implanted zinc oxide N:ZnO

The small mass loss of the $CaSc_2O_4$ sample in Fig. 2.6 was close to the noise level of this measurement. We can increase the relative accuracy often by increasing the sample mass m_s over the range 10...100 mg, which is typical for simultaneous DTA(DSC)/TG measurements, because the absolute noise level is often almost constant for a given experimental setup and usually does not depend on the sample mass. This is possible especially with TG sample holders like shown in Fig. 2.2 (d), which carry only one large crucible with 1 ml or even higher volume. Then large $m_s > 1$ g are often possible.

If larger sample volumes are not available, then we can alternatively reduce the interfering influence of the sample atmosphere on the mass signal. Some TG devices use "counter pans" in a separate furnace to obtain a symmetrical gas flow

(Saadatkhah et al., 2020). As pointed out by Brown (2001), the performance of the measurement in an atmosphere with low mass density is an alternative. Figure 2.8 shows such TG measurement performed in flowing helium. The atomic mass of ^{4}He is one order of magnitude smaller than that of ^{40}Ar, and thus also the buoyancy of this gas. Besides, the lift force acting between flowing gas and sample holder shrinks with lower gas density.

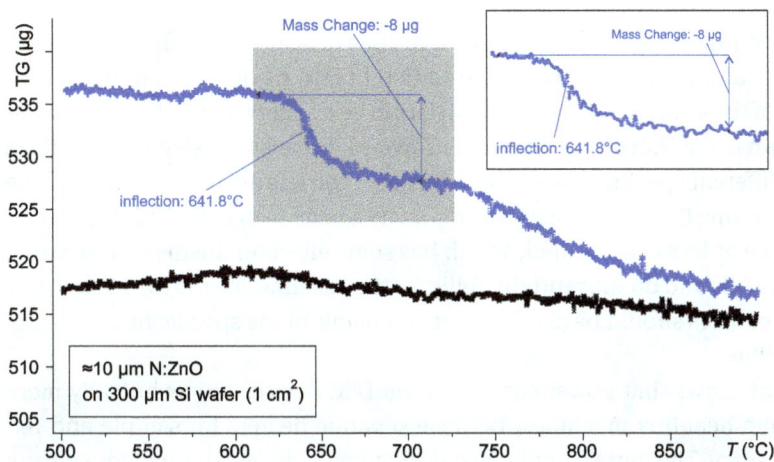

Figure 2.8: Thermogravimetry (heating with 10 K/min) of a nitrogen implanted ZnO layer on Si wafer with 300 μm thickness in He flow. Blue curve: first heating; gray curve: second heating.

In Fig. 2.8 the sample was a 10-μm N:ZnO layer, which was deposited on a silicon wafer. ZnO has a density of $\varrho = 5.61\,g/cm^3 = 5.61\,mg/mm^3$, and approximately $2\,cm^2$ of this wafer were available, which means that the N:ZnO layer had a mass of ca. $5.61 \cdot 0.01 \cdot 200\,mg \approx 11\,mg$. The sample was heated twice, and it is obvious that around 642 °C a mass loss of 8 μg occurs. The mass loss does not occur during the second heating, which proves that all implanted N atoms leave the sample during the first heating.

2.3 Differential scanning calorimetry (DSC)

2.3.1 DSC: technical details

The principles of two types of DSC devices are sketched in Figs. 2.1 (b) and (c). Another type, the Tian[3]–Calvet[4] calorimeter, will be only briefly mentioned here: Also,

---

3 Albert Tian (3 November 1880–18 July 1972).
4 Édouard Calvet (22 September 1895–2 April 1966).

there a symmetrical design with sample and reference is used, but with significantly larger volumes (Höhne et al., 1996). These large volumes also imply large time constants in the order of several ten minutes, which is very long compared to the time constants $\ll 1\,\text{min}$ (see Section 2.5.3) obtained with standard DTA or DSC carriers (see Section 2.5.3). The sample and reference are connected with a large number of thermocouples that work in series, which results in a strong signal with low noise. Such instruments are often used if high sensitivity is required, but lower time resolution is not impedimental.

It can be seen from Figs. 2.1 (a) and (b) that the construction of heat-flux differential scanning calorimeters is very similar to that of DTA devices. Often, just a DTA sample carrier of the kind shown in Fig. 2.2 (a) has to be exchanged by a DSC-type sample carrier shown in Fig. 2.2 (b). This is straightforward, because the electrical contacts are identical. Different specific types of heat-flux DSC carriers are often available: The Pt sheets surrounding the sensor faces (the right one can be seen in Fig. 2.2 (b)) may be constructed more or less complicated, which has some effect on the thermal sensitivity (see Section 2.5.2) and on the reproducibility of the baseline. This is the reason why special DSC-c_p carriers should be used for measurements of the specific heat capacity whenever possible.

Figure 2.1 (c) shows that power compensation DSC devices are technically more complicated than heat-flux machines, because separate heaters for sample and reference are necessary. The more complicated design limits the maximum temperature that can be reached by these devices to < 1000 °C. On the other side, the extremely small heaters in direct contact to sample and reference allow huge heating rates in the order of several 100 K/min and low time constants around 1 second. Also, the overlay of fast temperature fluctuations over a temperature program, like "StepScan" (Perkin[5]–Elmer[6]) or "temperature-modulated" DSC (NETZSCH, "TM-DSC"), can be realized better with low time constants. Schawe et al. (2006) demonstrated that stochastic temperature variations allow the determination of the specific heat capacity in one single measurement without further calibration. With "Fast Scanning Calorimetry" (FDSC), very small samples in the order of micrograms can be heated with rates up to several 1000 K/min (Quick et al., 2019; Schawe et al., 2020).

Such variations of DSC are often used if high sensitivity is required, e. g., for the detection of glass transitions (see Porter and Blum, 2000) or for measurements of the specific heat capacity $c_p(T)$. Already (Corbino, 1910, 1911)[7] exploited temperature variation for the measurement of thermal properties of materials by the "3ω method". For more DSC-related details, the reader is referred to Section 3.2.2 of this book and to the review by Gmelin (1997).

---

5 Richard Scott Perkin (1906–22 May 1969).

6 Charles Wesley Elmer (1872–7 December 1954).

7 Orso Mario Corbino (30 April 1876–23 January 1937).

2.3.2 DSC: examples

2.3.2.1 Heat of fusion

In a power compensation DSC, the signal has the unit mW (or mW/mg if scaled by the sample mass) and corresponds to the electrical power P that is needed to keep the sample temperature T_s the same as the reference temperature T_r. In contrast, in a heat flow calorimeter the temperature difference $T_s - T_r$ is measured by thermocouples in units of μV or $\mu V/mg$. However, the typically known sensitivity K_s of the sample carrier (see Section 2.5.2) allows us to convert the DSC signal of heat flow devices in mW.

In the same way like described there, also DTA devices can be calibrated for sensitivity. Such calibrations are often useful because they allow a better quantitative comparison of thermal effects that occur at different temperatures. The integral $W = \int P \, dt$, and hence the area under a DSC/DTA peak, corresponds to the total heat energy W, which is exchanged during a thermal event that produces the peak. In the case of melting, this is the heat of fusion, ΔH_f.

The blue curve in Fig. 2.9 shows the DTA melting peak of calcium fluoride on a $\mu V/mg$ scale. One problem is obvious there: Even if the baseline outside the peak is straight, it can be situated at different levels. One possible reason for such behavior is, e. g., different thermal contact between sample and crucible. This happens often if powder samples are molten, and the liquid sample exchanges heat better than the initial powder. Different specific heat capacity of liquid and solid phase can be another reason, because a higher c_p shifts the curve to the endothermal direction (see Section 2.3.2.2).

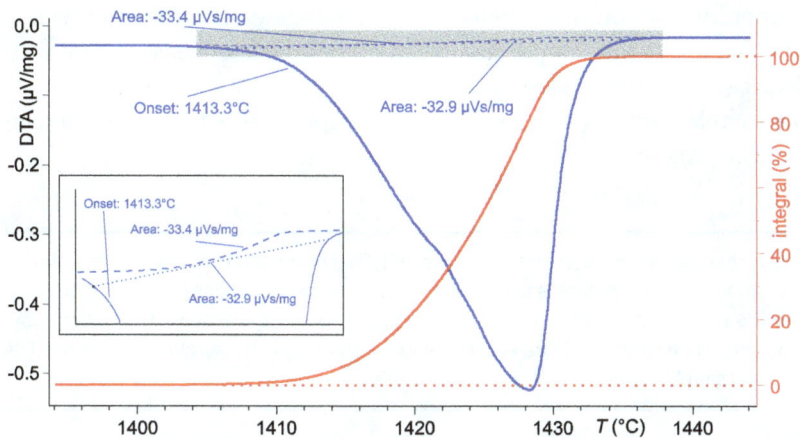

Figure 2.9: Blue curve: DTA melting peak of CaF_2, measured with a NETZSCH STA 449C "Jupiter". Pt crucibles, $\dot{T}$ = 10 K/min, flowing Ar atmosphere, m_s = 222.76 mg. The insert shows two possible constructions for baselines: a dotted straight line and a dashed line that starts tangential from both sides at different levels. In between the position of the baseline is proportional to the passed share of the peak, expressed by the red integral curve.

For the determination of the peak area, the position of the base line inside the peak must be known. Only if the DSC/DTA curve on both sides is at the same level, a straight line is the optimum choice. If not, often a "tangential base line" is preferred. In Fig. 2.9 a straight (linear) and a tangential base line are compared for the melting peak of CaF_2. In this case, where the DTA signal at both sides of the peak is similar, the areas over the linear base line ($-32.9\,\mu Vs/mg$) is not much different from the area over the tangential base line ($-33.4\,\mu Vs/mg$). Usually, the difference is larger if the level of the baseline on both sides is more different. In some cases, alternative definitions of a baseline must be used. An example is shown in Fig. 3.14, where cooling curves start partially inside the crystallization peak. Then, of course, the baseline position at the high temperature side is not available. Hence a horizontal, left started baseline was used, which terminates at the intersection with the DTA curve at the high-T side.

---

i If possible, the area of DSC or DTA peaks should be calculated over a tangential base line. Such baseline starts straight from both sides parallel to the thermal signal outside the peak. In the crossover region the baseline position is calculated in iterations from the share of the peak that is already passed, which correspond to the red curve in Fig. 2.9.

2.3.2.2 Specific heat capacity

The specific heat capacity as a function of temperature $c_p(T)$ can be measured with standard DSC devices by the comparison of three DSC curves:

$DSC_{bas}(T)$ (the baseline) is obtained from a heating measurement with empty crucibles for reference and sample.

$DSC_{ref}(T)$ (the reference curve) is obtained with a known reference sample (often, a sapphire disc, but other reference materials are also possible), which is placed inside the sample crucible after the baseline measurement with identical temperature program.

$DSC_s(T)$ (the sample curve) is obtained with the sample, again with identical temperature program.

---

i Sapphire in its closest sense is a mineral that consist of α-Al_2O_3 (corundum) with coloring trace impurities like Fe, Ti, Cr, or V. Natural sapphire is usually blue, but other colors are possible. If Cr^{3+} prevails as impurity, then the color of (colorless) corundum changes to red, and the mineral is called ruby. The denomination corundum would be more exactly, because the colorless (pure) α-Al_2O_3 is used as reference; but instead sapphire prevails.

For these three measurements, the sensitivity $K_s(T)$ (2.10) is not relevant; this means that a sensitivity calibration is not necessary. It is important that all three measurements are performed under identical conditions. Therefore it is useful to start them not at room temperature, because this is not well defined. Rather, it is recommended to begin with slow heating (e. g., 5 K/min) to some isothermal segment (e. g., 40,°C) where the program waits for ca. 30 min. This guarantees that all three heating runs (basis, reference, sample) start from identical conditions, independent on environmental conditions that may change with time. For the subsequent heating segment used to determine c_p, the heating rate should not be too low; 10 K/min are possible but often 20 K/min gives more accurate results. The reason is that the evaluation relies on the shift of the DSC curve to the endothermal direction, resulting from the thermal load of the reference or the sample, respectively. This can be seen from the DSC curves in Fig. 2.10, where the black $DSC_{bas}(T)$ curve is on top, and the red $DSC_{ref}(T)$ and the blue $DSC_s(T)$ curves are shifted to the endothermal direction. This shift becomes larger for larger $\dot{T}$ because then the same amount of enthalphy is exchanged in a shorter period of time. If the three measurements are finished, then $c_p(T)$ (in J/(g K)) can be calculated by the formula

$$c_p(T) = \frac{m_{ref}}{m_s} \times \frac{DSC_s(T) - DSC_{bas}(T)}{DSC_{ref}(T) - DSC_{bas}(T)} \times c_{p,ref}, \tag{2.5}$$

where m_{ref} and m_s are the masses of the reference and the sample, and $c_{p,ref}$ is the specific heat capacity function of the reference material, which must be known. (Usually, the evaluation according (2.5) is performed with the analysis software that comes with the DSC device, and then the $c_p(T)$ functions for reference materials are available from this software.)

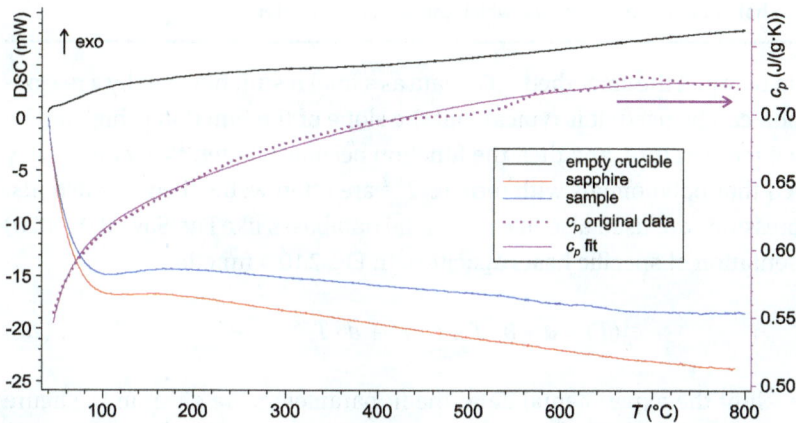

Figure 2.10: Measurement of the specific heat capacity of a MgGa$_2$O$_4$ crystal (63.72 mg) with sapphire (84.62 mg) as reference. The measurement was performed by dynamic DSC with $\dot{T} = 20$ K/min in a NETZSCH STA449C. $c_p(T)$ data are fitted to function (2.6) with parameters $a = 0.596730, b = 3.271283 \times 10^{-4}, c = -2.1081862 \times 10^{-7}, d = -136.8806$ (in J/(g K)).

The dotted curve in Fig. 2.10 shows the results of such analysis for a crystal of the wide bandgap semiconducting oxide MgGa$_2$O$_4$, which was grown by Galazka et al. (2015).

Kondrat'eva et al. (2016) published more recent DSC data for this material, which are in excellent agreement with Fig. 2.10 for low T and by ca. 4 % higher at the highest T for this measurement.

For every DSC measurement and especially for the determination of c_p, the crucibles should be covered with lids. Preferably, metal crucibles (e. g., platinum) should be used. Al_2O_3 crucibles and lids are not well suited especially at high temperatures, because they are transparent for infrared radiation, which leads to heat energy loss.

As pointed out before, it is often recommended to use high heating rates for c_p measurements, because then the thermal loads of sample and reference, and hence the differences in the numerator and the denominator of equation (2.5), are larger, which reduces experimental error. For the same reason, the masses m_s and m_{ref} should be not too small. It is recommended to perform smoothing of the DSC curves prior to the c_p calculation, because $c_p(T)$ functions are usually smooth, except near second-order phase transitions (see Section 1.4.2.2 and Fig. 1.3). Besides, several subsequent heating cycles can be performed and then averaged. Here it turns out that occasionally the first curve of such series deviates significantly from the following ones, and then this curve should be disregarded.

DIN 51007/ASTM E1269/ISO 11357 recommend after the heating segment a second isothermal segment, which should last at least 10 minutes. The reason is that the (constant) baseline may reside on different levels in both isothermal segments, which can then be corrected.

It is possible to take just the smoothed $c_p(T)$ data as a final result, but usually a reasonable fit function can be used. It is typical that the slope of the function is high at low T around room temperature, and then the function becomes flatter. Maier and Kelley (1932)[8] showed that polynomials with term $\propto T^{-2}$ are often well suited for such fits, and such expressions are used also in commercial databases like FactSage 8.0 (2020) for the representation of specific heat capacities. In Fig. 2.10 a function

$$c_p(T) = a + b \cdot T + c \cdot T^2 + d \cdot T^{-2} \tag{2.6}$$

was used to well fit the experimental data; the fit parameters are given in the figure caption.

For a power compensated differential scanning calorimeter, Rudtsch (2002) determined the relative uncertainty of $c_p(T)$ as 1.5 %. For a heat-flux DSC device, Luisi (2014) reported slightly higher uncertainties ranging from 2.1 % (at 400 °C) to 4.8 % (at 1200 °C).

---

8 Kenneth Keith Kelley (16 December 1901–28 September 1991).

An alternative method for the determination of c_p works in isothermal segments with an overlay of periodic T variations (periods around 1 minute and amplitudes of several 0.1 K are typical). Also, from such time modulated DSC (TMDSC) measurements with a sample, compared to an identical measurement with a standard, the specific heat capacity can be derived. TMSDC measurements are especially useful for sensitive materials that cannot be heated over an extended temperature range, such as plastics. We will discuss details in Section 3.2.2.

2.4 Evolved gas analysis (EGA)

2.4.1 EGA: technical details

A significant negative TG step (see Section 2.2) occurs only if parts of the sample are evaporating. If the measurement is performed in a gas flow, then we can lead this flow completely or partially to a gas analyzer, such as an infrared spectrometer (IR) or a quadrupole mass spectrometer (QMS). Such procedures are called "Evolved Gas Analysis" (EGA). Both analyzing techniques have different benefits and drawbacks:

Quadrupole Mass Spectrometer. Work under vacuum ($\approx 10^{-5}$ mbar = 1 mPa), which makes high-vacuum equipment necessary. QMS measures the ratio between mass and charge (m/z) of molecules or molecule parts. Molecule parts or "fragments" are often formed if gas molecules entering the mass spectrometer are ionized. The few examples in Table 2.1 where extracted from a report by Mohler et al. (1952);[9] the strongest signal is always set to 100 %, and rather than the m/z value the corresponding chemical fragment is given. Such "fragmentation patterns" can be found for many inorganic and organic molecules, e. g., in the "NIST Chemistry WebBook" (NIST, 2020). This database takes into account not only chemical fragmentation, e. g., $H_2O \rightarrow OH + H$, which means that for water, besides $m/z = 18$, additional signals must be expected at least for $m/z = 17$ and $m/z = 1$. Additionally, the natural abundance of isotopes is accounted for. (See, e. g., the zinc isotopes in Fig. 3.9.) Principally, all atoms or molecules can be detected, but their ionization strength in the "ion source" is different; accordingly, quantitative measurements are not straightforward. Most devices offer two detection modes of the QMS: In the SCAN mode a certain m/z range is scanned again and again. This mode has the advantage that no unexpected species can be missed out, but, on the other side, the sensitivity is not very high in this mode, depending on noise level. Besides, if signals of the gas flow (Ar, N_2, ...) inside the device are included in the scanning range, then these very strong peaks (e. g., $m/z = 40$ for $^{40}Ar^+$, 20 for $^{40}Ar^{2+}$, 36 for $^{36}Ar^+$ (0.34 % abundance), 38 for $^{38}Ar^+$ (0.06 % abundance)) can be disturbing. Here helium can be a good alternative rinsing gas, because it gives only one

---

9 Fred Loomis Mohler (23 August 1893–2 December 1974).

Table 2.1: Relative strength of QMS signals for different fluorocarbons as measured with an ionization voltage of 70 volts. C_2F_4 shows additionally small signals for C_2 ($m/z = 24$), C_2F ($m/z = 43$), and C_2F_2 ($m/z = 62$).

Molecule	C	F	CF	CF_2	CF_3	CF_4	C_2F_3	C_2F_4	C_2F_5	C_2F_6
CF_4	7.8	6.7	4.9	11.8	100	0	0	0	0	0
C_2F_6	1.5	1.2	18.3	10.1	100	0	0	0.6	41.3	0.2
C_2F_4	12.6	2.8	100	29.5	2.8	0	63.1	33.8	0	0

strong signal at $m/z = 4$. In the MID mode ("Multiple Ion Detection") a predefined choice of m/z values is followed almost continuously by the QMS. Here a higher sensitivity is possible, but unexpected species can be missed out.

The low pressure inside the QMS makes the detection of species with small partial pressure easier, compared to IR spectrometers.

Infrared Spectrometer. Work under atmospheric pressure, like usually the thermal analyzer, is more affordable than QMS devices. IR spectrometers well discriminate organic and inorganic molecules by measuring their bond vibrations. However, they cannot measure single atoms, e. g., of evaporating elements or noble gases. Also, molecules of two identical atoms (N_2, O_2, Cl_2) are not IR active. Like for mass spectra, reference data can be found in "NIST Chemistry WebBook" (NIST, 2020).

For both types of analyzers, the gas has to be moved from the place where species are evaporating (this is typically the sample crucible inside the thermoanalytic device) to the analyzer itself. Usually, this is done by one of the following coupling systems:

Capillary. The gas outlet of the thermal analyzer is connected by a thin valve (steel, fused silica) to the analyzer. This valve ("capillary") has to be thin to reduce the dead gas volume responsible for a time shift between species emanation and registration. Besides, a large dead volume results in smeared tails of the intensity vs. time signal. Capillary coupling works very well for the registration of gases (H_2O, CO_2, SO_2, NO/NO_2,...). However, it may be critical for the registration of species that evaporate at high T but have low vapor pressure at low T. Often, the capillary is heated to a few hundred degrees to reduce condensation of such species.

Skimmer. A two-stage pressure reduction system that is often used for the coupling of the atmospheric pressure of the thermal analyzer to the high vacuum inside a QMS. It consists of two subsequent orifices with several 10 μm diameters holding an "intermediate vacuum" between them. This intermediate vacuum is of order 100 Pa and is regulated in such a way that the pressure inside the QMS is constant. The skimmer system has a small volume and is held almost at the sample temperature. Consequently, condensation of substances such as volatile metals, halides, oxides, or "big" organic molecules is prohibited or at least reduced. Skimmers are made of Al_2O_3 ceramic ($T_{max} \approx 1600\,°C$) or vitreous carbon ($T_{max} \approx 2000\,°C$).

2.4.2 EGA: examples

2.4.2.1 DTA/TG/QMS of calcium oxalate $Ca(COO)_2 \cdot H_2O$

Already in Section 2.1.2.2 the thermal decomposition of calcium oxalate hydrate (whewellite[10]) was described as a process with three subsequent steps where (1) water, (2) carbon monoxide, and (3) carbon dioxide are released from the sample. In the previous Fig. 2.5, this decomposition was performed in Ar and O_2 flows. Figure 2.11 shows the decomposition of $Ca(COO)_2 \cdot H_2O$ again in flowing Ar, but with simultaneous analysis of the emanating gas with a QMS, which was coupled to the STA with a skimmer. In this case the whole system of skimmer and protective tube between sample holder and furnace consisted of vitreous carbon. This carbon environment in flowing Ar creates strongly reducing conditions for the sample, which we will discuss in more detail in Section 2.6.2.

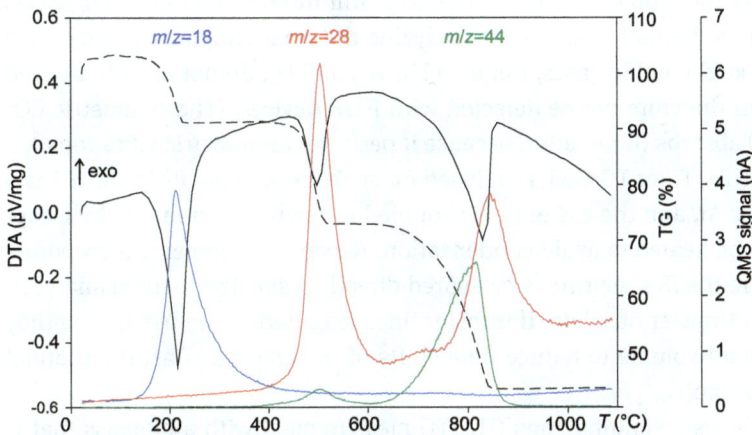

Figure 2.11: Simultaneous DTA (black full line) and TG (black dashed line) of 32.71 mg calcium oxalate hydrate $Ca(COO)_2 \cdot H_2O$ in Ar flow with QMS MID detection (Multiple Ion Detection; see Section 2.4.1) of the emanating gas. $\dot{T}$ = 10 K/min. QMS signals m/z = 18 (H_2O), m/z = 28 (CO), and m/z = 44 (CO_2) are shown. See Fig. 2.5 for comparison.

Figure 2.11 shows the TG curve (dashed line) and the QMS detector signals for H_2O, CO, and CO_2. The first TG step results from the release of water, and indeed the m/z = 18 signal has a strong maximum there. The second step, resulting from the release of carbon monoxide, is associated with the expected peak for the QMS signal m/z = 28. However, also a smaller peak for the m/z = 44 signal (carbon dioxide) appears there,

---

10 William Whewell (24 May 1794–6 March 1866).

which can be explained by the Boudouard[11] reaction

$$2\,CO \rightleftharpoons CO_2 + C, \tag{2.7}$$

which brings CO and CO_2 always in equilibrium. The third TG step due to the calcination of $CaCO_3$ under the release of CO_2 shows, as expected, an even stronger peak for the green $m/z = 44$ curve, but here the peak of the red $m/z = 28$ curve is comparably strong, resulting from the Boudouard reaction. The CO signal is enhanced partially by the excess of C, which is present from the vitreous carbon skimmer. If the same experiment would be performed with an Al_2O_3 skimmer or with a capillary coupling between STA device and QMS, then we could expect a weaker CO signal with this third TG step.

2.4.2.2 TG/FTIR of biomass

Instead of a mass spectrometer, a Fourier[12]-transform infrared spectrometer (FTIR) can be coupled to a thermal analyzer for analyzing the emanating gas. Symmetrical gas molecules, like the noble gases, but also N_2, H_2, and O_2, do not absorb infrared radiation and can therefore not be detected with FTIR devices. (The symmetric CO_2 molecule O=C=O absorbs IR radiation because it performs asymmetric vibrations.)

Usually, the DTA/TG or TG device is rinsed for such measurements by an IR inactive gas, like N_2 or Ar, and the gas outlet is coupled to the IR spectrometer by a capillary, which can be heated to avoid condensation. Newer developments incorporate the "gas cell" were the IR spectrum is measured directly inside the thermal analyzer, which makes gas transfer obsolete. If transfer lines (capillaries) are used, then they should have a small volume to reduce time shifts of IR signals against the thermal processes in the sample.

Figure 2.12 shows a simultaneous DTA/TG measurement with a biomass that is degraded and burned in flowing air. The initial small TG step $\Delta m/m < 10\,\%$ results obviously from humidity of the organic material. The degradation itself takes place from ca. 300...500°C in the second and third TG steps. Both are separated by a kink in the TG curve, which is connected with a dip between two exothermal DTA peaks.

During the TGA-FTIR measurements, spectral data are repeatedly collected. It is natural that at strong TG steps, where much gaseous products are released from the sample, also the FTIR spectra show strong overall absorbance. This can be used to build a Gram[13]–Schmidt[14] reconstruction, which is shown for the same measurement

---

11 Octave Leopold Boudouard (10 May 1872–15 December 1923).

12 Jean-Baptiste Joseph Fourier (21 March 1768–16 May 1830).

13 Jørgen Pedersen Gram (27 May 1850–29 April 1916).

14 Erhard Schmidt (13 January 1876–6 December 1959).

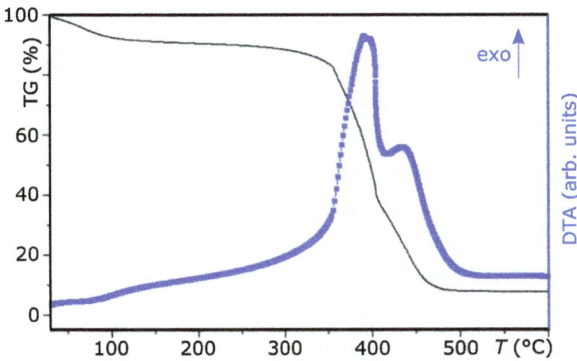

Figure 2.12: DTA/TG curves of an activated carbon sample that was produced from a biomass. Measurement in air flow. Figure adapted from de Oliveira et al. (2017) under CC BY 4.0. See also Fig. 2.13.

in Fig. 2.13. Each point of such a plot corresponds to the total IR absorbance of the evolved components in the measured spectral range.

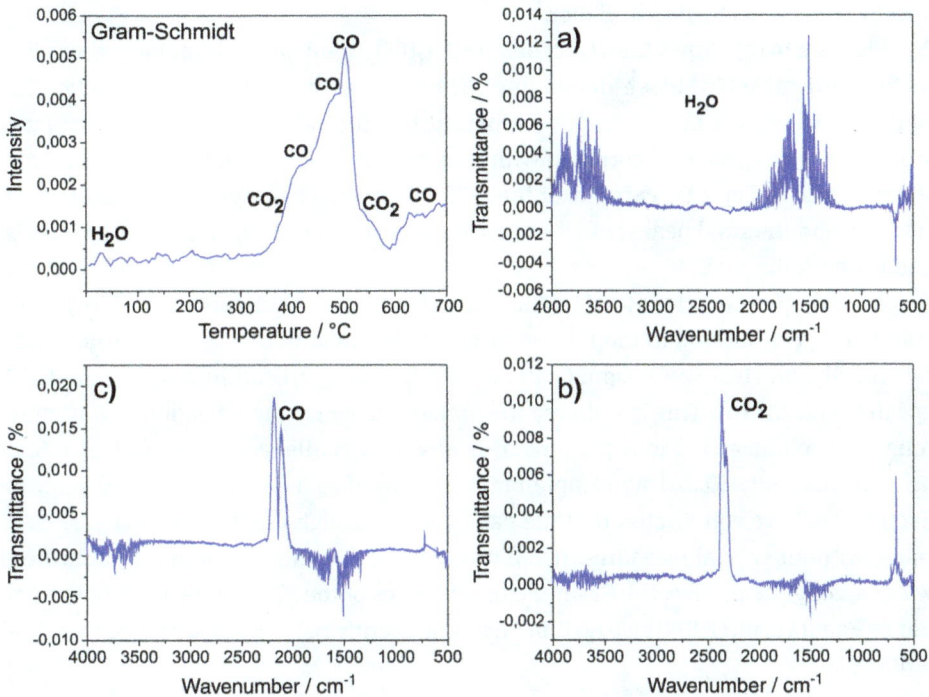

Figure 2.13: Gram–Schmidt curve for the same material that was used in Fig. 2.12, together with FTIR spectra of the main gaseous products. Figure adapted from de Oliveira et al. (2017) under CC BY 4.0.

2.4.2.3 TG/FTIR of lanthanum carbonate hydrate $La_2(CO_3)_3 \cdot nH_2O$

Not only the oxides of metals from the first and second main groups of the periodic system posses strongly (Lewis[15]) basic properties; the basicity of the rare-earth (RE) metal oxides is on a comparable level. This basicity leads to significant affinity to Lewis acids like carbon dioxide under the formation of carbonates. Besides, they tend to absorb water; Bernal et al. (2004) report a broad variety of intermediate phases with partially noninteger stoichiometry in the systems $RE_2O_3–H_2O–CO_2$.

To study the basicity, Maitra (1990) performed simultaneous DTA-TG measurements of the rare-earth carbonates $RE_2(CO_3)_2$ in a Rigaku Thermoflex PTC-10A device with heating rate of 10 K/min up to 1250 °C in Pt crucibles. Measurements were performed in flowing N_2 and flowing CO_2. The $RE_2(CO_3)_2$ were obtained preliminary by precipitation from aqueous $RE(NO_3)_3$ solutions with Na_2CO_3. The sequence in decreasing basicity found by Maitra (1990) is La > Pr ≈ Nd > Sm > Gd ≈ Eu > Tb ≈ Ho ≈ Er > Dy ≈ Tm ≈ Yb ≈ Lu > Ce.

Paama et al. (2003) investigated the emanation of gaseous species from commercial RE carbonate hydrates $RE_2(CO_3)_3 \cdot nH_2O$ (RE = La, Eu, Sm) with a thermogravimetric analyzer TGA-7 (Perkin-Elmer) that was coupled via steel capillary to a FTIR spectrometer, also by Perkin-Elmer.

Figure 2.14 (a) shows the TG results for $La_2(CO_3)_3 \cdot nH_2O$ (full line) together with its first derivative DTG (dash-dotted line). The calculation of the DTG is often helpful to decide where the mass loss per temperature unit (and hence time) has maxima, which can be expected to correlate with maxima for specific emanating gases. Even if not shown in Fig. 2.14, extrema of the DTG curve often also correspond to extrema (usually endothermal peaks) of DTA curves, because the release of gaseous species is usually endothermal.

A small TG effect near 100 °C (which can be seen more clearly in the DTG) is attributed by the authors to the release of water adsorbed to the sample furnace. Unfortunately, no clear water signal can be seen in this temperature range in the FTIR spectra, Fig. 2.14 (b). This is not very uncommon, because the detection of a ubiquitous molecule like H_2O in a gas stream is always a challenge. We should take into account that only 10...20 mg sample mass were used in a gas flow of 80 ml/min. In Section 2.6.2, we will discuss that the gas flowing through the thermal analyzer contains, without special measures, usually several ppm (parts per million) impurities, water among them. Also, the Gram–Schmidt curves of the FTIR data shown in the paper by Paama et al. (2003) give no hint on a significantly rising release of species below ≈ 400 °C.

---

15 Gilbert Newton Lewis (23 October 1875–23 March 1946).

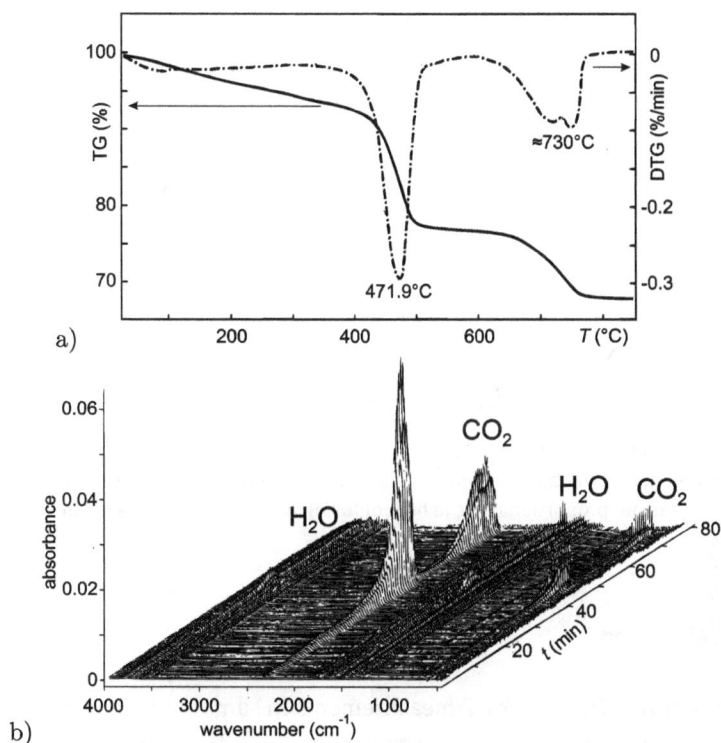

Figure 2.14: Thermogravimetry of commercial lanthanum carbonate hydrate (Fluka, 99.9%) with simultaneous EGA by FTIR spectrometry. (a) TG and DTG measured with a Perkin–Elmer TGA-7 analyzer, $\dot{T}$ = 10 K/min in 80 ml/min flowing N_2. (b) Simultaneous FTIR spectra vs. time, measured with a Perkin-Elmer System 2000. The end of the time scale (80 min) corresponds to 800 °C. Figures adapted with permission from Paama et al. (2003).

2.5 Calibration of thermoanalytic devices

The calibration of thermoanalytic devices, and especially of sample holders for DTA, DSC or TG measurements described in the previous Sections 2.1–2.3, has different purposes:
- It may help to improve the accuracy of measured data (e. g., T).
- It can be the precondition for determining the "sensitivity" of a sample carrier, which allows us to obtain heats of transformation ΔH from measured peak areas.
- It can improve the sharpness of peaks by numerical desmearing of DSC curves.

At least, temperature calibration of DTA and DSC sample carriers (Sec. 2.5.1) is always highly recommended. Calibration of heat flow DSCs for sensitivity (Sec. 2.5.2) is mandatory. Thermal resistance calibrations (Sec. 2.5.3) can usually be regarded as optional. Often, these calibrations can be performed on the basis of the same measurements, like shown in Fig. 2.15.

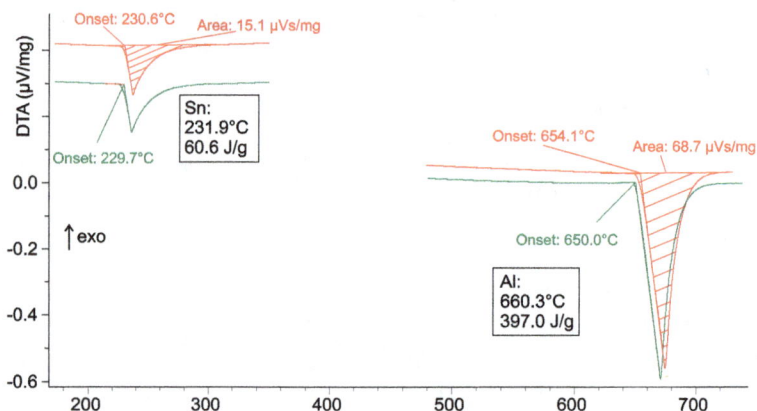

Figure 2.15: Calibration measurements of a DTA/TG sample holder with melting tin and aluminum, heating rate $\dot{T} = 10\ \text{K/min}$. Red: first heatings; green: second heatings. Onsets are the experimental melting points T_{exp}; peak areas are proportional to the heat of fusion ΔH_{f}. Literature data given in the boxes.

2.5.1 Temperature calibration

Thermocouples (see Section 1.3) used for T measurements in sample carriers are not always identically manufactured, and the thermovoltages may be slightly different from the data given in Table 1.4. Besides, aging may occur especially after chemical contamination. The influence of different crucible types, of atmosphere, and of heating rate are other possible origins for slightly faulty T determination with the thermoanalytic equipment.

Consequently, new thermocouples should always be calibrated to account for such incorrect temperature measurements. In the ideal case, several calibration substances (for some examples, see Table 2.2) are measured under the same conditions as the samples under investigation. These calibration substances undergo at known temperatures a well-known thermal effect (melting, phase transition), which is measured. The difference between the measured and expected transformation temperatures gives the experimental error, which is used to correct the results obtained with further samples. Fortunately, the influence of some parameters such as atmosphere and even crucible material is usually small compared to the influence of new (or aging) thermocouples. Thus a few calibration runs can serve for all measurement tasks.

If the calibration substance is placed in the crucible, then it has usually an arbitrary "as supplied" shape: wires or spheres are typical for metals, powders for inorganic compounds. If the calibration sample melts and solidifies subsequently, then it usually assumes a geometry defined by the crucible bottom. This often results in better thermal contact between crucible and sample and, consequently, in a slightly different shape of the peak, like shown in Fig. 2.15 for grains of tin and aluminum. It

Table 2.2: Data of some calibration substances used for DTA and DSC. ($BaCO_3$ shows the second and third transitions at higher T, which are not well suited for calibration because they are very close to the decomposition temperature to BaO and CO_2.)

Substance	transformation	T (°C)	ΔH (J/g)
Hg	melting	−38.8	11.44
H_2O	melting	0.0	333.4
Ga	melting	29.8	80.0
KNO_3	phase transition	128.7	50.0
In	melting	156.6	28.6
Sn	melting	231.9	60.6
Zn	melting	419.6	107.5
K_2CrO_4	phase transition	668.0	37.0
$BaCO_3$	phase transition	808.0	94.9
Ag	melting	961.8	104.6
Au	melting	1064.2	63.7
Ni	melting	1455.0	290.4
Pd	melting	1554.8	157.3
Pt	melting	1768.0	165.00
Rh	melting	1962.0	≈199
Ir	melting	2446.0	≈107

is recommended to use second heating curves for the calibration if during the subsequent thermal analysis runs, also repeated heating/cooling cycles are performed. For expensive but long-term stable calibration substances like gold or palladium, it is possible to store them inside their crucibles for reusing them in future calibration runs.

Usually, the control software of thermal analyzers offers the possibility to create "calibration files" by entering experimental melting points T_{exp} for a series of at least 2–3 calibration substances in the kind of Table 2.3. The T_{exp} are compared with the nominal values T_{nom}, and a (possibly weighted) fitted polynomial is used to approach the differences. Such a polynomial (2.8) is shown together with calibration points in Fig. 2.16.

$$\Delta T = c_0 + c_1 T_{exp} + c_2 T_{exp}^2 \tag{2.8}$$

The conditions of calibration should be as similar as possible to the real measurements that will be performed on their basis. Unfortunately, this recommendation can sometimes not be fulfilled completely: Melting of calibration metals in metallic crucibles (e. g., Au in a Pt crucible) is sometimes impossible and often results in alloying and destruction. Besides, some other calibration metals are subject of oxidation if heated in an atmosphere containing O_2; sometimes even oxygen traces from leaks or the gas flowing through the equipment are sufficient to prevent reliable measurements.

An unorthodox example for the simultaneous calibration of temperature and sensitivity was demonstrated in Fig. 1.52. The calibration substance, in this case Y_2O_3, was

Table 2.3: Example of a temperature calibration table with mathematical weight put on each point.

Substance	T_{nom} (°C)	T_{exp} (°C)	Weight	T_{corr} (°C)
dummy	0	1	1.000	0.0
Sn	231.9	232.6	0.500	231.9
Zn	419.6	419.2	1.000	419.2
Al	660.3	660.0	1.000	661.3
Au	1064.2	1057.8	1.000	1062.5
Ni	1455.0	1449.8	0.700	1459.3
Pd	1554.8	1542.0	1.000	1552.8

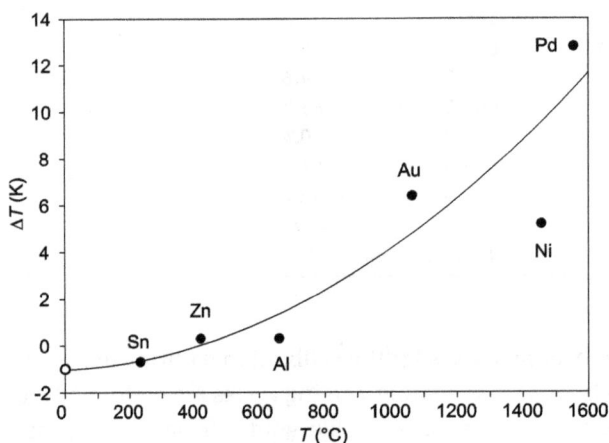

Figure 2.16: Calibration curve for a DTA/TG sample carrier with 6 calibrated points and a fitted correction polynomial (2.8) with parameters $c_0 = -1.0385$, $c_1 = 5.45 \times 10^{-4}$, $c_2 = 4.621 \times 10^{-6}$. The "dummy" point at 0 °C has no practical relevance and leads just to smoother fitting for low T.

placed during a measurement in the reference crucible, and the sample in the sample crucible. Then, of course, melting of the reference substance produces a DTA peak in the exothermal direction.

2.5.2 Sensitivity calibration

Let us assume that during a DTA or DSC measurement, the DTA or DSC curve that is plotted according to the ASTM norm (see info box on p. 2) bends upward. This indicates that T_s rises slightly faster than T_r, indicating an exothermal process in the sample. Vice versa, a downward bend indicates an endothermal process, like the melting peaks in Fig. 2.15. It turns out that the deviation from the undisturbed "baseline" outside the melting peak is proportional to the thermal power $W(t)$ (in mW) that flows in this moment into the sample. This power flows over a certain time period, and the area under the peak (in J)

$$A = \int W(T)\,dt \qquad (2.9)$$

is proportional to the consumed heat energy Q, which in this case is the heat of fusion ΔH_f.

Estimate the lengths of the melting peaks in Fig. 2.15 on a time scale!　　　　　?

However, only for power compensation DSC instruments (Fig. 2.1(c)), A and Q are identical. For DTA and heat flow DSC instruments, the measured signal is the thermovoltage ΔU (in μV), which results from the temperature difference between **S** and **R**. With data like those given in Table 1.4, ΔU can be converted into a temperature difference ΔT (in K), but the direct determination of Q is not possible. Rather, for this conversion, the factor

$$K_s(T) = \frac{A}{Q} \qquad (2.10)$$

is required, which is called the sensitivity of the thermoanalytic setup; it has the dimension [V/W] and describes the efficiency of the setup in creating a DTA signal (in μV) from a given thermal power W. As indicated in equation (2.10), K_s is a function of temperature and drops usually for higher T for the following reasons (Fig. 2.17):
- To create a DSC (or DTA) signal in the sensor, heat must flow between the sensor and the substance through the crucible. However, the thermal conductivity drops for most materials for higher T, approximately above the Debye[16] temperature.
- Total electromagnetic radiation, including infrared radiation, rises with T following the T^4 law. Consequently, then a significant share of Q is exchanged by radiation and does not contribute to the sensor signal.

It is convenient to determine several calibration points $K_s(T)$ with the same measurements that were already used for temperature calibration. In Fig. 2.15, besides the onsets (for T calibration), also the melting peak areas A were determined for both samples. The measured peak areas A (see, e. g., Table 2.4) are divided by literature data for the corresponding heats of transformation H, and by equation (2.10) we obtain K_s(exp) for the given T. A fit can be performed by useful functions of the kind

$$K_s = (c_2 + c_3 z + c_4 z^2 + c_5 z^3)\exp[-z^2]$$
$$\left(z = \frac{T - c_0}{c_1}\right). \qquad (2.11)$$

---

16 Peter Debye (24 March 1884–2 November 1966).

Table 2.4: Example of a sensitivity calibration table for a DSC sample carrier with mathematical weight put on each point. The fit parameters for this calibration with (2.11) are $c_0 = 20.0, c_1 = 1301.56, c_2 = 1.0585, c_3 = -0.59454, c_4 = -0.30376, c_5 = 0.70568$.

Substance	T (°C)	H (J/g)	A (µVs/mg)	K_s(exp) (mV/W)	Weight	K_s(fit) (mV/W)
dummy	20.0	−100.0	−105.00	1.050	1	1.058
In	156.6	−28.6	−28.33	0.991	1	0.983
Zn	419.6	−107.5	−87.50	0.814	1	0.790
Al	660.3	−397.0	−224.8	0.566	1	0.610
Au	1064.2	−63.7	−27.05	0.425	1	0.394
Ni	1455.0	−290.4	−83.18	0.286	1	0.290
Pd	1554.8	−157.3	−41.52	0.264	1	0.272

(The "dummy" value in Table 2.4 guarantees a monotonous falling function $K_s(T)$ but is not of great relevance, because reliable enthalpy measurements are only possible if the heating rate $\dot{T}$ has reached a constant value, which for DSCs starting from room temperature is hardly possible below ca. 100–120 °C.)

> **!** DSC devices have not only benefits, because the more sophisticated construction (partially radiation shields, larger contact area between flat crucible bottom and sensor) makes them also more delicate with respect to chemical or mechanical damage. Another point is a typically smaller sample volume inside DSC crucibles. This is beneficial in terms of peak sharpness (see Section 2.5.3), but if the components of the sample tend to evaporate, then the concentration shift inside small DSC crucibles is often larger than with larger DTA crucibles.

Figure 2.17 shows on top the DSC sensitivity function $K_s(T)$ obtained from the data in Table 2.4 together with $K_s(T)$ for a DTA sample holder calibrated under identical experimental conditions. The significantly higher sensitivity of the DSC holder is obvious, which is a severe benefit if small effects have to be analyzed. Due to the significantly higher sensitivity of DSC carriers, they are preferred if the magnitude of thermal effects (ΔH, c_p) is under investigation.

Usually, DSC devices reach around 1400 ... 1600 °C their maximum working temperature, because then thermal radiation becomes too strong, and the sophisticated construction of DSC carriers becomes unstable. Beyond, DTA has to be used. It becomes even more difficult to detect small thermal effects for higher T, because the negative sensitivity slope from Fig. 2.17 continues. This is even worse if the maximum working temperature of sample holders with Pt/Rh-based thermocouples (type B, $\approx$ 1750 ... 1800 °C, Table 1.4) is exceeded. Then tungsten-based sample holders with W/Re alloy thermocouples are used. Table 1.4 shows that their thermopower is much lower than for Pt/Rh sensors, which leads to even lower sensitivity. Besides, reasonable calibration for T and K_s is there almost impossible: tungsten-based sample holders are preferably used with W crucibles, because the standard crucible material Al_2O_3 melts already at 2054 °C. However, all certified calibration materials for high

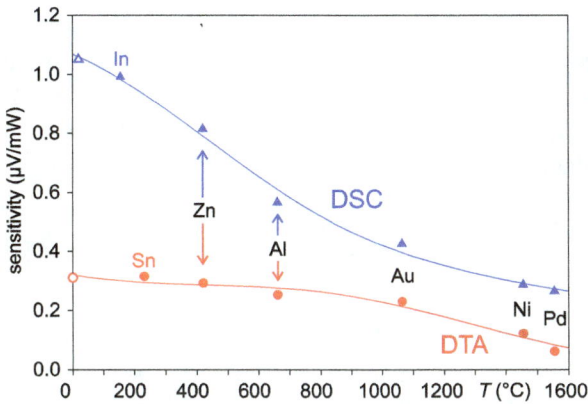

Figure 2.17: Sensitivity calibration curves of DSC and DTA sample holders in an identical thermal analyzer (NETZSCH STA 449 "F3" with rhodium furnace, Al_2O_3 crucibles, Ar atmosphere, $\dot{T}$ = 10 K/min).

$T \geq 800\,°C$ are metals that can hardly be molten inside a metal crucible, because alloying may occur.

Figure 2.18 demonstrates the c_p method, which is often a good alternative for sensitivity calibration. It relies on subsequent heating runs without sample (gives the "baseline" of the device) and with a suitable and well-defined standard sample. This can, e. g., be a slice of sapphire (crystalline α-Al_2O_3, more exactly called corundum). In Fig. 2.18, baseline measurement and sample measurement were performed four times to confirm reproducibility (which is obviously given). Then we can take the average of the four curves for further calculations.

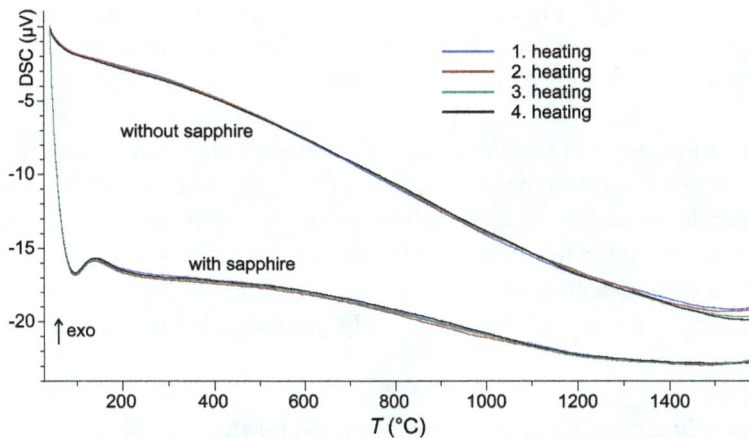

Figure 2.18: The c_p method for sensitivity calibration $K_s(T)$. Top: Four successive DSC curves without sample for the determination of the baseline. Bottom: Four successive DSC measurements with the same experimental setup under identical conditions but with a sapphire slice as sample.

The sample measurements ("with sapphire") are shifted to the endothermal direction, compared to the baseline. This results from the additional thermal load on the sample side. The mass of the standard sample $m^{(st)}$, the specific heat capacity of the standard $c_p^{(st)}$ and the heating rate $\dot{T} = \partial T/\partial t$ of the measurements are known, and we obtain the sensitivity curve from the calculation

$$K_s(T) = \frac{|DSC^{(st)}(T) - DSC^{(basis)}(T)|}{m^{(st)} c_p^{(st)}(T)\dot{T}}. \tag{2.12}$$

($DSC^{(st)}(T)$ and $DSC^{(basis)}(T)$ are the DSC curves with or without the standard sample). The c_p method for sensitivity calibration has the advantage that it can be performed with all crucibles and up to very high $T \approx 1600\,°C$ without the risk of damage.

! Instead of real calibration measurements, we can perform in the high-T range beyond 2000 °C occasional runs with melting Al_2O_3 inside W crucibles. For Al_2O_3, $T_f = 2054\,°C$ and $\Delta H_f = 118.41\,kJ/mol =$ 1161 J/g are reliable literature data from FactSage 8.0 (2020).

In the ideal case, all calibration measurements should be performed under the same experimental conditions (crucible, heating rate, atmosphere, lid or no lid) like the subsequent measurements with substances under investigation. With the arguments given above, like partial incompatibility of Pt crucibles with molten metals, this is sometimes impossible. Different possibilities can often solve this problem, with not too severe impact on the functions $K_s(T)$:

-   Calibration is performed with metals in Al_2O_3 crucibles, even if samples like some molten oxides or halides must be measured in Pt crucibles.
-   It is possible to perform the T calibration with uncritical Al_2O_3 crucibles, because T calibration depends not very much on the crucible material. Sensitivity calibration can independently be performed in Pt crucibles with the c_p method described above Fig. 2.18, e. g., with a crystalline Al_2O_3 plate ("sapphire") as a calibration sample.
-   Calibration is performed in Pt crucibles, but a thin inner Al_2O_3 "liner" avoids direct contact between crucible and sample. Also, a thin layer of Al_2O_3 powder can be used to separate the molten metal from the crucible, but with care.
-   As a general rule, we can consider to protect DSC sample holders from sticking together with metallic crucibles by placing a thin ($\ll$ 1 mm) flat liner or washer made of an almost inert material (often, again Al_2O_3 ceramic) below the crucible bottom.

Figure 2.19 compares functions $K_s(T)$ that were measured for the same DSC sample holder with type S thermocouples (see Table 1.4) in a NETZSCH STA 449 "F3". Only the blue curve was measured with metallic calibration samples that were molten in Al_2O_3 crucibles, like demonstrated in Table 2.4. The other three curves were obtained

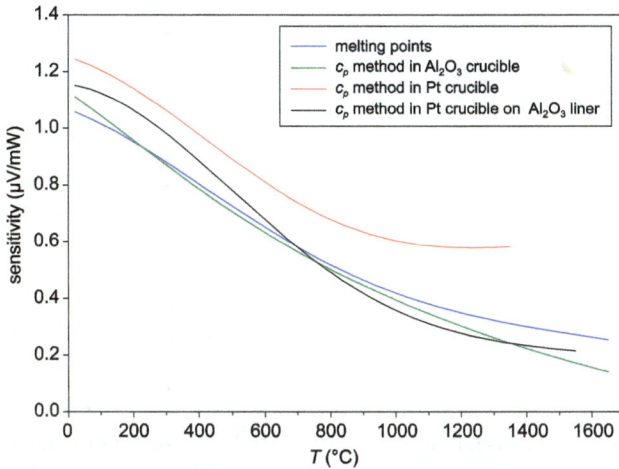

Figure 2.19: Four calibration curves for the same DSC sample holder obtained with identical heating rates of 10 K/min in argon flow and with different types of crucibles and determination (see the text).

with the c_p method under slightly different conditions, like indicated in the figure legend. The blue $K_s(T)$ curve obtained from melting points is similar to the green curve obtained using the c_p method with the same crucibles.

It is interesting to note that the red $K_s(T)$ curve, obtained by the c_p method in Pt crucibles, is always higher; here, obviously, the opaqueness of Pt for infrared radiation is beneficial, because heat losses by infrared radiation are omitted. As mentioned above, such arrangement is somewhat risky, because during long-term measurements sticking of the crucible on the sample holder may occur. An additional thin (0.2 mm) Al_2O_3 washer between crucible can prevent sticking but also slightly reduces the sensitivity to similar values like for the other arrangements (black curve).

2.5.3 Thermal resistance calibration

A change of the sample mass during thermal analysis can be immediately detected, because the mechanical connection from the sample through crucible and sample holder to the balance is rigid. Figure 2.20 shows that the situation is different for thermal signals. If a thermal event produces or consumes heat at a time t, then this heat energy must diffuse through parts of the sample (at least, for heat produced in the center) through the crucible and (at least, for heat flow DSC's) through the "disk" to the sample thermocouple. Heat conduction is a slow process, and thus the time constant τ for the delay when the heat arrives at the sensor can easily be in the order of several 10 s up to 1 min.

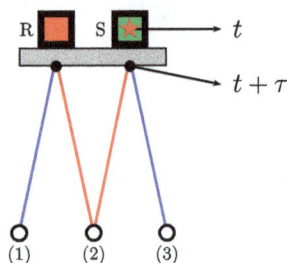

Figure 2.20: If at a time t a thermal event happens inside the sample, then the signal appears at the sample thermocouple at a later time $t + \tau$. The time constant leads to a convolution (smearing) of the DSC signal.

Of course, even a sharp thermal pulse from inside the sample Φ^{sample} would not arrive after the time lag τ as a sharp pulse at the sensor. Rather, the peak is smeared out by the convolution integral

$$\Phi^{\text{meas}} = \Phi^{\text{sample}} * g = \int_0^t \Phi^{\text{sample}} g(t - t') \, \mathrm{d}\, t', \tag{2.13}$$

where g is a device-dependent function, and $*$ is the convolution operator. For a detailed description how the convolution and the reverse process deconvolution can be quantitatively described, we refer, e. g., to Höhne et al. (1996), Moukhina and Kaisersberger (2009). However, subroutines to perform a calibration routine with the effect that the deconvolution of experimental DSC curves can be automatically performed in subsequent measurements are implemented in some control software products for thermoanalytic devices. Here we will demonstrate the effect of thermal resistance and the resulting convolution of DSC and DTA signals with a practical example.

Figure 2.21 shows the heating and subsequent cooling of an Au sample. The onset of the melting peak is very close to the expected $T_{\text{f}} = 1064.2\,°\text{C}$ from Table 2.2. However, during cooling, crystallization does not occur at this equilibrium temperature; rather, we can see "supercooling". Supercooling is not very seldom, because the solidification is a process that requires the formation of crystal seeds, which again requires some activation energy. If the liquid sample finally crystallizes at 1039.4 °C, then it is already significantly out of equilibrium. Consequently, it "jumps" almost immediately into its solid equilibrium state, which we can see from the almost vertical rise of the crystallization peak.

The same crystallization (cooling) peak is shown in Fig. 2.22 on a time scale. The heat of crystallization that was the origin of this peak was produced inside the sample almost as a heat flash T_{s}. Nevertheless, the DTA signal smoothly returns to the baseline. Such a natural cooling curve can be described by a single exponential, or for more complicated situations, with double exponential functions of the

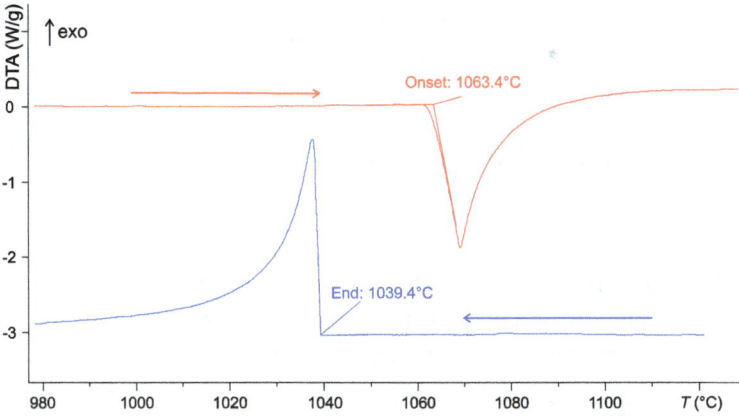

Figure 2.21: DTA heating and cooling curves of a gold sample ($\approx$ 50 mg) in a NETZSCH STA 409 with T and sensitivity calibrated DTA/TG sample carrier type S. Heating/cooling rate was ±20 K/min.

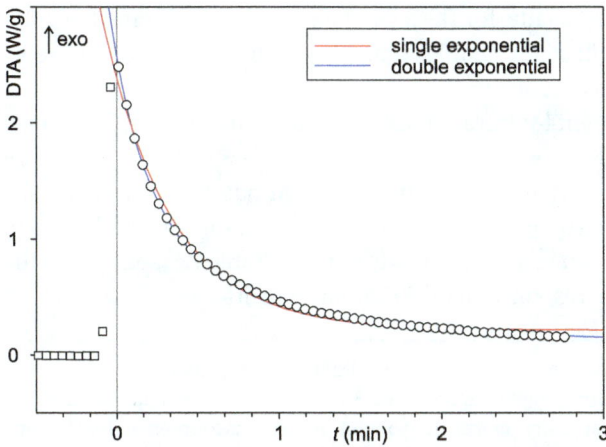

Figure 2.22: The cooling peak from Fig. 2.21 on a time scale (dots) with single exponential, eq. (2.14), $y_0 = 0.210, a = 2.1671, b = 2.3616$ and double exponential, eq. (2.15), $y_0 = 1.235\times10^{-5}, a = 1.289, b = 5.006, c = 1.150, d = 1.237$ fits to its decay.

kind

$$f(t) = y_0 + a \exp(-bt), \tag{2.14}$$

$$f'(t) = y_0 + a \exp(-bt) + c \exp(-dt), \tag{2.15}$$

where the parameters b and d represent inverse time constants. The red and blue fit curves in Fig. 2.22 show that already a single exponential fit gives a good approximation for the experimental data, and from the fit we obtain the relation $\tau = b^{-1} = 0.4234$ min ≈ 25 s. For special applications and extreme thermal resolution, DSC sensors can be constructed with silver parts. Then time constants $\tau < 1$ s can be reached, because Ag has the highest thermal conductivity among all metals.

2.6 Technical conditions for thermal analysis

2.6.1 Crucibles for thermal analysis

Samples for DTA, DSC, and TG measurements have to be placed inside crucibles. The choice of crucible materials and shapes is huge for comparably low temperatures $T_{max} \lessapprox 500$ °C, and then often metal crucibles made of aluminum or sometimes Al_2O_3-based ceramics are chosen. (However, the latter can be also used for significantly higher $T_{max} \leq 1650$ °C.) Al crucibles have several advantages:

- Al crucibles are cheap, which means that they have not to be necessarily cleaned after the measurement. This is important if, e. g., the sample melts or bakes so tight to the crucible bottom that it cannot be easily removed.
- Many Al crucible types (especially, for DSC) can be closed with clamped or cold-welded lids, which is beneficial for samples that are sensitive against constituents of ambient air, like oxygen or moisture.
- Al crucibles are intransparent for infrared radiation, in contrast to Al_2O_3 ceramic (see the notice box below).
- Al crucibles can be made very light, because the metal has a low mass density (2.7 g/cm^3) and can be formed into thin crucible walls. This, together with its high thermal conductivity (> 3 times higher than for Pt; see Table 2.6), enhances the sensitivity and reduces the response time of thermal sensors.

! At least for all DSC measurements, covering the crucible by a lid is recommended, especially if high accuracy is needed like for c_p measurements (see Section 2.3.2.2). Metallic lids should be used for such purposes because Al_2O_3 ceramics are almost transparent in the near infrared spectral region (see Krell et al., 2003), which is the maximum of thermal emission for typical temperature regions of thermal analysis. Also through Al_2O_3 crucible walls heat losses by radiation occur, especially for high $T \gg 1000$ °C, but also then metallic lids (e. g., from Pt) are recommended.

The crucibles have to fulfill the following requirements, which significantly restrict the choice of materials:

1. It is indispensable that the crucible withstands the maximum temperature T_{max} of the measurement. T_{max} must be significantly lower than the melting temperature T_f of the crucible material, because most materials become softer and start to "creep" already several 100 K below T_f.
2. The crucible must be stable also under the often elevated temperatures of the measurement against the atmosphere in the thermal analyzer. As an example, graphite crucibles are quickly oxidized if heated in an atmosphere that contains oxygen. (We discuss this topic in more detail in Section 2.6.2.)

3. The crucible must not react with the sample. For instance, it is often dangerous to heat metal samples in metal crucibles such as platinum. Already below the T_f of crucible metal and sample, both can interdiffuse and form alloys (compounds or solid solutions). Especially, platinum is very sensitive with respect to some non-metallic elements: Under reducing conditions, it forms eutectics with low T_{eut}. Lupton et al. (1995) report some low melting eutectics for several systems: Pt–As (597 °C), Pt–Si (830 °C), Pt–P (588 °C), Pt–S (1240 °C). Under oxidizing conditions, these "Pt poisons" (also Bi, Sb, and some other elements belong to this group) are converted to uncritical arsenates, silicates, or other salts.

Figure 2.23 shows stability curves for several crucible materials with respect to temperature and oxygen. The latter might either be an intentional component of the gas flow supplied during thermal analysis, or it can be present as an impurity in the "inert" gas like Ar or N_2. The curves for Cu and Ni end at a circle symbol that marks the melting point T_f of the metal. (The melting points for the other crucible materials are out of the scale.) The diamonds below T_f mark the upper limit where crucibles of the corresponding material can be used. A broader choice of crucible materials can be found in Table 2.5.

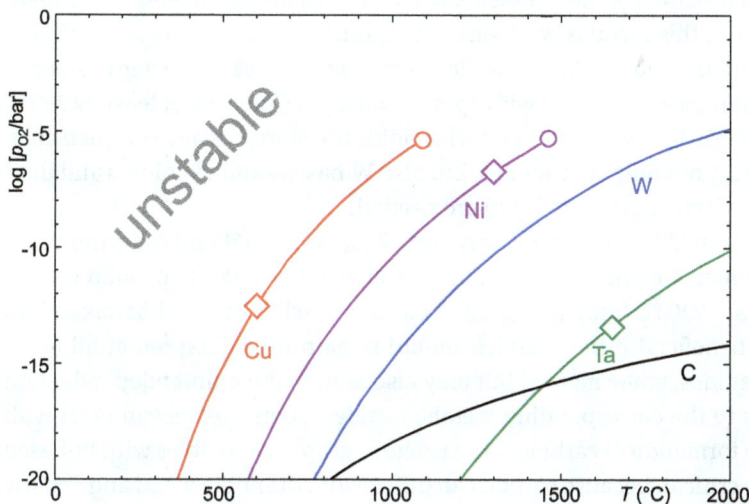

Figure 2.23: Stability limits with respect to atmospheres containing O_2 of several crucible materials for thermal analysis. Above the curves, crucible materials can be destroyed by oxidation. Circles mark melting points, and beyond the diamond symbols, crucibles tend to become soft and mechanically unstable.

Not solely T is a parameter that restricts the conditions where materials can be used, e. g., for crucibles; rather, it is a combination of T and the surrounding gas, which is represented here by its oxygen partial pressure p_{O_2}. Figure 2.23 is a predominance diagram; we will discuss in more detail this type of phase diagrams in Section 2.8.3. For each element, the lines separate a phase field where the element itself is stable (bottom right) from phase fields where one or more oxides of this element are stable (top left).

Certainly, thermodynamic equilibrium is not always reached; this is the reason why all materials shown in Fig. 2.23 can exist metastable at room temperature in ambient air. However, already then these materials are usually covered by a thin oxide layer. The reaction rate becomes significantly larger upon heating, and at sufficiently high T, equilibrium is usually reached. It is not possible to define an upper temperature limit for the metastability of materials, which is generally valid. Rather, such a limit depends on the chemical environment and especially on the grain size and hence also on the surface quality. As an example, iron is obviously metastable under ambient conditions for centuries, because we find it in old buildings and weapons. Nevertheless, fine-grained iron oxidizes in air spontaneously under self-heating (it is "pyrophoric"; see Galwey and Gray (1972)[17,18]).

All crucible materials shown in Fig. 2.23 are stable only under very low $p_{O_2} \lessapprox 10^{-5}$ bar, which means that the atmosphere must be virtually free of oxygen, unless T is so low that the crucible is still metastable. Tungsten crucibles can be used up to 2400 °C, which makes them almost indispensable for DTA measurements at the highest T. Unfortunately, this metal is very sensitive against oxidation. Consequently, the atmosphere during thermal analysis should be as "clean" (free of oxygen) as possible. Repeated pumping and refilling with "pure" noble gas (Ar or He, at least 99.999 % purity, better 99.9999 %) is recommended. This holds the more, because for such measurements, certainly not only W crucibles, but also W-based sample holders and thermocouples (e. g., of type D; see Table 1.4) are needed.

Also graphite crucibles withstand very high $T_{max} \approx 2400$ °Cand have the benefit that they are much cheaper. However, the extremely low stability of solid carbon against oxygen (at 1500 °C, only $p_{O_2} \lessapprox 10^{-14}$ bar are possible) should be taken into account. This is beneficial if the samples should remain free of oxygen at all (e. g., halides, chalcogenides, some metals) but may also lead to the unintended reduction of oxide samples to the corresponding metals. Besides, some metals can react with carbon under the formation of carbides. The usage of graphite crucibles with Pt-based thermocouples is critical, because carbon can diffuse into the noble metal and destroy it, or at least it can interfere with the thermocouple and change its Seebeck coefficient (see Section 1.3.1). If such combinations of Pt with C are necessary, then between subsequent measurements the sample holder should be heated frequently to $\gg 1000$ °C in flowing oxygen or air, which can help to restore the initial quality of the thermo-

---

[17] Andrew Knox Galwey (born 13 March 1933).
[18] Peter Gray (25 August 1926–7 June 2012).

Table 2.5: Some crucible materials for thermal analysis. Remark: Alloying Pt by typically 5% Ir, Au, or Rh strengthens the material, without significantly changing its melting temperature.

Material	T_{max} (°C)	Remarks
different steels	200–1000	also sealed for high pressures
Titanium	≈500	also sealed for high pressures
Aluminum	≈600	especially for DSC with organics
Silver	700	very quick response, stability against strong bases
Copper	700	for oxidation induction time (OIT) measurements
Gold	≈900	high chemical stability
Nickel	≈1300	stable against alkaline melts
Pt	1600	stable against most atmospheres and samples
Pt/Ir, Pt/Au, Pt/Rh	1650	stable against most atmospheres and samples
Tantalum	1650	stable against N_2 and most nitrides
Tungsten	2400	sensitive against O_2
Alumina ceramic	1650–1700	stable against most substances including metals
Duran glass	≈550	can be sealed as ampoule
Quartz glass	≈1100	can be sealed as ampoule
MgO ceramic	≈2100	stable against metals and oxide melts with CuO
Graphite	2400	sensitive against O_2 and some metals
Zirconia ceramic	2000	stable against most metals

couples. Combinations of graphite crucibles with W-based sample holders and thermocouples are not critical.

For all conditions shown in Fig. 2.23, in a Pt-O_2 predominance diagram, only Pt metal is stable, and different oxides such as PtO and PtO_2 are formed only under lower T or higher p_{O_2} (see, e. g., van Spronsen et al., 2017). Platinum metal melts at 1768 °C (Table 2.2) but becomes above 1600...1650 °C so soft that it cannot be used as crucible material. Alloying Pt with ca. 5 % Rh or Ir hardens the material. Alloying with Au can additionally reduce wetting of crucible walls by some melts. Besides its stability against oxygen, platinum proves to be very resistent against halogens, most salt melts, and strong acids.

Platinum and its alloys with other noble metals possess high stability with respect to temperature and many substances. The most important threats for Pt are: (1) "Pt poisons" like C, S, or P, as mentioned above; (2) very reducing conditions; (3) melts of other metals. Especially, a combination of these influences should be avoided.

As insert for metallic crucibles, to protect them from critical samples (e. g., during calibration measurements, because most calibration substances are metals; see Table 2.2), ceramic "liners" are offered. Ceramic "washers" between the flat bottom of a DSC crucible and the DSC sensor (see Fig. 2.2 (b)) help to prevent sticking of Pt crucibles to the sensor.

The thermal conductivity λ of the crucible material is another parameter, which can become relevant if the time constant of the measurement system is important (see Section 2.5.3). Typically, λ for metals is larger than for ceramics, but exact values depend also on the materials pretreatment (e. g., cold working), which influences the defect content, and also on minor components. Alloying usually reduces λ. The black and red DTA curves in Fig. 2.24 (b) were measured under identical conditions; just with the difference that for the black curve, a ceramic Al_2O_3 crucible was used, and for the red one, a Pt/Ir crucible. The latter metallic material gives sharper peaks and a better separation of both endothermal effects, because λ for platinum is more than

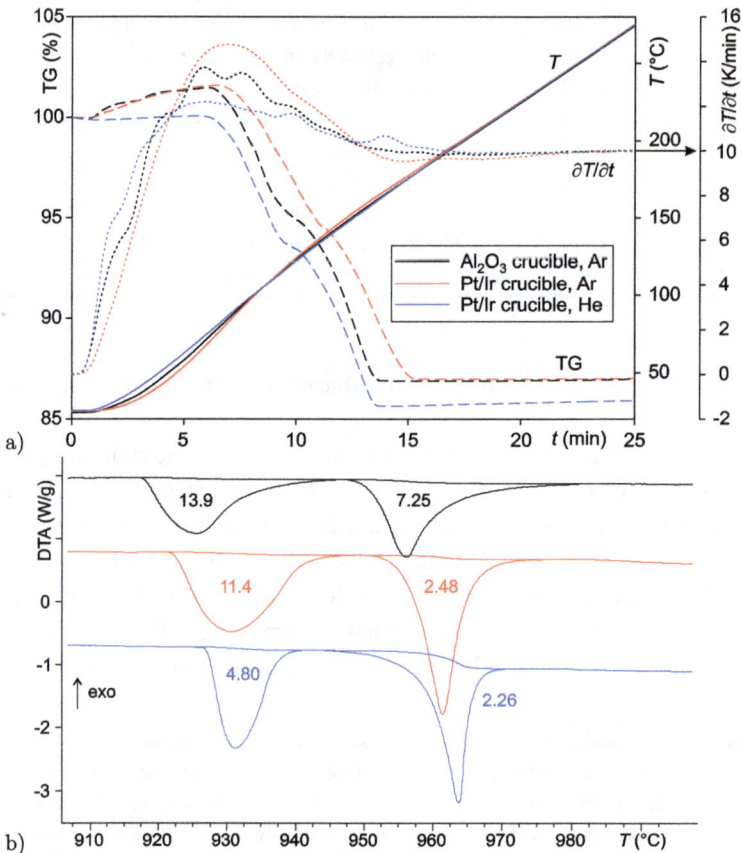

Figure 2.24: DTA/TG measurements of $BaCl_2 \cdot 2H_2O$ in a NETZSCH STA409 (type S sample holder, identical heating rates $\dot{T}$ = 10 K/min, identical gas flow rates 40 ml/min, but with different crucible/gas combinations). Sample masses: black 48.40 mg, red 43.61 mg, blue 43.53 mg. (a) First 25 minutes of the measurements, TG, T, and $\dot{T} = \partial T/\partial t$ are shown vs. time t. (b) DTA peaks related to the phase transition (Hull et al., 2011) and subsequent melting. Ratio of the peak width ΔT (in Kelvin) and the peak height (in W/g) is given as a parameter for every peak (both values determined with sigmoidal baseline). Compare also with Fig. 3.10.

twice as high as for the ceramic. The improvement is significant especially for the second peak, where the sample becomes liquid, and heat transport is mainly determined by the thermal conductivity of the crucible itself. For the first peak, the difference of the peak width/hight ratio is not so strong because here the sample consists still of solid grains. Then thermal transport is significantly influenced by the conductivity of the gas atmosphere between the grains. The big difference for this peak appears between the last two DTA curves, where Ar (red) is exchanged with He (blue). In the next section, we will discuss that the latter can transport heat much faster, which results in a smaller phase transition peak.

The DTA peak widths in Fig. 2.24 (b) are given on the level of 37 % of the peak maximum. Why? [?]

The temperature program of the thermal analyzer is usually controlled by a thermocouple situated in the vicinity of the heater element to ensure quick response of the heating power to the temperature via a PID or PI (proportional–integral–differential) circuit. Often, a second "cascaded" PID or PI circuit is added to gain even faster response, and this second circuit can be controlled by the reference thermocouple (see, e. g., Yang et al., 2019).

Figure 2.24 (a) shows the initial phases of three TG measurements under almost identical conditions on a time scale. Besides the TG signal (dashed lines), also T (full lines) and $\dot{T}$ (dotted lines) are shown. A heating rate $\dot{T}$ = 10 K/min was set for all measurements, and, as usual, some transient period of ca. 15 min is needed before a constant rate is reached. The measurements in Ar show the largest "overshoot" of $\dot{T}$ near t = 7 min (maximum value there $\geq$ 14 K/min). Best response is observed if the Pt/Ir crucible is used in a He atmosphere, because then thermal transport is enhanced further.

2.6.2 Atmosphere

Thermal analysis is usually performed in a gas atmosphere, which flows through the device, and often this gas has the ambient pressure of 1 bar. Only seldom measurements are performed in vacuum, because a gas flow has usually several benefits:
-   It supports heat transport from the heater element to the sample holder and sample.
-   The gas itself can be a reaction partner of a process under investigation, e. g., oxygen in a combustion process like described in Section 2.4.2.2.
-   The gas flow helps to "rinse out" species that are emanating from the sample, like water (as reaction product or adsorbed humidity), CO_2 (from carbonates), or SO_2 (from sulfates).

– From the chemical point of view, the measurement in a well-defined gas atmosphere is often preferred over the conditions given in a vaguely defined vacuum,
because this is never "empty". Rather, it will contain some residual gas, which
can be air (slightly oxidative, resulting from O_2) or vacuum pump oil (reductive,
resulting from organics).

If measurements are performed under very low pressure p, it may happen that the
mean free path of gas molecules approaches the geometrical dimensions of the thermal analyzer. This means that a free path is at standard conditions for most gases in
the order of 100 nm and approaches for $p = 0.1\,Pa = 10^{-3}$ mbar the order of 10 cm. If
under such conditions, gas molecules emanate from the sample, then they can produce a recoil on the sample holder, which is interpreted by a TG system as a temporarily increased mass. Bertram and Klimm (2004) compared TG curves for $SrCO_3$, which
were measured in Ar flow ($p = 1$ bar) and in vacuum ($p \approx 0.06$ Pa), respectively. The
results are shown in Fig. 2.25. Only in the latter case a positive TG signal from initial
100 % to > 130 % suggested an increased mass, before finally the TG signal dropped to
the expected level resulting from the loss of CO_2. Under vacuum, the $SrCO_3$ decomposition is finished at significantly lower T, because the equilibrium of this calcination
is shifted to the side of the products (increase of volume).

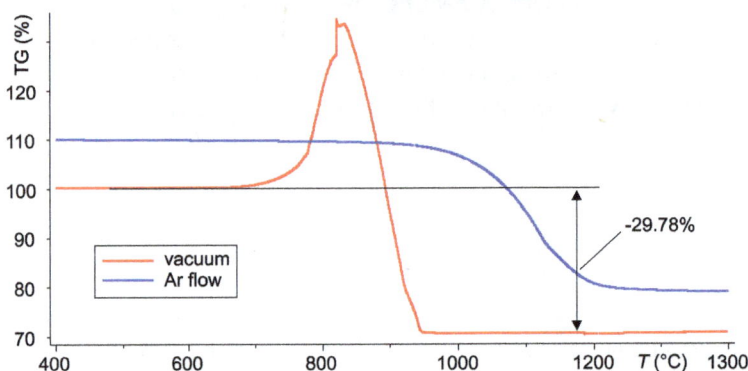

Figure 2.25: TG of ca. 40 mg $SrCO_3$ in vacuum ($p \approx 0.06$ Pa) or in an Ar flow of 40 ml/min ($p = 1$ bar,
curve shifted by +10%), respectively. Amended from Bertram and Klimm (2004) with permission
from Elsevier.

If no specific chemical reaction with the sample is intended, then often flows of argon
or nitrogen are considered to be "inert" and hence are often used during thermal analysis. However, the term "inert" should be handled with care because trace impurities
of commercial gases and leaks of the thermal analyzer always lead to background concentrations of impurities like oxygen or water in the order of at least several parts per
million (ppm).

For measurements in vacuum, special attention must be paid to the temperature controller! A control thermocouple, which is located close to the heater unit, will give correct information. In contrast, if the sample thermocouple is additionally used in a cascaded controller (e. g., "sample temperature control", STC), then damage of the furnace can occur, because the heating energy cannot reach the sample thermocouple by conduction. This situation can be risky, because the T controller can overheat the furnace. It is highly recommended to rely only on the furnace thermocouple for vacuum measurements.

Klimm and Ganschow (2005) investigated the melting behavior of FeO created in situ by heating iron(II) oxalate dihydrate $Fe(COO)_2 \cdot 2H_2O$ in 5N Ar (99.999 % purity). Similarly to calcium oxalate hydrate, which was described in Section 2.1.2.2, this material decomposes into a multistage process finally until 450 °C to iron(II) oxide FeO (wüstite,[19] see Section 3.1.3). Upon further heating, without interruption of the Ar flow, the sample showed the melting peak of FeO near 1400 °C. After this, again without interruption of the Ar flow, the sample was repeatedly cycled between $T_{max} \approx 1600$ °C and room temperature. In these subsequent heating runs the melting peak became broader and deformed and gradually shifted to higher T. After several heating/cooling cycles, the peak remained stable at ≈1560 °C. This temperature corresponds to the melting point of magnetite Fe_3O_4. The apparent but unexpected perception: 5N Ar oxidized FeO to Fe_3O_4!

This observation can be understood in terms the $Fe-O_2$ predominance diagram in Fig. C.24. Klimm et al. (2015) gave the following values for the equilibrium p_{O_2} at the three corners of the stability field of FeO, from top to bottom: 2.5×10^{-6} bar, 7.8×10^{-11} bar, 1×10^{-26} bar. These values are so low that they could not be reached in the thermal analyzer and with the 5N Ar flow. Consequently, the excess oxygen shifted the system composition to the stability field of magnetite, which is an Fe(II,III) oxide.

In Fig. C.24 the $p_{O_2}(T)$ curve for CO_2 is above the FeO stability field, and the curve for CO is below it. Klimm and Ganschow (2005) succeeded with a suitable gas mixture (85 % Ar, 10 % CO_2, 5 % CO), which was calculated to have a $p_{O_2}(T)$ dependency running through the FeO stability field. Indeed, FeO proved to be stable in this "reactive atmosphere" and repeated DTA runs always showed the melting peak at 1400 °C.

For contemporary thermal analyzers, gas flow rates in the order 20...50 ml/min are typical: closer to the lower limit if no emanating species are expected, which have to be rinsed out, and closer to the upper limit in the opposite case. It should be taken into account that a too high flow rate can have a negative effect on the homogeneity of the temperature field inside the device. On the other side, often, somewhat higher flow rates in the order of 100 ml/min are used if devices for evolved gas analysis (EGA) are coupled to the thermal analyzer. (See also Section 2.4.1.) Then it is desirable to correlate TG steps, which are due to the release of specific species from the sample

---

19 Fritz Wüst (8 July 1860–20 March 1938).

with corresponding signals in the EGA system. If, however, the gas flow is too low, then the dwell time inside the thermal analyzer is large, which can smear the signals in the EGA unit.

Special care must be taken if the gas analyzer is coupled via a skimmer: The skimmer splits the incoming gas flow into one part, which is pumped through the EGA unit, and a second part, which leaves the thermal analyzer through the normal exhaust. If the incoming flow is smaller than the flow through the EGA unit, then the difference can be sucked from the exhaust backward into the system. In the best case, this disturbs just the EGA measurement. In the worst case, a carbon skimmer can be destroyed by air. Figure 2.26 shows that the viscosity of gases increases significantly with T. Hence the gas flow through the EGA unit will usually become smaller upon heating. This means that if no air is sucked into the analyzer at the beginning of a measurement, this will remain so if the skimmer system is heated.

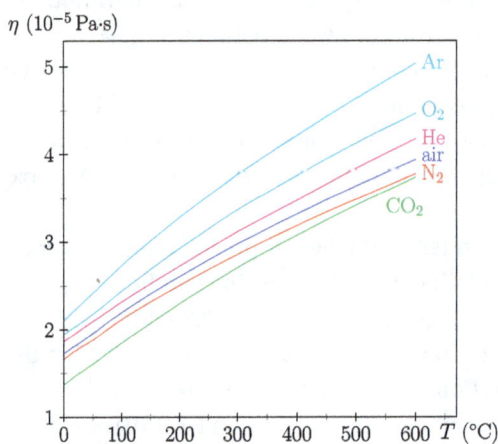

Figure 2.26: Dynamic viscosity η of several gases as a function of temperature. Data from The Engineering Toolbox (2021).

Table 2.6 shows the mass density and the thermal conductivity of some gases, which are often used for rinsing thermal analyzers, the viscosity of these gases as function of T can be found in Fig. 2.26. Under the typical conditions of thermal analysis, gases can be considered as almost ideal, and the mass density reduces with T,

$$\varrho(T) \propto \frac{1}{T} \tag{2.16}$$

(Charles[20] law). The thermal conductivity of gases rises significantly with T; for air, the ratio of λ for T is $0\,°C : 500\,°C : 1000\,°C : 1500\,°C \approx 1 : 2.4 : 3.3 : 4.1$. Nevertheless,

---

20 Jacques Alexandre César Charles (12 November 1746–7 April 1823).

Table 2.6: Mass density ϱ for several gases at a pressure p = 1 atm and thermal conductivity λ of gases plus some crucible materials (data for 20…25 °C from The Engineering Toolbox, 2021).

Gas	Ar	N$_2$	O$_2$	Air	He	CO$_2$		
ϱ (kg/m^3)	1.661	1.165	1.331	1.205	0.1664	1.842		
λ (W/(m K))	0.016	0.024	0.024	0.0262	0.142	0.0146		
Solid	**Al**	**Pt**	**Steel**	**Ag**	**Ni**	**W**	**Al$_2$O$_3$**	**Graphite**
λ (W/(m K))	236	71	≈20…60	428	94	182	≈30	≈170

according the Stefan[21]–Boltzmann[22] law, the energy exchange by radiation rises much stronger with T^4 and becomes dominating typically above 1500 °C.

Figure 2.24 demonstrates that from the technical point of view, helium is often a very good choice for rinsing thermal analyzers: ϱ is very low, which reduces disturbing buoyancy effects, and λ is high, which improves temperature control of the device and often results in faster attainment of a linear $T(t)$ function. Another positive aspect is the small atomic mass of 4, which may be beneficial for QMS analysis of emanation gas in the SCAN mode; see Section 2.4.1. Unfortunately, helium is expensive, and consequently often other gases are used.

The density of gas mixtures is the weighed average of the components, but not the thermal conductivity. See, e. g., Lindsay and Bromley (1950)[23] and the data for N$_2$, O$_2$, and air in Table 2.6.

As an example for the immense influence that the atmosphere can have on the results of thermal analysis, the reader is referred to Fig. 2.5, where the DTA peak, which is connected with the second TG step (release of CO from calcium oxalate), is reversed from endothermal (if measured in Ar) to exothermal (if measured in O$_2$).

However, not only pure oxygen (or air) can act as oxidizing agents. As pointed out above for FeO, already the trace impurities of commercial gases, like O$_2$ or H$_2$O, can lead to unexpected and undesired reactions with the sample. This holds, e. g., for many metals that can be converted to oxides. Many halides, especially fluorides, can react already with traces of moisture (below 1 ppm) and/or oxygen in the gas flow to oxides or oxyfluorides.

Already Abell et al. (1976) tried to reduce the degree of unintended hydrolysis during DTA measurements with fluorides with a static atmosphere; in Section 3.3.3, we discuss this topic in more detail. Also, DTA devices for the highest T > 2000 °C are often closed systems, because then the introduction of oxygen is limited. Getter elements are another possibility to significantly reduce the concentration of hazardous gases near the sample. Such "oxygen traps" (some of them are also capturing moisture

---

21 Jožef Štefan (24 March 1835–7 January 1893).

22 Ludwig Eduard Boltzmann (20 February 1844–5 September 1906).

23 LeRoy Alton Bromley (30 September 1919–24 February 2004).

and/or sulfur) can be installed in the gas line that delivers the flow to the thermal analyzer; or they can be placed in situ in the vicinity of the sample. They are commercially available from different suppliers.

In the latter case the getter element works at high T close to the sample temperature, and it must be ensured that the getter itself (which is usually a metal or an alloy) cannot evaporate and pollute the interior of the thermal analyzer, including the sample. Figure C.17 shows an Ellingham-type predominance diagram of the system $Zr–O_2$ and demonstrates that zirconium has very favorable properties. The melting point of Zr is rather high (T_f = 1855 °C), which guarantees that the solid getter elements remain mechanically stable to very high T_{max} ≈ 1650 °C. Even close to T_f, the phase boundary β-Zr/ZrO_2 is located not much above 10^{-20} bar O_2! A very low p_{O_2} level can be obtained, because the rate of oxygen absorption of Zr-based getter elements at elevated T is high. Another beneficial aspect of Zr getters is indicated by the green dashed line in Fig. C.17. This curve marks the (p_{O_2}, T) limits where the sum $\sum p_i$ of the partial pressures p_i of all species containing Zr exceeds 10^{-6} bar. Below this limit, evaporation and transport of zirconium can be regarded insignificant. The high affinity of Zr to oxygen also reduces the water content to very low levels.

Zr has not only a strong affinity to O_2, but also to N_2, which is demonstrated in Fig. 2.27. The equilibrium partial pressure at 1600 °C is ca. 10^{-21} bar for O_2 and ca. 10^{-11} bar for N_2, which means that also nitrogen is trapped from the gas (however, less efficiently). Of course, for this reason, oxygen cannot be removed from a nitrogen flow by a Zr-based getter.

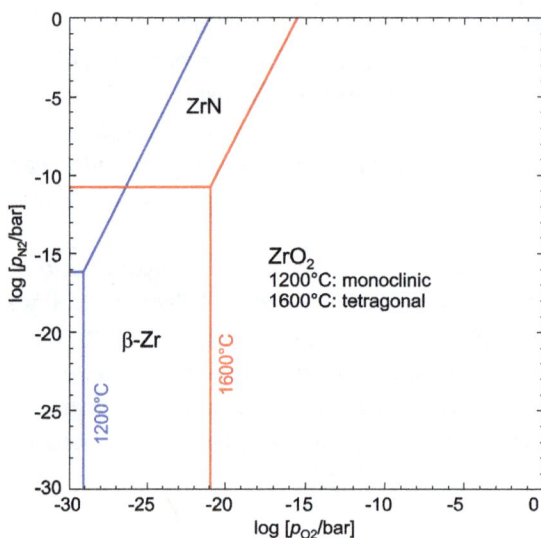

Figure 2.27: The stability field of β-Zr in the system $Zr–O_2–N_2$ at 1200 °C (blue) and 1600 °C (red) is restricted to very low partial pressures for both gases, especially to low p_{O_2}.

It should be noted that also titanium is a good getter metal (Miyazawa et al., 2019 and Fig. C.16).

2.6.3 Sample mass

It is obvious that the magnitude of a thermal event in the sample, e. g., the consumed amount of heat during melting, scales proportionally with the sample mass m_s. Depending on the sensitivity $K_s(T)$ of the thermoanalytic device (see Section 2.5.2) and on the accuracy and noise level of the voltmeters used to register the thermoelectric voltages, with modern DSC devices, $m_s \approx 5 \ldots 20$ mg or even less can be used. For DTA, $m_s \approx 20 \ldots 60$ mg is more typical. This is significantly better than in historic devices, which required $m_s \gg 100$ mg (Aharoni et al., 1961)[24] or even several grams.

If the scientific problem under investigation allows it, then a small sample mass should be preferred. **!**

Figure 2.28 shows a series of DTA melting peaks of zinc. All measurement conditions were identical for these curves, except the sample mass that was increased stepwise from initial $m_s = 8.83$ mg (bottom; see label at the curve) to finally 338.39 mg. All curves were obtained during a second heating, which means that the sample was premolten in a first heating and hence formed one homogeneous body.

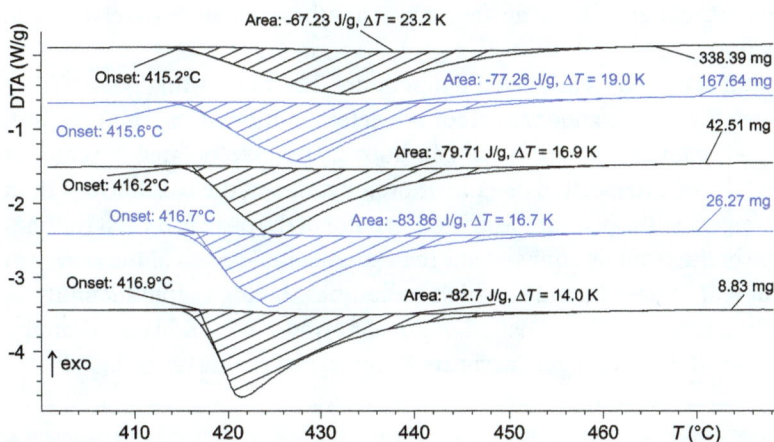

Figure 2.28: DTA heating curves ($\dot{T} = 10$ K/min) with zinc samples of different mass in flowing Ar (NETZSCH STA 409 with T and sensitivity calibrated DTA/TG sample carrier type S). ΔT is the peak width at $e^{-1} \approx 37\%$ of its height, determined with asymptotic baselines.

---

24 Amikam Aharoni (5 August 1929–21 January 2002).

It is typical that the first (falling) part of the melting peak of a substance that melts congruently (this means at a clearly defined melting point T_f like all chemical elements; see Section 1.4.2) is fairly good described by a straight line, except the parts very close to its beginning and near the crossover to the rising part after the melting peak minimum. This nearly linear behavior results from the limited heat flux into the sample crucible, which is restricted by the thermal conductivity of the system. From equation (1.1) it is obvious that the heat energy $Q(t)$ introduced into the sample would rise for short peaks (where the temperature difference between sample and crucible environment can be considered almost constant) linearly with t, and due to the constant heating rate $\dot{T}$, also linearly with T, which is the temperature at the reference crucible. Then Q is consumed by the melting process as long as solid sample remains. The required amount of heat is $m_s \cdot \Delta H_f$, where ΔH_f is the heat of fusion in J/g. (For very large m_s, the difference between onset and peak minimum scales weaker because the temperature difference between sample and crucible outside cannot be regarded constant.)

---

i In modern thermal analyzers, T is always measured near the reference, because only this way the requirement of a constant heating rate can be fulfilled.

Rather than plotting the measured DTA signal (scaled either in μV or after sensitivity calibration in mW) versus T or (less frequently) t, it is sometimes useful to divide it by m_s (scaling then in μV/mg or W/g), which makes measurements with different masses better comparable. This is very helpful, e. g., during the experimental determination of phase diagrams (see Section 3.3), if one starts with a comparably small sample and adds subsequently growing amounts of another component to change the sample composition.

For events that start at a clearly fixed temperature, like the melting point T_f, the "extrapolated onset" is a very good model for the determination of this temperature, even under severely changed experimental conditions. This extrapolated onset (or in brief just "onset") is the intersection of a linear prolongation of the baseline left from the peak and a tangent to the peak at the inflection point of the peak in its first (falling) part. Irrespective of the severely changed m_s, the experimental onsets of the curves in Fig. 2.28 are different by less than 2 K, which underlines the high reproducibility of this determination. (The real T_f = 419.6 °C of Zn is, however, ca. 3 K higher than the experimental onsets due to a not very accurate T calibration in this range.)

⚡ The onset of a melting peak gives a very good approximation of the melting point of a pure substance. The peak temperature, where the DTA or DSC signal reaches its minimum, depends on several experimental parameters and is usually shifted significantly toward higher T for larger m_s.

If high caloric accuracy is required, then DSC is usually superior to DTA. Indeed, the range of peak areas in Fig. 2.28 is ≈ 67 . . . 84 J/g, which is significantly lower than the literature value 107.5 J/g (see Table 2.2). Nevertheless, for many purposes, only relative trends are required, and then the reproducibility of a DTA sample carrier might be sufficient.

The rising parts of the melting peaks become broader, because the melting process of larger samples requires more heat and, consequently, more time. This is the main reason why the peaks in total become broader. It is usual to measure the peak width in thermal analysis at the level of e^{-1} of its minimum or maximum. The FWHM ("full width at half maximum") values which are used, e. g., in optical spectroscopy or X-ray diffractometry, are not used in thermal analysis.

2.6.4 Temperature change rate

Thermal analysis is mainly performed during heating "segments" or "ramps" and less frequently during cooling or under isothermal conditions. For heating and cooling segments, the positive or negative heating rate is an important parameter, and practically quite often a rate of $\partial T/\partial t = \dot{T} = \beta = \pm 10\,\text{K/min}$ is chosen as a compromise between the drawbacks of higher or lower rates.

Figure 2.29 shows subsequent heating curves of the same Au sample under identical experimental conditions; $\dot{T}$ is just changed from 20 K/min to very low 0.5 K/min. Like in the previous Fig. 2.28, it turns out that the onset temperatures scatter not too much, by just 4.4 K at this (compared to Zn from the previous figure) high $T_f = 1064.2\,°\text{C}$ (see Table 2.2). Obviously, the higher heating rates delivered more accurate results. As described in Section 2.5.1, it should be taken into account that also the T calibration is performed with a certain rate, in this case the typical 10 K/min.

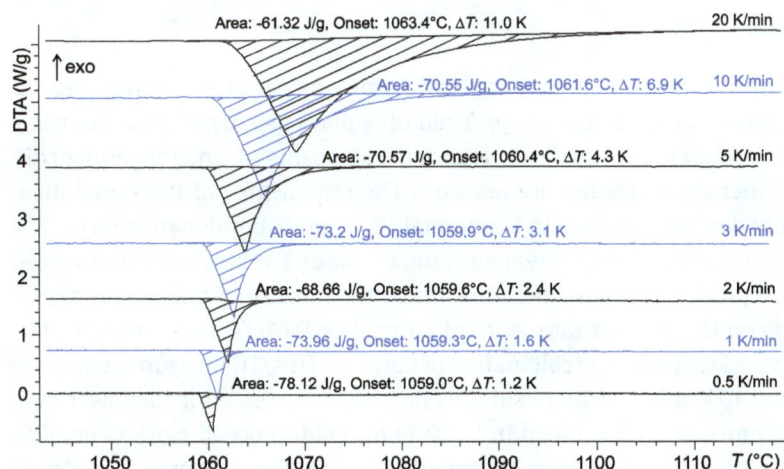

Figure 2.29: DTA heating curves obtained with 39.26 mg Au in flowing Ar with different rates (NET-ZSCH STA 409 with T and sensitivity calibrated DTA/TG sample carrier type S). ΔT is the peak width at $e^{-1} \approx 37\%$ of its height.

The peak area represents the heat of fusion, which, according to Table 2.2, is $\Delta H_f =$ 63.7 J/g. The experimental values are not very distant from this literature value, but with better accuracy for the higher heating rates. It seems surprising that visually the peaks for low rates are much smaller than for high rates, but nevertheless calculated areas are similar. This is a direct effect of the different rates: even if the consumed heat energy Q is always the same, it flows at lower rate over a longer period of time t into the sample. From equation (1.1) it follows that t and W (consequently, also the DTA signal) are indirect proportional to each other.

The most significant difference is shown for the peak width ΔT, because the required heat $Q = m_s \cdot \Delta H_f$ flows into the sample during a certain time t, which depends only weakly on $\dot{T}$. As shown in Fig. 2.22, also the return of the DTA signal to the baseline requires time in the order of minutes. However, for a given period of time, T (measured at the reference!) changes as $\propto \dot{T}$, resulting in visually strong DTA or DSC peaks, which extend over a wide T range for high rates.

! Heating/cooling rates $\dot{T} = \pm 10$ K/min are often a good choice for DTA and DSC measurements and often represent a reasonable compromise between total measurement time, magnitude of the thermal signal, and resolution between neighboring thermal events. Sometimes deviations are required: If the sensitivity of the sensor is small (often, for very high T; see Section 2.5.2), then weak effects sometimes cannot be seen, as a result of noise. Under such conditions, higher rates (around 2000 °C often 20 K/min) can be useful. Vice versa, the separation of near subsequent effects is often impossible if a too high rate and/or a too high sample mass are used. Then a small sample should be measured with low rate in a very sensitive DSC.

2.7 Kinetics

2.7.1 Introduction

Homogeneous transformations of first or second order appear at a specific temperature or over a specific temperature range. Then time plays usually no role. For some other processes, such as solid–solid reactions or decompositions, time-dependent diffusion steps or thermal activation are relevant. Then the degree of the transformation significantly depends on time in isothermal (T = const.) or dynamic (e. g., $\dot{T}$ = dT/dt = const.) measurements (T – temperature, t – time). For kinetically influenced processes, it is typical that the temperature where the transformation occurs significantly depends on the heating rate, e. g., of a dynamic DTA/TG measurement. Figure 2.30 demonstrates this for the calcination of $CaCO_3$ in DTA/TG measurements with $\dot{T}$ = 2 K/min and 10 K/min. Decomposition is finished at lower T for the lower rate. Simple melting processes, like shown in Fig. 2.29 for gold, proceed close to equilibrium, without an influence of kinetics. Consequently, there the measured T_f is almost independent of $\dot{T}$.

The numerical treatment is based usually on the degree of conversion α that starts from 0 before the beginning and finally reaches 1 after the reaction is finished. $\alpha(T)$

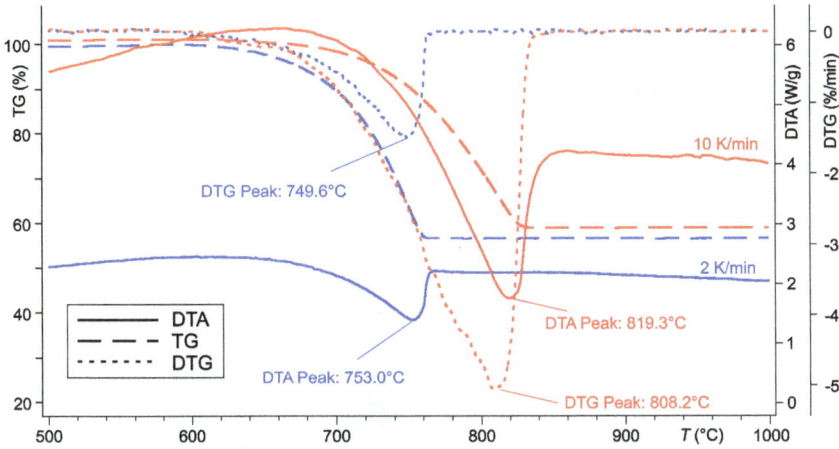

Figure 2.30: DTA/TG measurements of CaCO$_3$ (m_s = 11...14 mg) with two heating rates in a NETZSCH STA 409C (Ar flow of 80 ml/min in open Al$_2$O$_3$ crucibles). DTG ist the first derivative of the TG curve.

can be determined either from DTA (DSC) signals as the share of the partial peak area $A'(T)/A$ (A – full peak area). An often better alternative (if applicable) is the determination from the share of the mass loss TG$'(T)$ compared to the total mass loss TG. The latter method is more accurate because TG signals are registered immediately, whereas DTA effects always show a time lag resulting from limited thermal conductivity. Typically, time constants of DTA sample carriers are in the order of a few 10 seconds. Both methods are for the thermal decomposition of CaCO$_3$ demonstrated in Fig. 2.31.

Guldberg and Waage (1879)[25,26] proposed the concept of "Massenwirkung" (mass action) for the description of chemical equilibrium. Starting from this, the rate equation

$$\frac{d\alpha}{dt} = k(T)f(\alpha) \tag{2.17}$$

says that the rate of reaction is proportional to some function $f(\alpha)$ of the progress of conversion. For thermal analysis with constant heating rate $\beta = \dot{T}$ = const., the term dα/dt can be replaced by βdα/dT. The proportionality factor k in equation (2.17) depends on T, and for this dependence, Arrhenius (1889)[27] introduced the expression

$$k(T) = Z \exp\left[\frac{-E_a}{RT}\right], \tag{2.18}$$

---

25 Cato Maximilian Guldberg (11 August 1836–14 January 1902).

26 Peter Waage (29 June 1833–13 January 1900).

27 Svante August Arrhenius (19 February 1859–2 October 1927).

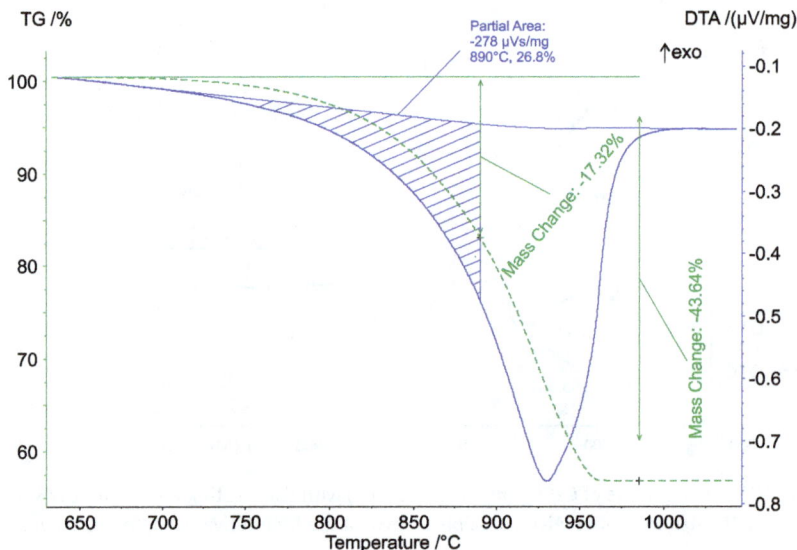

Figure 2.31: DTA (blue full line) and TG (green dashed line) for the thermal decomposition of $CaCO_3$ with 10 K/min heating rate in flowing Ar. At 890 °C the "partial area" of the DTA peak is $\alpha = 0.268$, whereas already $\alpha' = 17.32/43.64 = 0.3969$ of the TG step is reached.

which is now called the Arrhenius equation with activation energy E_a and preexponential factor Z. Substituting equation (2.18) into (2.17) and taking the logarithm gives

$$\ln \frac{d\alpha}{dt} = Zf(\alpha) - \frac{E_a}{R} \frac{1}{T}. \tag{2.19}$$

For thermoanalytic measurements with constant heating rate, a plot of $\ln[d\alpha/dT]$ versus reciprocal temperature should then give a straight line of slope $-E_a/R$. For the thermal decomposition ("calcination") of $CaCO_3$, such a plot is shown in Fig. 2.32. Plots of this kind are useful to check the consistency of experimental data prior to further kinetic analysis.

Probably, the first paper on isothermal kinetic studies with thermobalances was published 1925 by Akahira;[28] for a historical overview, see Ozawa et al. (1995).

2.7.2 Rate laws

A large number of rate laws (rate equations $f(\alpha)$; see eq. (2.17)) can be used to describe thermoanalytic data. This includes not only single-step reactions Ⓐ → Ⓑ, but

---

28 Takeo Akahira (1895–1981).

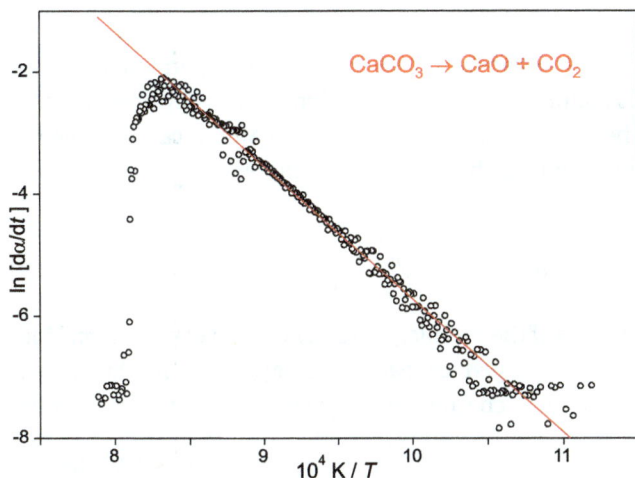

Figure 2.32: Evaluation of the TG signal for the decomposition of $CaCO_3$ from Fig. 2.31 according to eq. (2.19).

also two-step reactions with an intermediate product of the kind Ⓐ → Ⓑ → Ⓒ and concurring reactions. Here we give only a short overview of possible rate laws, and in the following, we will present a few methods for numerical analysis of thermodynamic data. For a more sophisticated analysis, we recommend the application of special software.

2.7.2.1 Reaction of nth order

Different assumptions can be made for the function $f(\alpha)$ in equation (2.17), e. g.,

$$f(\alpha) = (1 - \alpha)^n, \tag{2.20}$$

which describes a reaction of nth order. This was described by Borchardt and Daniels (1957).[29,30] The reaction order can often (but not always) be deduced from stoichiometry factors of the underlying chemical reaction. In the case of a simple decomposition, we often find $n = 1$.

Zero-order kinetics is always an artifact of the conditions under which the reaction is carried out, e. g., restricted contact of reaction partners in catalytic reactions at surfaces. Then the concentration of educts and products changes linearly with time. A fractional reaction order often indicates a chain reaction or another complex mechanism.

---

29 Hans J. Borchardt (≈1930–17 November 2015).

30 Farrington Daniels (8 March 1889–23 June 1972).

2.7.2.2 Autocatalysis

Prout and Tompkins (1944)[31] investigated the kinetics of potassium permanganate decomposition, which proceeds mainly according to equation (1.39). They observed that the reaction is accelerating because the products (especially MnO_2) catalyze the decomposition process. Such behavior can be described by the Prout–Tompkins equations of the kind

$$f(\alpha) = (1 - \alpha)^n p^m, \tag{2.21}$$

which includes the concentration p of the product, possibly with an exponent m. Modifications of equation (2.21) were developed for reactions comprising a marked induction period (Prout and Brown, 1965). Such functions give sigmoidal $\alpha(t)$ curves under isothermal conditions.

2.7.2.3 Johnson–Mehl–Avrami–Kolmogorov model

After publications by Kolmogorov[32] and Mehl,[33] together with his PhD student Johnson,[34] Avrami (1939, 1940)[35] published a kinetic model that takes into account rates of seed formation and the problem of overlapping parts of a sample that are already reacted. This model allows a description of isothermal data for reactions Ⓐ → Ⓑ with inhomogeneous nucleation and is now called the Johnson–Mehl–Avrami–Kolmogorov (JMAK) equation. There it is assumed that initially nuclei of B are produced with constant rate, $n \propto t$. Besides, the radius of every seed grows as $r \propto t$, and hence its volume as $V \propto t^3$. Consequently, in the beginning of the reaction, we find the rate $d\alpha/dt \propto t^4$. Later, the growing seeds get in contact (see Fig. 2.33), and the rate drops. The whole process can be described by the sigmoidal function

$$\alpha(t) = 1 - \exp[-kt^n], \tag{2.22}$$

where α is again the degree of conversion.

Only in the case described above, with sporadic random nucleation and homogeneous growth in three directions, we have $n = 4$ in equation (2.22). n may be smaller if either nucleation appears only predetermined at seeds or if growth of the new phase proceeds only as two-dimensional (e. g., at grain boundaries or during epitaxy) or one-dimensional (e. g., along dislocations). An overview is given in Table 2.7, and some functions $\alpha(t)$ are shown in Fig. 2.34.

31 Frederick Clifford Tompkins (29 August 1910–5 November 1995).

32 Andrey Nikolaevich Kolmogorov (25 April 1903–20 October 1987).

33 Robert Franklin Mehl (30 March 1898–29 January 1976).

34 William Austin Johnson (born 9 November 1913).

35 Mael Avrami Melvin, born as Moshe Yoel Avrami (1913–1 October 2014).

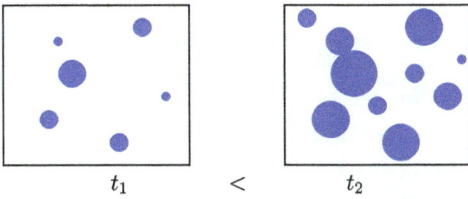

Figure 2.33: Avrami kinetics: after an initial phase (time t_1) with unlimited nucleation and growth, grains of the new phase B (circles) get in contact at t_2, resulting finally in saturation.

Table 2.7: Avrami exponents n of equation (2.22) for different growth mechanisms of phase B.

Growth mechanism	n (random nucleation)	n (seeded nucleation)
volume	4	3
planar	3	2
linear	2	1

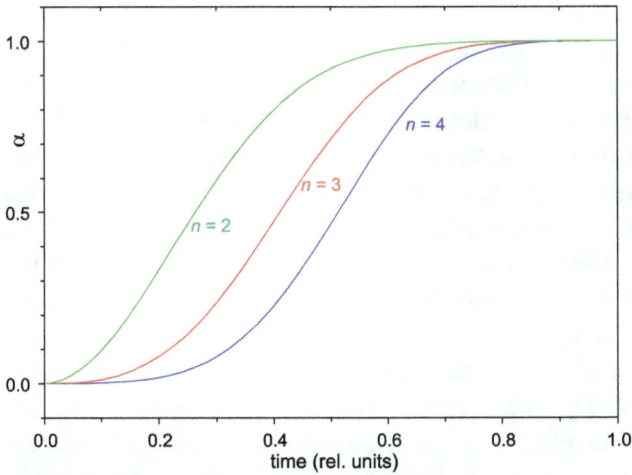

Figure 2.34: Avrami kinetics for random nucleation and volume ($n = 4$), planar ($n = 3$) or linear growth ($n = 2$), see Table 2.7.

From (2.22) we can derive

$$\ln[-\ln(1-\alpha)] = \ln k + n \ln t, \tag{2.23}$$

and, consequently, it is useful to plot $\ln[-\ln(1-\alpha)] = \ln\ln[1/(1-\alpha)]$ versus $\ln t$, because such plots are linear with slope n.

2.7.3 Methods of kinetic analysis

Most suppliers of thermoanalytic devices offer, besides the essential control and analysis software, software packages for kinetic analysis. Software from other companies is also available (AKTS, 2021). The application of such commercial computer programs is highly recommended if kinetic investigations are a regular task, if complicated mechanisms (e. g., multistep reactions) occur, or if numerical forecasts have to be performed for conditions that could not be applied during the thermoanalytic measurements, such as drug stability in hot and humid environments.

Vyazovkin et al. (2011) published recommendations of the Kinetics Committee of the International Confederation for Thermal Analysis and Calorimetry (ICTAC) for some technical conditions for the performance of kinetic measurements by DSC, DTA, and TG. Isothermal measurements and measurements with constant heating rate β are taken into account. The authors point out that both temperature regimes have their benefits and drawbacks. However, at least measurements with β = const. allow us to include the situation of zero extent conversion (α = 0) in the subsequent numerical evaluation.

Many evaluation methods are "isoconversional", which means that the conversion rate $d\alpha/dT$ for some arbitrary constant extent of conversion α is only a function of T. Typically, we have to perform a series of ca. 3...5 measurements with different temperature programs; this could be either runs with different heating rate or several isothermal measurements at different T. Without assuming any specific reaction model, these methods allow us to use equation (2.19). For this reason, these methods are often called "model free" and allow to derive the activation energy $E_a(\alpha)$ for different phases of the process $0 \leq \alpha \leq 1$ from plots $\ln[d\alpha/dT]$ vs. T^{-1}. Significant variations of $E_a(\alpha)$ indicate multistep kinetics.

As an example, below we show DTA/TG measurements with ZnS powder heated with different rates in an Ar/O_2 flow. Figure 2.35 (a) shows the DTA results, and Fig. 2.35 (b) shows the TG results that were obtained simultaneously. All curves are baseline corrected and will be used later for some kinetic evaluations. (See also Section 3.1.4 for measurements under oxygen-free conditions.)

For all DTA curves, a pronounced exothermal effect is obvious. This can be explained by the sum equation

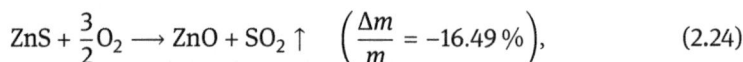

$$ZnS + \frac{3}{2}O_2 \longrightarrow ZnO + SO_2 \uparrow \quad \left(\frac{\Delta m}{m} = -16.49\,\% \right), \tag{2.24}$$

which describes the netto reaction. The higher ΔH_f of ZnO (−350 kJ/mol) compared to ZnS (−205 kJ/mol) is overcompensated by the large enthalpy production of burning sulfur to SO_2 (ΔH_f = −297 kJ/mol under standard conditions), resulting in the exothermal netto reaction (2.24). However, the slower measurements with $\dot{T} = \beta \leq 1.5\,K/min$

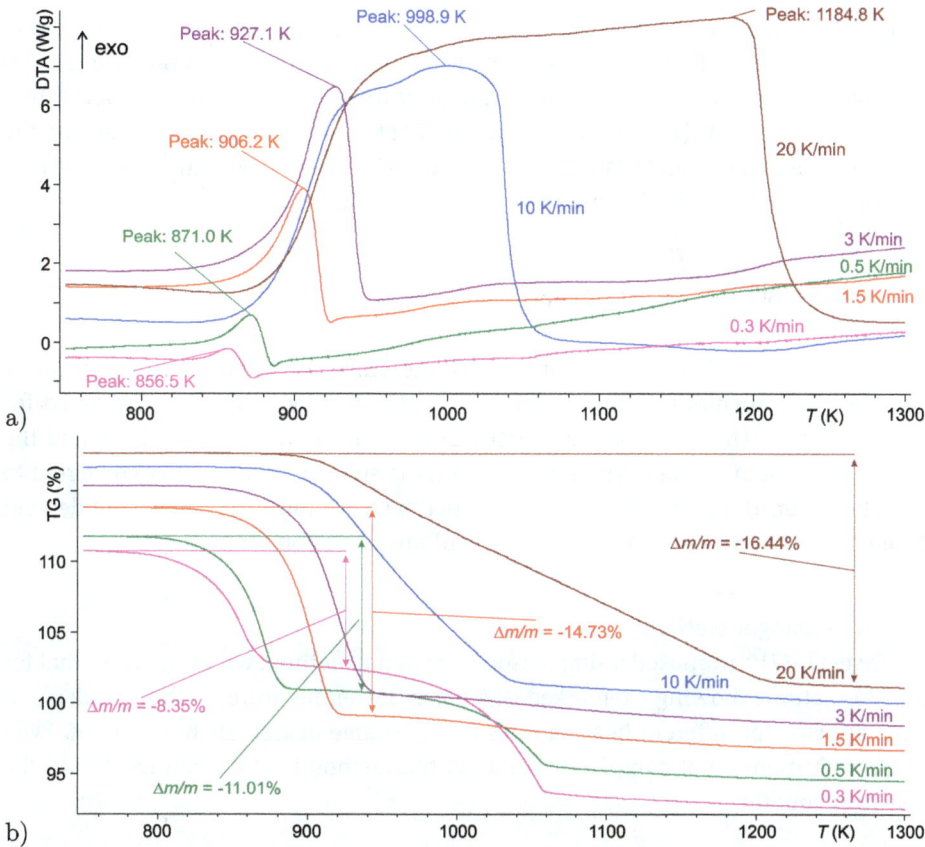

Figure 2.35: Simultaneous DTA/TG of ZnS powder (m_s = 20 ... 25 mg) in a gas flow of 50 ml/min Ar + 50 ml/min O_2 with rates in the range $\dot{T} = \beta$ = 0.3 ... 20 K/min. DTA curves are shown in the top panel, and TG in the bottom. Heating rates are written to the curves.

demonstrate that the process is more complicated: The initial exothermal peak is followed by a small endothermal effect. This hints clearly on a more complicated multistep mechanism.

The multistep behavior becomes even more obvious from the TG curves in Fig. 2.35 (b), which possess two very pronounced steps for the lower rates. Indeed, Sofekun and Doraiswamy (1996) found the phase $ZnO(ZnSO_4)_2 = 2\,ZnSO_4 \cdot ZnO$ during roasting under suitable conditions and developed a kinetic model for the roasting process; see also Shamsuddin (2016). A hypothetical chemical equation that converts ZnS completely to the oxysulfate would be

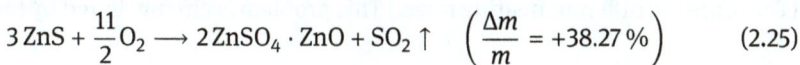

$$3\,ZnS + \frac{11}{2}O_2 \longrightarrow 2\,ZnSO_4 \cdot ZnO + SO_2 \uparrow \quad \left(\frac{\Delta m}{m} = +38.27\,\%\right) \qquad (2.25)$$

and resulted in an increase of the sample mass by 38.3 %, opposite to the netto reaction (2.24), which reduces the mass by 16.5 %. In the experimental data in Fig. 2.35 (b) the first mass loss step grows with the heating rate from 8.35 % (0.3 K/min) over 11.01 % (0.5 K/min) to 14.73 % (1.5 K/min). For even higher rates, this step is less clear, and the whole process finally shows a single TG step of –16.44 % at 20 K/min. This value is in good agreement with the $\Delta m/m$ expected from equation (2.24).

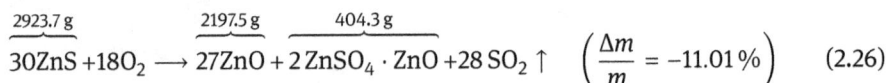

$$\underbrace{30ZnS}_{2923.7\,g} + 18O_2 \longrightarrow \underbrace{27ZnO}_{2197.5\,g} + \underbrace{2\,ZnSO_4 \cdot ZnO}_{404.3\,g} + 28\,SO_2 \uparrow \quad \left(\frac{\Delta m}{m} = -11.01\,\%\right) \quad (2.26)$$

In equation (2.26) the reactions (2.24) and (2.25) are combined in such away that the experimentally found $\Delta m/m$ = –11.01 % results, which was obtained for β = 0.5 K/min. Then 90 % of the educt ZnS are converted directly to ZnO, and the rest in a first step to the oxysulfate. The latter is subsequently further converted to ZnO. However, the rates of both reaction paths (2.24) and (2.25) obviously depend differently on T, which suppresses the oxysulfate formation for high β.

2.7.3.1 Kissinger method

Kissinger (1957)[36] proposed a simple isoconversional method, which assumes that for transformations showing a remarked DTA peak, the temperature T_p of the peak maximum appears for different heating rates β at the same degree of conversion α. Even if this condition is not completely fulfilled, the method is still often used, and the Kissinger equation

$$\ln\left[\frac{\beta}{T_p^2}\right] = \ln\left[\frac{ZR}{E_a}\right] - \frac{E_a}{RT_p} \quad (2.27)$$

says that plots of $\ln[\beta/T_p^2]$ vs. $1/T_p$ give straight lines of slope $-E_a/R$. The intercept additionally allows to calculate the preexponential factor Z introduced in equation (2.18). Blaine and Kissinger (2012) gave an interesting report on the history of this method, its limitations, its author, and further developments.

Figure 2.36 shows a Kissinger plot of the DTA data for ZnS roasting. It turns out that only the T_p values for the lower heating rates $\beta \leq 3$ K/min can be fitted well by a straight line. This is a hint that under those low heating rates, the reaction mechanism is different from that for the higher rates $\beta \geq 10$ K/min.

2.7.3.2 Kissinger–Akahira–Sunose (KAS) method

As mentioned above, the condition α = const. at $T_{p,i}$ for all β_i is often not fulfilled. (The β_i are the different heating rates.) This problem is circumvented by the Kissinger–

---

36 Homer Everett Kissinger (29 August 1923–7 July 2020).

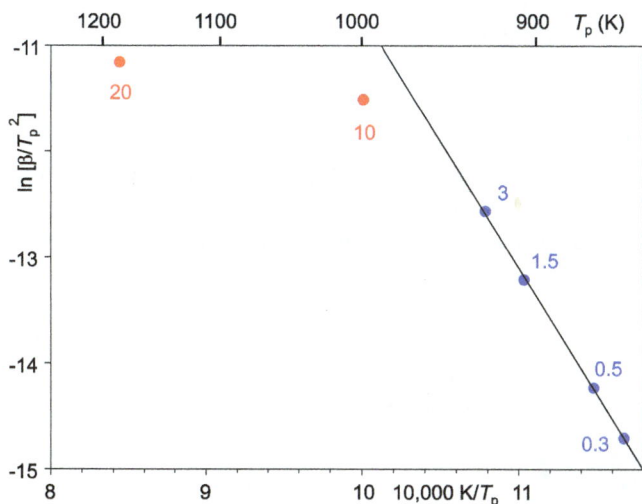

Figure 2.36: Kissinger plot for the roasting of ZnS with DTA peak data taken from Fig. 2.35 (a) (heating rates given as labels). The slope of the linear fit for $\beta \leq 3\,K/min$ gives $E_a = 199\,kJ/mol$.

Akahira–Sunose (KAS) equation

$$\ln\left[\frac{\beta_i}{T_{\alpha,i}^2}\right] = \text{const.} - \frac{E_\alpha}{RT_{\alpha,i}}, \tag{2.28}$$

where for multiple conversions $0 < \alpha_i < 1$, the corresponding $T_{\alpha,i}$ are used. For reactions with a mass change, the α_i are obtained easily from the TG signal. If the DTA signals are used, then the α_i correspond to the "partial peak area", which is demonstrated, e. g., in Fig. 2.31. However, it should be noted that especially for too large samples and high heating rates, the α_i for a given T can be significantly different, as a result of the limited rate of thermal conductivity from the sample to the sensor. A change of $E_a(\alpha)$ during the reaction progress, which can be detected by KAS, indicates a change of the reaction mechanism.

Both the original Kissinger and KAS methods are often used for the kinetic analysis of a variety of processes: Palomares-Sanchez et al. (2018) analyzed the growth kinetics of In_2O_3 thin films by KAS and compared the results with alternative methods of analysis. Shi and Tu (1989) investigated the kinetics of oxygen exchange in $YBa_2Cu_3O_{7-\delta}$ high-T superconductors by KAS. The thermal decomposition of a composed organic material (rice straw) was investigated by Kongkaew et al. (2015). For Kissinger plots, the temperature of maximum mass loss rate, which is the peak in DTG curves (first derivative of the TG) was used. They found similar results: ca. 182...222 kJ/mol for KAS and 173 kJ/mol for the original Kissinger analysis.

2.7.3.3 Ozawa–Flynn–Wall (OFW) method

The method of Ozawa (1965)[37] and Flynn and Wall (1966)[38,39] obtains E_a from plots of $\ln(\beta)$ vs. reciprocal temperature for different values $0 < \alpha < 1$. These plots give almost straight lines of slope $\approx -1.05E_a/R$.

OFW belongs to the group of isoconversional methods. If kinetic parameters like E_a that describe the reaction vary in the course of the process, then this can be interpreted in terms of a multistep reaction mechanism (Venkatesh et al., 2013). However, the method tends to fail if the types of concurring reactions are very different.

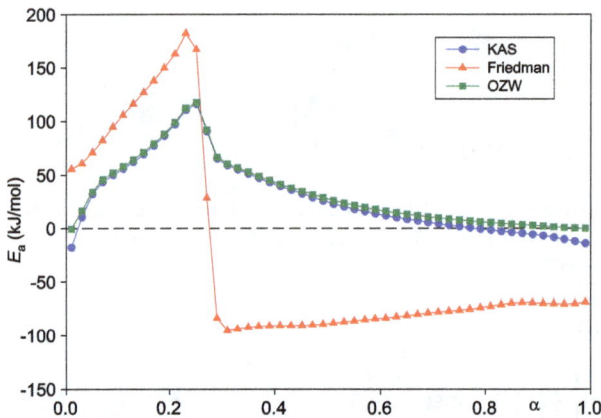

Figure 2.37: Kinetic analysis of ZnS roasting TG data from Fig. 2.35 (b) by different methods. Friedman results (see Section 2.7.3.4) are unrealistic because this method is not suitable for parallel reactions that occur during this process.

Bianchi et al. (2008) used different kinetic analysis methods, including OFW and Avrami, for investigating DSC data on crosslinking reactions of polymers. In this study, different methods proved to be complemental, and the results agreed well.

2.7.3.4 Friedman analysis

Friedman (1964) demonstrated that plots

$$\ln\left[\frac{d\alpha}{dt}\right] = \ln\left[\beta\frac{d\alpha}{dT}\right] \quad \text{vs.} \quad T^{-1} \tag{2.29}$$

give straight lines of slope $-E_a/R$. The Friedman analysis is an isoconversional method. The determination of E_a (and also of the preexponential factor) uses points

---

37 Takeo Ozawa (14 February 1932–2 October 2012).

38 Joseph Henry Flynn (28 October 1922–24 April 2011).

39 Leo A. Wall (ca. 1919–20 September 1972).

at identical conversion (e. g., $\alpha = 0.01, 0.02, \ldots, 0.99$) from the measurements at different heating rates or different isothermal conditions. There is no assumption on the reaction type. It can be used for multistep reactions. However, these reactions should be independent and should not run in parallel. Figure 2.37 compares the analytic procedures of Friedman with that of Kissinger–Akahira–Sunose and Ozawa–Flynn–Wall. KAS and OZW can handle the parallel reactions (2.24) and (2.25) satisfactory; Friedman fails here. It is a benefit, however, that Friedman analysis can be performed with dynamic and isothermal measurements.

The thermal decomposition of calcium oxalate hydrate was already discussed in Section 2.1.2.2 and is a useful test for thermal analyzers and kinetic analysis routines. Freeman and Carroll (1958) used this material for the introduction of their kinetic analysis method, which allows the derivation of kinetic parameters from a single dynamic measurement. With their method, which is not presented here in detail, the authors obtained

- Step 1 (release of H_2O): $E_a = 92\,kJ/mol$,
- Step 2 (release of CO): $E_a = 310\,kJ/mol$,
- Step 3 (release of CO_2): $E_a = 163\,kJ/mol$.

In the following, experimental data from TG measurements with different heating rates will be analyzed with the "Kinetics Neo" software (NETZSCH, 2021). Commercial software is helpful for such extensive numerical calculations, but not mandatory.

The dots in Fig. 2.38 represent experimental $\alpha(T)$ data obtained from three measurements of $Ca(COO)_2 \cdot H_2O$ with heating rates from 1 to 10 K/min. All measurements are baseline corrected. As expected, the TG steps shift to higher T for higher $\dot{T}$.

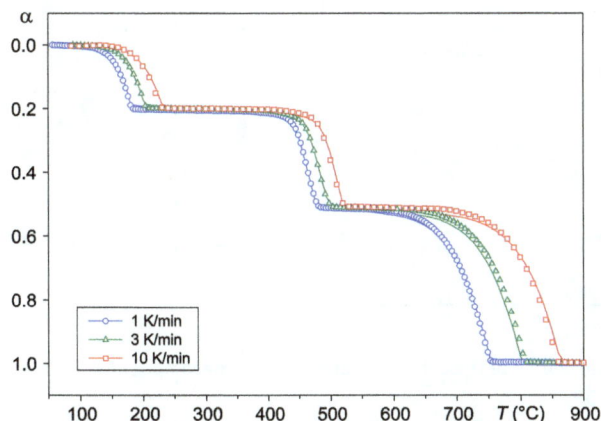

Figure 2.38: Conversion α of calcium oxalate hydrate obtained from TG measurements with three different heating rates (points = experimental values). The subsequent steps result from the release of H_2O, CO, and CO_2, respectively. Measurements with $m_s = 20 \ldots 22$ mg were performed in Ar flow with a NETZSCH STA 449C "F3".

Figure 2.39 shows Friedman plots (eq. (2.29)) for the experimental points from Fig. 2.38. Again, the three measurements are discriminated by different colors, and for every measurement, dots were calculated in steps α_i = (0.01, 0.02, ..., 0.99). For better visibility, subsequent dots are connected by colored lines.

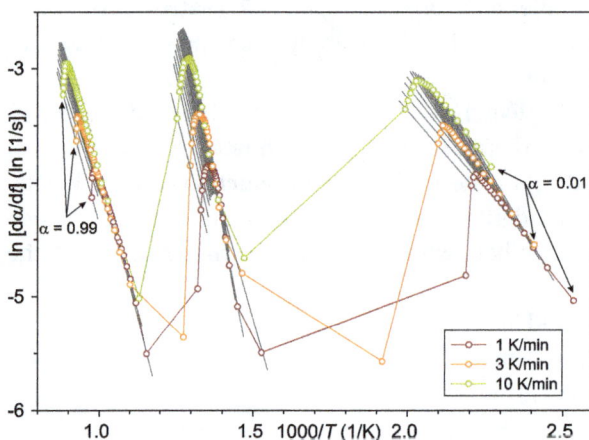

Figure 2.39: Friedman analysis for the calcium oxalate hydrate decomposition data from Fig. 2.38, calculation for $0.01 \leq \alpha \leq 0.99$ in steps of 0.01. The gray linear fits connect points of identical α, and their slope $-E_a/R$ is shown in Fig. 2.40.

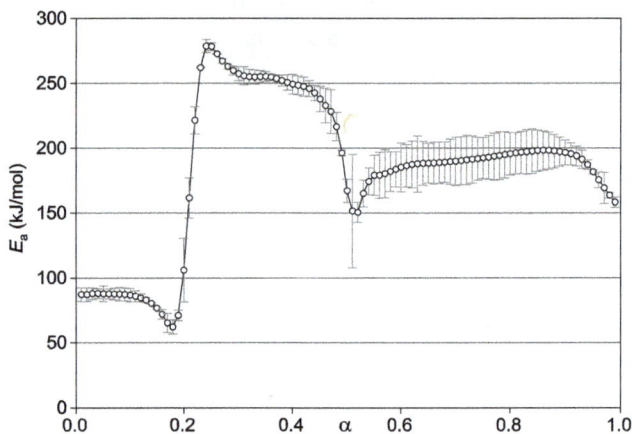

Figure 2.40: Activation energy E_a in the course of the calcium oxalate hydrate decomposition, calculated from the Friedman plot in Fig. 2.39.

For the three measurements, the α_i ($i = 1, 2, ...$) for every single i can be well fitted by the gray lines, and these gray lines are grouped into three sets corresponding to the TG steps in Fig. 2.38. The slopes $-E_a/R$ of the gray lines within each set differ not much,

and the calculated E_a for the whole decomposition reaction is shown in Fig. 2.40. For the different decomposition steps, we subsequently have
- Step 1 (release of H_2O): $E_a \approx 85\,kJ/mol$,
- Step 2 (release of CO): $E_a \approx 260\,kJ/mol$,
- Step 3 (release of CO_2): $E_a \approx 180 \ldots 200\,kJ/mol$,

which is in only fairly good agreement with the results by Freeman and Carroll (1958) given above. The remaining differences should not be overestimated and are often found if kinetic analysis is performed with different analysis methods and if different experimental raw data are used. The full lines in Fig. 2.38 are calculated functions $\alpha(T)$, based on the calculated values E_a given above.

2.7.3.5 Coats–Redfern method

Coats and Redfern (1964)[40] showed that the plots

$$\log_{10}\left[\frac{1 - (1 - \alpha)^{1-n}}{T^2(1-n)}\right] \quad \text{vs.} \quad \frac{1}{T} \quad (n \neq 1), \tag{2.30}$$

$$\log_{10}\left[\frac{-\log_{10}(1-\alpha)}{T^2}\right] \quad \text{vs.} \quad \frac{1}{T} \quad (n = 1) \tag{2.31}$$

are straight lines of slope $-E_a/\ln[10]R$. ($\ln[10] \approx 2.30$). In a subsequent paper (Coats and Redfern, 1965) it was shown that for the initial phase of a reaction, say $\alpha \leq 0.3$, the expressions from equations (2.30) and (2.31) can be simplified to plots of

$$\ln(\alpha/T^2) \quad \text{vs.} \quad \frac{1}{T}, \tag{2.32}$$

which should be straight lines with slope $-E_a/R$.

As usual, the functions $\alpha(T)$ calculated from thermoanalytic measurements are the starting point for Coats–Redfern analysis. In contrast to some other kinetic analysis methods, one single dynamic measurement is sufficient for Coats–Redfern analysis. Nevertheless, Fig. 2.41 shows $\alpha(T)$ data obtained from two TG-measurements of ZnS roasting, leading finally to the formation of ZnO as expressed by the sum equation (2.24). The different reaction paths for the low and high heating rates are obvious and were already discussed in the beginning of Section 2.7.3. Therefore the following Coats–Redfern analysis is restricted only on the initial phase of this reaction, where the intermediate byproduct $ZnO(ZnSO_4)_2$ is not yet formed.

The ZnS roasting reaction is of (almost) first order, because O_2 prevails in the gas flow in excess, and equation (2.31) can be used for analysis. The correspondingly calculated dots are shown in the top part of Fig. 2.42 for both heating rates. Apparently,

---

40 John P. Redfern (1933–2019).

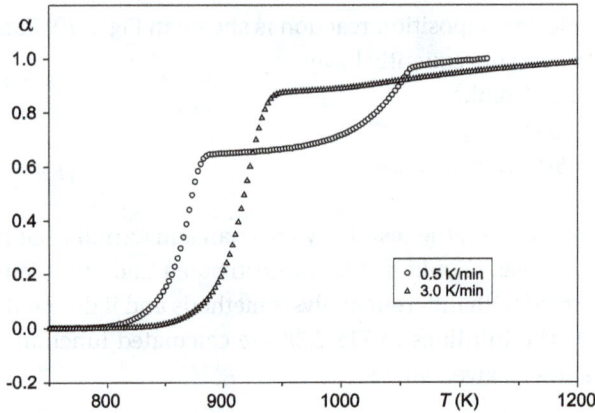

Figure 2.41: Reacted fraction $\alpha(T)$ for ZnS heated in Ar/O_2 with $\dot{T}$ = 0.5 K/min (circles) or 3 K/min (triangles). The data were calculated from the corresponding TG curves in Fig. 2.35 (b).

their slopes are identical. From the linear fit through the β = 0.5 K/min points the activation energy E_a = 339 kJ/mol is obtained for the initial phase of ZnS roasting.

Figure 2.42 analyzes only the initial phase of the roasting process, and for the corresponding small $\alpha \leq 0.3$, Coats and Redfern (1965) proposed the simpler plots (2.32). Those plots are steeper by the factor $\ln[10] \approx 2.30$, and the calculated E_a = 321 kJ/mol is similar to the value reported before for the detailed analysis.

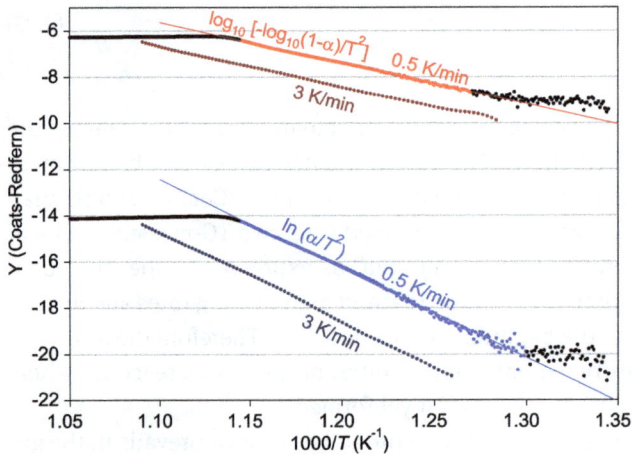

Figure 2.42: Analysis of the $\alpha(T)$ curves from Fig. 2.41 by the Coats and Redfern (1964) analysis from eq. (2.31) (top) and by the simplified formula reported by Coats and Redfern (1965) analysis from eq. (2.32) (bottom). The slope of the linear fit in the bottom part is $-E_a/R$ = $-38,588$, resulting in E_a = 321 kJ/mol. The slope of the linear fit in the top part is $-E_a/\ln[10]R$ = $-17,694$, resulting in E_a = 339 kJ/mol.

2.8 Thermodynamic modeling

Often, the results of thermal analysis, e. g., for the determination of a phase diagram, are presented "as measured". As an example, Fig. 2.43 shows experimental points for the eutectic system NaCl–CsCl measured by DTA. These points were connected by more or less arbitrary lines; just for the eutectic, a linear fit was performed. Three problems are obvious:

1. The (green) eutectic line rises monotonously, which must be an experimental error, because the eutectic temperature in a simple eutectic system is constant (see Section 1.5.1.3).
2. The (blue) line indicating the solid state transition of CsCl from its low-T CsCl-structure to the high-T NaCl structure (see Fig. 1.17 (c) and (d)) shows an upward bend in the middle, which is also unphysical.
3. The (red) liquidus curves intersect slightly above the green eutectic line.

It is a disadvantage of the color lines in Fig. 2.43 that they present the experimental data without taking into account basic rules of thermodynamics. Besides, additional thermodynamic data, like heat of fusion ΔH_f, c_p, and enthalpy, either from other measurements or from the literature, cannot be used easily for the construction of the phase diagram. "Thermodynamic assessment" with the CALPHAD method ("CALculation of PHAse Diagram" — see, e. g., Chang et al., 2004) is a modern and successful tool to overcome such problems.

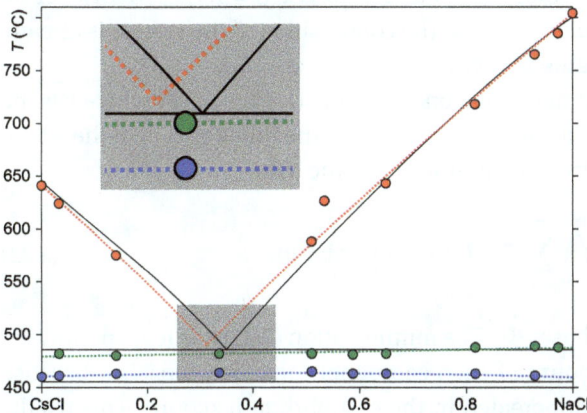

Figure 2.43: Experimental points (DTA heating curves) with 11 compositions x NaCl + (1 − x) CsCl ($0 \leq x \leq 1$). The dashed color lines are "guides to the eye" connecting the experimental points fairly good (simple linear fit for the eutectic (green) line). The black lines result from a thermodynamic assessment of the system.

The basic idea behind such calculations is that experimentally evident phase boundaries are not fitted, e. g., by the Schröder–van Laar equations (1.35), (1.36), (1.37),

(1.38). Instead, the numerical Gibbs energy functions $G(T, x)$ for all phases are adjusted in such a way that all experimental data are optimally adapted. Several commercial and freeware software packages are available that can perform this task, and it is beyond the scope of this book to present them here in detail. The black phase boundaries in Fig. 2.43 result from such thermodynamic assessment and are a reasonable fit of the experimental data without contradiction to thermodynamic laws.

Here we will not give a detailed introduction into the technical details of numerical modeling software packages. In the framework of this book, exclusively FactSage 8.0 (2020) or its predecessor versions, including ChemSage (GTT Technologies, Herzogenrath, Germany), were used. Thermo-Calc (2020) and Pandat (2020) are alternatives. For solid–gas equilibria, the freeware TraGMin 5.1 (2008) can be used.

! Whenever useful and possible, experimental findings of thermal analysis should be discussed in terms of an appropriate thermodynamic model, rather than being presented in its original form.

2.8.1 Calculation of thermodynamic equilibrium

The Gibbs energy G introduced in Section 1.2.3 has a minimum under equilibrium conditions. With a few exceptions, thermal analysis is performed under isobar conditions (p = const., typically, p = 1 bar), and G depends only on the temperature T and the composition of the system. The composition of a system with C components is given by $C - 1$ concentration data, preferably as molar fractions x_j ($j = 1, 2, \ldots, C - 1$). For binary systems, the index j is obsolete because the composition of the system is already defined by one concentration value $x_1 = x$.

Usually, the system contains more than one phase Φ_i ($i = 1, 2, \ldots, P$; where P is the number of coexisting phases). Then for the binary case, the Gibbs energy of the whole system, which has to be minimized at equilibrium, is the sum

$$G(T, x) = \sum_{i=1}^{P} G^{\Phi_i}(T, x) \longrightarrow \text{minimum} \tag{2.33}$$

of the Gibbs energies of the P phases Φ_i. The minimization (2.33) has to be performed in such a way that the total amount of each component remains constant, because they can neither be destroyed nor created in the equilibrium reactions. This results in constraints for the amounts for the different phases that can be formed from the given amount of each component. The thermodynamic simulation programs mentioned above are able to perform this optimization and can calculate equilibrium states of systems this way. This includes the calculation of the enthalpy difference that occurs if the external conditions (such as T and p) are changed. It will be shown in Section 2.8.2.1 that this enthalpy difference corresponds to the heat flow between a thermoanalytic sample and its environment and hence to the DTA or DSC signal.

2.8.1.1 Data for stoichiometric phases

As pointed out in Section 1.2.3.1, for "pure" phases with fixed stoichiometry, the x_i are no free parameters, and G is only a function of T. (As mentioned above, the optional dependence on p can be neglected for the conditions of thermal analysis.) In some printed collections, e. g., Barin (1995), tables for thermodynamic data such as G, H, S, c_p can be found for many substances. Such tables are valuable resources for calculations "by hand", but analytical expressions are preferred. This was initially done by simple power series with positive exponents, but already Maier and Kelley (1932)[41] pointed out that often at least one negative exponent is useful, because this way unphysical maxima can be avoided. Rather, these authors proposed expressions of the type

$$c_p(T) = a + bT + cT^{-2}, \tag{2.34}$$

where just three constants a, b, c are sufficient to represent $c_p(T)$ data over an extended temperature range. Functions similar to (2.34) can represent very well the typical $c_p(T)$ dependence, which is often square-root like at low T and almost linear at high T (see Fig. 1.3). Fit functions of type (2.34) were developed further by Shomate (1954)[42] and are now (often with more than three coefficients) called the Shomate equations. In web resources like NIST (2020) and in databases that come with thermodynamic simulation software thermodynamic data are stored this way.

For ambient pressure $p = 1$ bar, the Gibbs energy of a stoichiometric phase can be calculated from $c_p(T)$ by (Harvey et al., 2020)

$$G^{\Phi}(T) = H^{\Phi}(T_{\text{ref}}) + \int_{T_{\text{ref}}}^{T} c_p \, dT - T \left[S^{\Phi}(T_{\text{ref}}) + \int_{T_{\text{ref}}}^{T} \frac{c_p}{T} \, dT \right], \tag{2.35}$$

where $H^{\Phi}(T_{\text{ref}})$ is the standard enthalpy of formation, which is typically negative, because just the release of H^{Φ} is the reason why the phase Φ is formed from its constituents. The first two terms represent $H^{\Phi}(T)$, as shown in equation (1.19), and the term in brackets is $S^{\Phi}(T)$; see equation (1.24). As an example, the reader is referred to İlhan et al. (2017), who derived the thermodynamic data $\Delta H = H(T) - H(T_{\text{ref}})$ and $\Delta S = S(T) - S(T_{\text{ref}})$ for ceramic $BaTa_2O_6$ samples from $c_p(T)$ data measured by dynamic DSC. In their paper, an additional quadratic term was added to the fit function (2.34), which results in expressions of the type

$$c_p(T) = a + bT + cT^2 + dT^{-2} \tag{2.36}$$

---

41 Kenneth Keith Kelley (16 December 1901–28 September 1991).
42 Charles Howard Shomate (28 January 1915–28 January 1997).

with four fit parameters a, b, c, d. The maximum deviation of a few single experimental points from their fit equation was in the order of 2 % but was typically much smaller. Besides, İlhan et al. (2017) compared their experimental $c_p(T)$ data with theoretical values derived using the Neumann–Kopp rule (1.7) with the binary oxides BaO and Ta_2O_5 as basis. It turned out that the Neumann–Kopp values are very close to the experimental results only between room temperature (were measurements started) and $\approx$ 800 K; otherwise, Neumann–Kopp data are systematically below measured data. The difference reaches $\approx$ 8 % at the maximum temperature of this investigation, which is 1300 K. (The difference near room temperature was only 1.4 %.)

---

i The first integral in equation (2.35) is the enthalpy increment ΔH with respect to the reference temperature (usually, T_{ref} = 298 K). The second integral in equation (2.35) is the entropy increment ΔS with respect to T_{ref}. If the function $c_p(T)$ is given in analytic form of the kind shown in equation (2.34), e. g., from fitted experimental data, then both increments can be derived from it.

By definition, for chemical elements under standard conditions, we have $H^\Phi(T_{ref}) = 0$. Leitner et al. (2010)[43] showed that for the case that data for chemical compounds cannot be found in the literature, the Neumann-Kopp rule often gives a good approximation for $H^\Phi(T_{ref})$ for most substances. However, Schick et al. (2019)[44] pointed out that an uncritical application of the Neumann–Kopp data can lead to errors.

The Neumann–Kopp rule was originally designed for the description of metallic alloys and often fairly well describes systems with mainly metallic bonding. However, deviations must be expected for systems where the bonding type of mixtures is significantly different from the bonding type(s) of the pure components. Thomas (2015) performed a thorough investigation of the system Li–Si, where the end members have metallic or covalent bonds, respectively. For intermediate compositions, as much as six lithium silicides with silicon concentrations $\approx 0.19 < x_{Si} \leq 0.5$ are reported. These compounds are Zintl[45] phases with partially ionic character; consequently, their $c_p(T)$ functions deviate significantly from the values obtained from data for Li and Si using the Neumann–Kopp rule.

Vibrations of the crystal lattice represent for most substances the major contributions to the specific heat capacity $c_p \approx c_V$, which was described in Section 1.1.3. Consequently, we can expect that the validity of the Neumann–Kopp rule relies on at least fairly similar interatomic forces between the atomic units of the components. We must expect severe deviations from the Neumann–Kopp rule for these compounds, because this is not the case for Li, Si, and different Li_xSi_y compounds.

---

43 Jindřich Leitner (born 11 September 1956).

44 Christoph Erich Georg Schick (born 16 March 1953).

45 Eduard Zintl (21 January 1898–17 January 1941).

Figure 2.44 (b) demonstrates this for the example $Li_{15}Si_4$, which has a cubic crystal structure, like its constituents Li and Si. The unit cell of lithium (Fig. 2.44 (a)) is body-centered cubic with metallic bonding. This means that all Li atoms are positively charged and are embedded in a "sea of delocalized electrons", which is responsible for undirected binding forces between the atoms. The crystal structure of silicon, in contrast, consists of a tetrahedral network with covalent Si—Si bonds. In the crystal structure of the intermediate $Li_{15}Si_4$, Si atoms are surrounded by 12 Li atoms in a slightly distorted cuboctahedral symmetry. Consequently, Si—Li binding forces prevail, which do not occur in both components. However, even for intermetallic compounds, significant deviations from the Neumann–Kopp rule sometimes occur. Silva et al. (2021) demonstrated this, e. g., for two iron-niobium compounds.

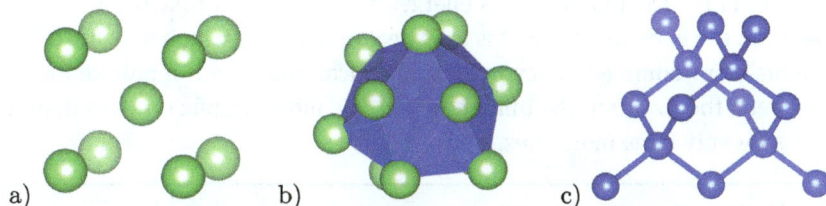

Figure 2.44: (a) 1 unit cell of Li (bcc structure, see Fig. 1.17, undirected metallic bonds, $F\bar{4}3m$). (b) distorted cuboctahedral [$SiLi_{12}$] building unit of $Li_{15}Si_4$ ($I\bar{4}3d$). (c) 4 [Si—Si_4] tetrahedra as building unit of Si crystals ($F\bar{4}3m$). Drawn with crystal structure data from Wyckoff (1963) and Zeilinger et al. (2013); green = Li, blue = Si.

For the estimation of thermodynamic data for chemical compounds from their constituents, it is preferred to obey similar atomic building units and binding forces for all phases. This means that, e. g., data for complex sulfides ABS_{x+y} should be calculated from simple sulfides AS_x and BS_y, rather than from the elements A, B, S. Whenever possible, also coordination numbers should be identical. **i**

Sometimes, it is possible to derive thermodynamic data for a complex compound ABX_{x+y}, not simply by adding the data from its constituents (X stands for the anion)

$$\underbrace{AX_x}_{\text{known}} + \underbrace{BX_y}_{\text{known}} \rightleftarrows \underbrace{ABX_{x+y}}_{\text{unknown}}, \tag{2.37}$$

but rather a formal exchange reaction can be defined to derive it from an isotype phase with similar ionic radii

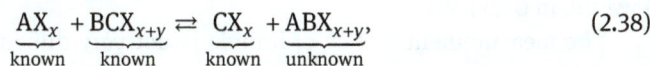

$$\underbrace{AX_x}_{\text{known}} + \underbrace{BCX_{x+y}}_{\text{known}} \rightleftarrows \underbrace{CX_x}_{\text{known}} + \underbrace{ABX_{x+y}}_{\text{unknown}}, \tag{2.38}$$

which ensures that all requirements from the information box above are fulfilled. An example is given in Section 3.3.5.

2.8.1.2 Data for mixture phases

In Section 1.2.3.2, it was shown that the Gibbs energy of mixtures is a sum (1.26) of three contributions:

– $G^0(T)$ is the weighed sum of the contributions of the pure components. For two components, this is the straight dashed line in Fig. 2.45, which connects the $G(T)$ values of the pure components; see equation (1.27).

– $G^{id}(T) = -TS^{id} \leq 0$ is the contribution of disorder, which reduces the Gibbs energy of (true, on an atomic scale) mixtures, compared to a pure macroscopic mixture with identical overall composition. S^{id} depends only on the number of equivalent "microstates" of the system ($\cong$ different possible arrangements of atoms). Hence it is only a function of the composition x but does not depend on the substances; see equation (1.29) and the red dotted curve in Fig. 2.45.

– $G^{ex}(T) = H^{ex}(T) - TS^{ex}(T)$ describes energetic (H^{ex}) and entropic (S^{ex}) interactions between the constituents. H^{ex} is the "heat of mixing", which appears if the components show some tendency to form chemical bonds or if repulsive forces occur between them. Often, the influence of H^{ex} is more significant than that of S^{ex}, unless for very large molecules.

---

i Deviations from ideal behavior tend to be small:

(1) at high T, because then interaction energies between atoms are often weaker, and the entropic contribution $G^{id} = -T\Delta S^{mix}$ becomes more important.

(2) in the liquid (or the more gaseous) state, because then stresses resulting from different sizes of atoms or molecules are less relevant.

(3) if chemically similar species are mixed, because then energetic interactions are usually weak.

(4) if the mixed species are of similar size; here it is less critical if a "small" atom replaces a "large" atom than vice versa.

Figure 2.45 shows for $T = 1200$ K (hence in the solid state) all three contributions for the system silicon–germanium, which form solid solutions $Ge_{1-x}Si_x$ for all concentrations $0 \leq x \leq 1$. It is obvious that already the sum $G^0 + G^{id}$ gives a fairly reasonable approximation for the total Gibbs energy G^Φ of the mixed crystal. The small contribution G^{ex} shifts G^Φ slightly upward. We can assume that the radius difference between Si and Ge atoms increases the inner energy of the solid phase, which results in a slight destabilization, compared to the ideal mixture. Such destabilizing effects can occur in both solid and liquid phases, but for one specific system, they are usually weaker in liquids, because there the atom positions are not fixed, in contrast to solids. If attractive forces between components A and B are stronger than the average of the interactions A—A and B—B, then H^{ex} becomes negative, and the function $G^\Phi(x)$ is slightly lower than $G^0(x) + G^{id}(x)$.

The measurement of H^{ex} of liquids is not very difficult because it can be performed in calorimeters by measuring the heat of mixing. Figure 2.46 shows experimental $H^{ex}(x)$ data for mixtures of chloroform with cyclohexane (blue curve) and ethanol

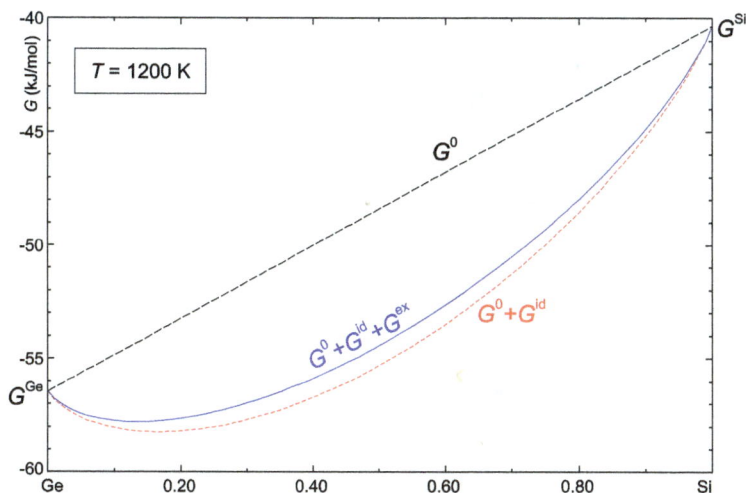

Figure 2.45: Functions $G(x)$ at $T = 1200\,\mathrm{K}$ for a mechanical mixture of silicon and germanium (G^0) compared to an ideal ($G^0 + G^{id}$) and a real ($G^0 + G^{id} + G^{ex}$) mixture (= solid solution).

(red curve). It is obvious that both curves start and end for the pure components at $H^{ex} = 0$, because there no interaction occurs. Cyclohexane shows with chloroform always positive (repulsive) interaction, with a maximum value around 640 J/mol at $x \approx 0.5$. For the interaction with ethanol, surprisingly, the interaction is exothermal ($H^{ex} < 0$) on the ethanol-rich side and endothermal for chloroform-rich mixtures.

Depending on the chemical nature of the systems, different models were developed for an analytical approximation of experimental functions $H^{ex}(x, T)$, and some of these models are implemented in the corresponding software packages mentioned above. Pelton (2019) gives a comprehensive introduction on this topic, and here we only present the following polynomial expression of the excess Gibbs free energy in binary systems with molar fractions of the components x_1, x_2:

$$G^{ex} = x_1 x_2 \sum_{j \geq 0} {}^j L_{12} (x_1 - x_2)^j, \qquad (2.39)$$

which is often useful. The coefficients ${}^j L_{12}$ are independent of composition but may be functions of temperature, e. g., ${}^j L_{12} = {}^j L_{12}^0 + {}^j L_{12}^1 \cdot T$.

The functions $(x_1 - x_2)^j$ are called the Redlich[46]–Kister polynomials (Redlich and Kister, 1948). The polynomials fulfill the condition $G^{ex} = 0$ for the pure components (no interactions), because x_1 and $x_2 = 1 - x_1$ are factors in equation (2.39); this holds

---

46 Otto Redlich (4 November 1896–14 August 1978).

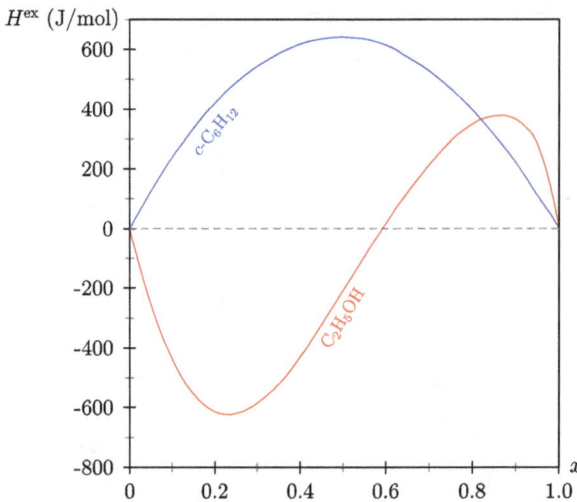

Figure 2.46: Experimental H^{ex} data for mixtures of chloroform (CH_3Cl) with cyclohexane (c-C_6H_{12}, blue curve, data from Nagata et al., 1980) and ethanol (C_2H_5OH, red curve, data from DDBST, 2021). $T = 298.15$ K, $x = 1$ is pure CH_3Cl.

for all j. For $j = 1$, an additional zero-crossing appears for $x_1 = x_2 = 0.5$, and G^{ex} has the opposite sign for $x \gtrless 0.5$, which allows a description of a behavior like shown for CH_3Cl–C_2H_5OH in Fig. 2.46. Figure 2.47 shows plots of the Redlich–Kister polynomials for $j = 0 \ldots 2$; it is obvious that a small contribution from $j = 2$ can adjust the antisymmetric contribution from $j = 1$ in such a way that different absolute values for $x \gtrless 0.5$ are possible. It turns out that higher contributions $j > 2$ are seldom needed for the description of real systems. Some combinations for the coefficients $^j L_{12}$ are as follows:

Ideal solution: $^j L_{12} = 0$ for all j, and hence $G^{ex} = 0$ (= Raoultian[47] solution).

Regular solution: Only $^0 L_{12} \neq 0$, and hence $G^{ex} = x_1 x_2 {}^0 L_{12}$ (black full line in Fig. 2.47).

Subregular solution: $^0 L_{12} \neq 0$, $^1 L_{12} \neq 0$, and $G^{ex} = x_1 x_2 ({}^0 L_{12} + {}^1 L_{12}(x_1 - x_2))$.

2.8.2 Simulation of thermoanalytic measurements

Harvey et al. (2020) presented examples where the calculated heat consumption during heating of an aluminum alloy is compared to a measured DSC curve or where the calculated mass loss during heating of the vitriol of copper ($CuSO_4 \cdot 5H_2O$) is compared to an experimental TG curve. It turns out that the calculations represent the ex-

---

47 François-Marie Raoult (10 May 1830–1 April 1901).

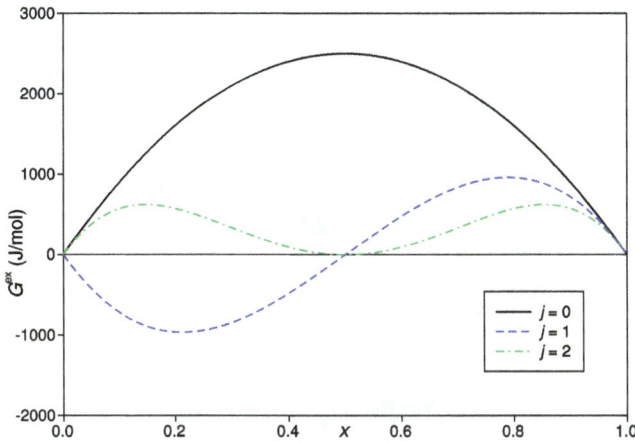

Figure 2.47: Contributions from Redlich–Kister polynoms (2.39) to G^{ex} for the degrees $j = 0, j = 1$, and $j = 2$. $L_j = 10,000$ was set for all j.

perimental results fairly well, but both calculations result in rather abrupt (for the TG curve step-like) curves, in contrast to the experimental results that are more smeared. It will be shown in the next section that the smearing of DTA/DSC curves can be accounted for using the thermal lag of the signals resulting from the finite rate of heat conduction. In contrast, mass changes during thermal analysis are measured almost immediately because they are directly registered by the balance. Nevertheless, also TG steps are usually not sharp, which will be shown in Section 2.8.2.2 for the thermal decomposition ("calcination") of $CaCO_3$ to CaO.

2.8.2.1 Simulation of DTA/DSC curves

The red curve in Fig. 2.48 represents DTA signals obtained upon heating of a pure barium chloride sample; it is from the same measurement shown as the uppermost curve of Fig. 3.30 (a) in Section 3.3.4. It is explained there that the first peak at 925.3 °C results from the transition between two modifications of $BaCl_2$, and the second peak with experimental onset at 958.5 °C marks the melting temperature.

Both processes are endothermal and result from a steplike enthalpy increment at the corresponding transition points, like shown for sulfur in Fig. 1.7. On the other side, the DTA or DSC signal is proportional to the difference of the heat flows to the sample and to the reference (see Section 2). For a power compensation device, the DSC signal is equivalent to the difference of the electrical heating power between sample and reference.

The power P is the time derivative of energy (here enthalpy H), but for a constant heating rate $\dot{T}$, the time derivative can be replaced by the temperature derivative

$$P = \frac{\partial H}{\partial t} = \dot{T}\frac{\partial H}{\partial T},$$

(2.40)

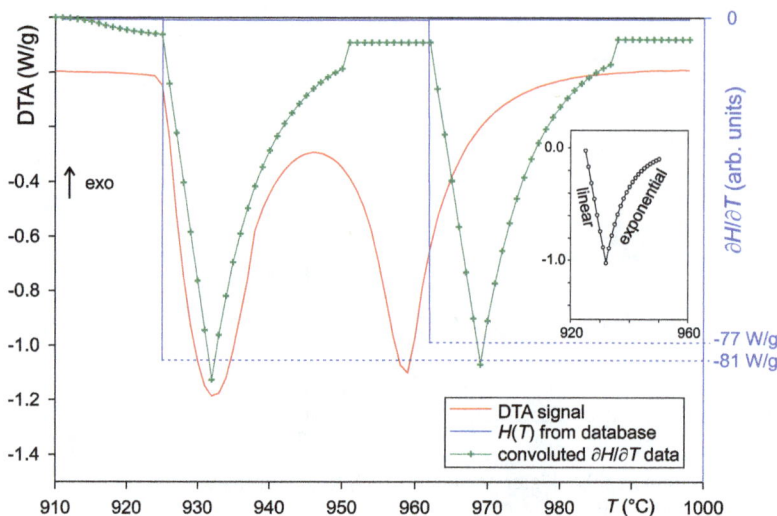

Figure 2.48: Experimental DTA data (red curve) for $BaCl_2$ compared with numerical $\partial H/\partial T$ data (blue curve) calculated from FactSage 8.0 (2020) data. The green dots and curve are the result of a numerical convolution of $\partial H/\partial T$ with the device function $g(T)$, which is shown in the insert.

and this temperature derivative of the function $H(T)$ for $BaCl_2$ is the blue function in Fig. 2.48. Except at first-order phase transitions, $H(T)$ changes only smoothly; consequently, $\partial H/\partial T$ almost vanishes on the scale of the diagram. It is not so at the transition points where $H(T)$ undergoes jumps with the height of the transition enthalpy ΔH. (The derivative was calculated numerically with 1 K steps, which results in finite spikes of height ΔH at the transition points.)

It was explained in Section 2.5.3 that peaks originating from a "sharp" first-order transition cannot be that sharp in the experimental DTA or DSC curve, because the heat flow rate to sample and reference is limited. If we assume the thermal resistance to be approximately constant, then the "transition rate" (e. g., for melting) is almost constant until the whole sample is molten, because heat flows with an almost constant rate into the sample. This results in a nearly linear slope of the first part of the melting peak. Afterward, the sample temperature returns back to the baseline with an exponential curve. (Actually, the heat flow rate is not exactly constant, because the furnace temperature continues to rise during melting. However, also the distance between the crucible surface and the still solid parts of the sample center grows.)

Such a shape can be observed for many experimental conditions if a pure substance is molten (see, e. g., Figs. 2.28 and 2.29) and can be considered as a "device function" of the corresponding experimental setup (furnace, sample holder, crucibles) with some typical time constants τ_i that describe the time delay of the peaks; see equations (2.14) and (2.15). For the setup used in Fig. 2.48, this device function is shown in the insert with a linear first (dropping) and an exponential second (rising) part of the peak. Also, there the dots mark 1 K steps, which means that the initial dropping part of the peak extends over 7 K. With $\dot{T}$ = 10 K/min, this corresponds to the time τ = 0.7 min = 42 s.

The expected DTA signal $\Phi^{meas}(T)$ is the convolution of the device function g with $\Phi^{sample}(T)$ in a manner described by equation (2.13). If $\Phi^{sample} = \dot{T}\partial H/\partial T$ and g is the device function, then the integration is performed over T. In experiments the functions are given numerically as time series, and the integral (2.13) is replaced by the sum

$$\Phi^{meas}(T) = \Phi^{sample}(T) * g(T) = \sum_{k=-\infty}^{\infty} \Phi^{sample}(k)g(T-k). \tag{2.41}$$

For long data series, calculation (2.41) is elaborate but can be performed with software packages like Mathematica© or Mathlab©. Besides, free internet resources like mymathtables (2020) can be used.

For BaCl$_2$, the convolution result (2.41) is shown as green dots and curve in Fig. 2.48. The shape of the first (phase transition) peak is fairly well reproduced by the convolution function, with the exception that the top of the DTA peak cannot be perfectly met by the simple function g that was used there. In contrast to the experimental DTA curve, both peaks are not overlapping in the convolution. This results from the fairly high T_f = 962 °C, assumed by FactSage 8.0 (2020), which results in a better separation of the peaks. However, it should be noted that not only in Fig. 2.48 a lower melting point (onset 958.5 °C there) was found: Also, in the literature, often values below 960 °C are reported. Koštenská (1976) observed T_f = 955 °C for BaCl$_2$, and she reports the phase transition at 920 °C.

The result of the calculation in Fig. 2.48 is comparatively obvious, because $\Phi^{sample}(T)$ is almost a delta function, and any signal $x(n)$ convolved with the delta function is left unchanged, $x(n) * \delta(n) = x(n)$ (Smith, 1997). In fact, the two green peaks are simple copies of the device function, which is shown in the insert.

Figure 2.49 shows an analogous treatment of data for one intermediate composition of the system LiCl–NaCl. For this composition, which is marked in Fig. 2.49 (b) by a red dashed line, the difference between T_{sol} and T_{liq} is 99 K. Correspondingly, the melting process itself extends over the rather wide T range between the sharp dips of the blue $\partial H/\partial T$ curve in Fig. 2.49 (a).

On the basis of the phase diagram in Fig. 2.49 (b), explain why the red DTA curve in Fig. 2.49 (a) has a "double peak" shape! Hint: Use the lever rule (Section 1.5.1.1)!

The convolution function $g(T)$ was determined here by melting an aluminum sample. The Al melting peak shown in the insert lies in the same T range like the melting peak of the (Li,Na)Cl sample. The shape of the Al melting peak is defined by the

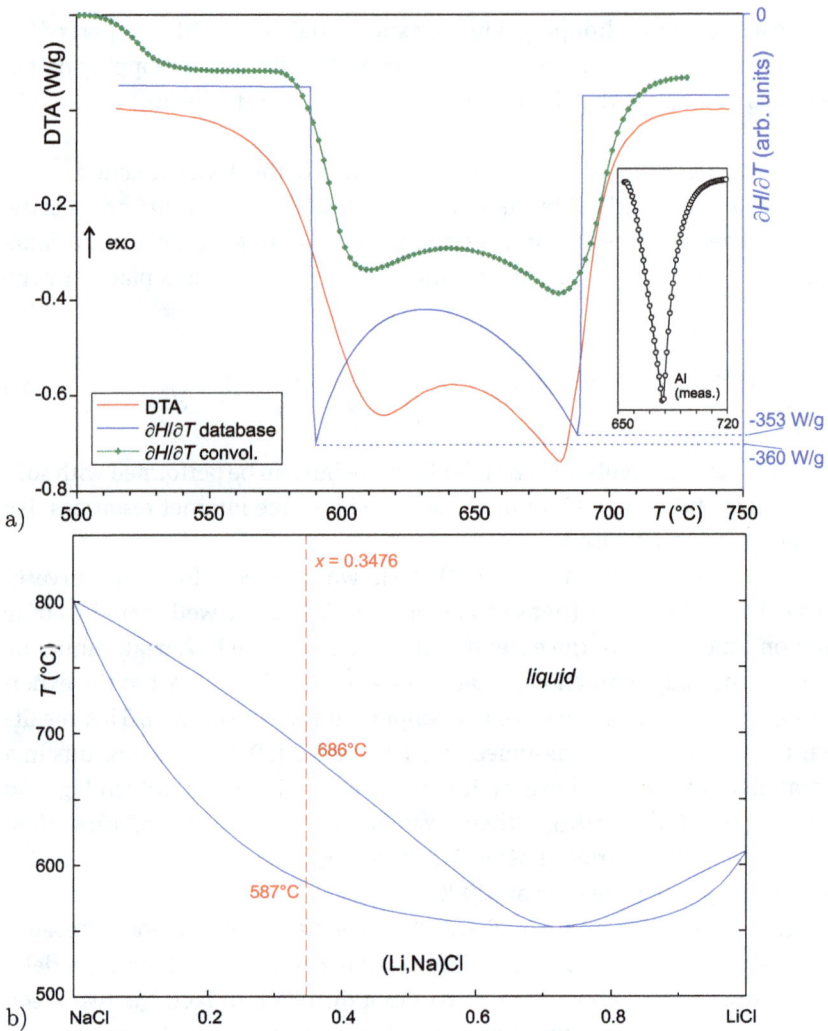

Figure 2.49: (a) Experimental DTA data (red curve) for a $Li_xNa_{1-x}Cl$ mixed crystal (x = 0.3476) compared to numerical $\partial H/\partial T$ data (blue curve) calculated from FactSage 8.0 (2020) data. The green dots and curve are the result of a numerical convolution (2.41) of $\partial H/\partial T$ with the experimental melting peak of Al (insert), which was used as the device function $g(T)$. (b) Calculated phase diagram LiCl–NaCl; the composition that was investigated in the upper panel is marked by the red dashed line.

function $g(T)$, which also smears $\Phi^{sample}(T)$(see eq. (2.41)), because Al melts sharp at 660.3 °C. The green curve is the result of the numerical convolution and is in reasonable agreement with the experimental DTA curve.

It is interesting to note that the first local minimum of $\partial H/\partial T$ near 590 °C (−360 W/g) is slightly deeper than the second minimum near 690 °C (−353 W/g). In contrast, for the simulated and measured DTA curves, the second minimum is deeper, which is a result of the delayed arrival of the enthalpy change at the DTA sensor.

2.8.2.2 Simulation of TG/EGA curves

As the result of thermodynamic equilibrium calculations, we obtain not only an energetic balance of the corresponding process, which can be used, e. g., for the simulation of a DTA curve. Moreover, also chemical conversions of the system are calculated. If the gas phase contributes to these reactions, then changes of the sample mass and hence TG effects have to be expected. Typical reasons for significant TG effects were mentioned in Section 2.2.

The bottom panel of Fig. 2.50 shows two subsequent DTA/TG heating curves of a commercial $CaCO_3$ powder sample, which is expected to decompose according equation (3.2) under the release of CO_2. This decomposition corresponds to the large TG steps $\approx 750\ldots905°C$. We will discuss further in Section 3.1.1 that a much smaller TG step $\approx 400\ldots416°C$ results from the decomposition of $Ca(OH)_2$ impurities. Both reactions are endothermal, which results in DTA peaks.

Changes of the sample mass act almost immediately on the balance that measures the TG signal; hence a convolution on the time scale like described in the previous Section 2.8.2.1 does not occur. Nevertheless, also TG steps are often smeared for the following reasons, which can often hardly be separated:
- The underlying chemical reaction can be slowed down by kinetic reasons, because an "activation barrier" must be overcome.
- In an equilibrium reaction like (3.2) the reaction rate is influenced by the concentration of the products, in this case, of CO_2 gas. Hence the efficiency of gas removal from the (solid or liquid) sample surface will influence the reaction rate measured by DTA and TG. Experimental details like the gas flow rate, usage of a lid, or the total pressure in the thermal analyzer can have a significant influence on the experiment.

The quantitative observation of such experimental details in calculations is challenging. Besides, it is significantly easier to calculate a thermodynamic equilibrium situation (given amount of substances under defined conditions T, p) than to calculate a process where T changes under permanent in- and outflow of rinsing gas. The top panel of Fig. 2.50 shows two calculation where $CaCO_3$ with 3 % $Ca(OH)_2$ added as "impurity" are heated in different amounts of the "rinsing gas" Ar. If m_{Ar} is the mass (in gram!) of argon and m_{gas} is the total mass of the gas phase, then $\Delta m = m_{gas} - m_{Ar}$ is the total mass of all gaseous reaction products (mainly H_2O and CO_2, plus byproducts) that emanate from the sample. This Δm is plotted as the "mass loss" in negative direction on the ordinate of Fig. 2.50 (top) for two different m_{Ar}, which are equivalent to 0.1 mol (3.99 g) or 0.5 mol (19.97 g), respectively. Both calculated $\Delta m(T)$ curves are similar to the experimental TG curve (blue dashed line in the bottom panel). As expected, the calculated TG steps occur slightly earlier for the larger amount of Ar.

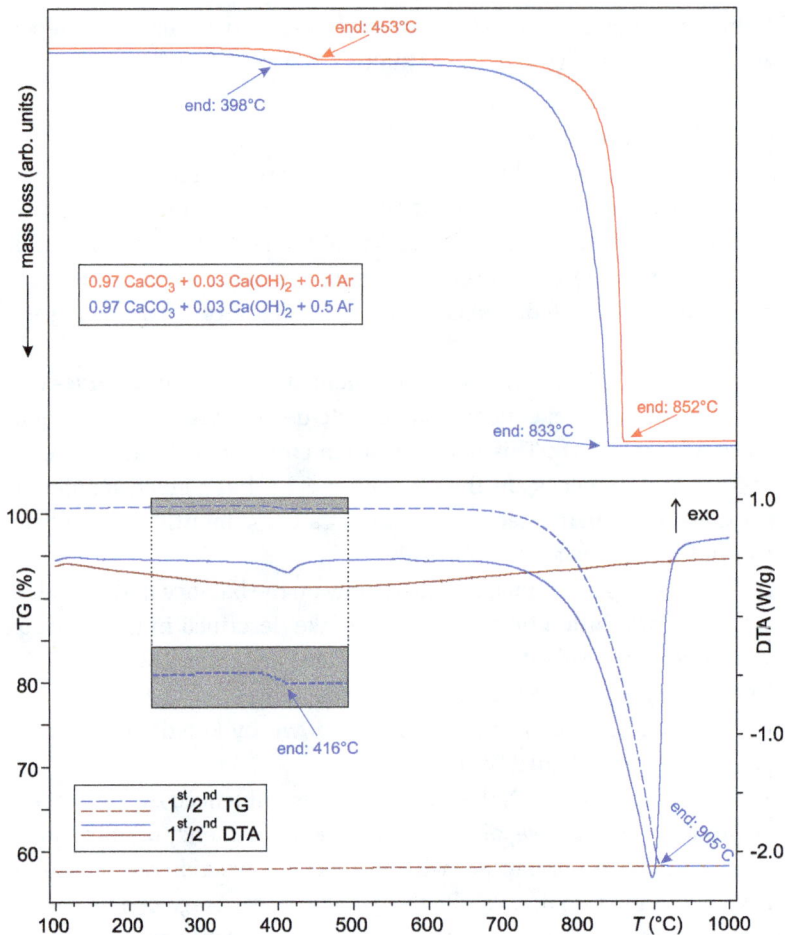

Figure 2.50: Top panel: Calculated mass loss of $CaCO_3$ with $Ca(OH)_2$ impurities, heated in different amounts of Ar. One curve is slightly shifted downward for better readability, calculations with FactSage 8.0 (2020). Bottom panel: DTA/TG measurement of commercial $CaCO_3$ (99.999% nominal purity) in flowing Ar, $\dot{T} = 10\,K/min$. First and second heatings run under identical conditions from 100 °C to 1000 °C. See also Fig. 3.1.

2.8.3 Predominance diagrams

If a metal Me can form oxides with different valence, e. g., m and $m + 1$, then a redox equilibrium reaction of the type

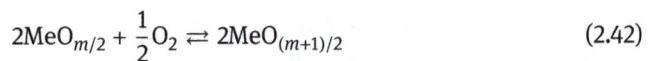

$$2MeO_{m/2} + \frac{1}{2}O_2 \rightleftarrows 2MeO_{(m+1)/2} \tag{2.42}$$

describes the mutual transformation of both oxides. The Gibbs energy change of reaction (2.42) can be written as

$$\Delta G^0 = -RT \ln K = -RT \ln\left(\frac{p^2_{(m+1)/2}}{p^2_{m/2} p^{1/2}_{O_2}} \right) \qquad (2.43)$$

with equilibrium constant K. Often, the partial pressures of the metal oxides can be neglected with respect to the oxygen partial pressure p_{O_2}, and (2.43) further simplifies to

$$\Delta G^0 = \Delta H^0 - T\Delta S^0 = RT \ln p_{O_2}. \qquad (2.44)$$

The plots $\Delta G^0 = RT \ln p_{O_2}$ vs. T are almost straight lines, because ΔH^0 and ΔS^0 are constant. Such plots were introduced by Ellingham (1944).[48] Pelton (1991) showed that a series of Ellingham plots (or Ellingham diagrams) $RT \ln p_{O_2}(T)$ for a multitude of oxidation states sets up a "predominance diagram". This is a special type of phase diagram where stability fields of different MeO_x are separated by (mainly) straight inclined lines. Transitions between different phases of one compound, such as melting, appear as vertical lines in the diagram.

Figure 2.51 (a) shows an Ellingham-type predominance diagram for the system $Mn-O_2$. Under the given conditions, MnO_2 has the highest valency +4; other oxides like $Mn_2^{7+}O_7$ are unstable (Glemser and Schröder, 1953)[49] and cannot be found in this equilibrium diagram. At ambient pressure, Mn assumes four different allotropic forms Mn(s), Mn(s2), Mn(s3), Mn(s4), which are separated by vertical lines (Hafner and Hobbs, 2003). Only the phase field of liquid manganese oxide MnO_x is separated by bent lines, which is typical for phases with variable stoichiometry in such diagrams.

Predominance diagrams can be constructed not only for systems metal (Me)–oxygen (O_2). Already in the original publication, Ellingham (1944) pointed out that a similar discussion can be also performed with sulfides instead of oxides. In fact, it is possible for every system with a less volatile chemical element in equilibrium with another chemical element of higher volatility. This means that "Me" in equation (2.42) can also stand, e. g., for C and that O_2 can be replaced by S_2, N_2, P_4, or halogens. For example, Klimm (2014) calculated the system $Al-N_2$, which contains only one binary compound AlN.

The Ellingham diagram in Fig. 2.51 (a) has the disadvantage that only the upper border $RT \ln p_{O_2}(T) = 0$ can be directly correlated to specific experimental conditions: this is pure oxygen at 1 bar pressure. For $p_{O_2} < 1$ bar, the atmospheric conditions cannot be read very easily from the diagram, because the ordinate values are also scaled

---

48 Harold Johann Thomas Ellingham (1897–1975).

49 Oskar Max Glemser (12 November 1911–5 January 2005).

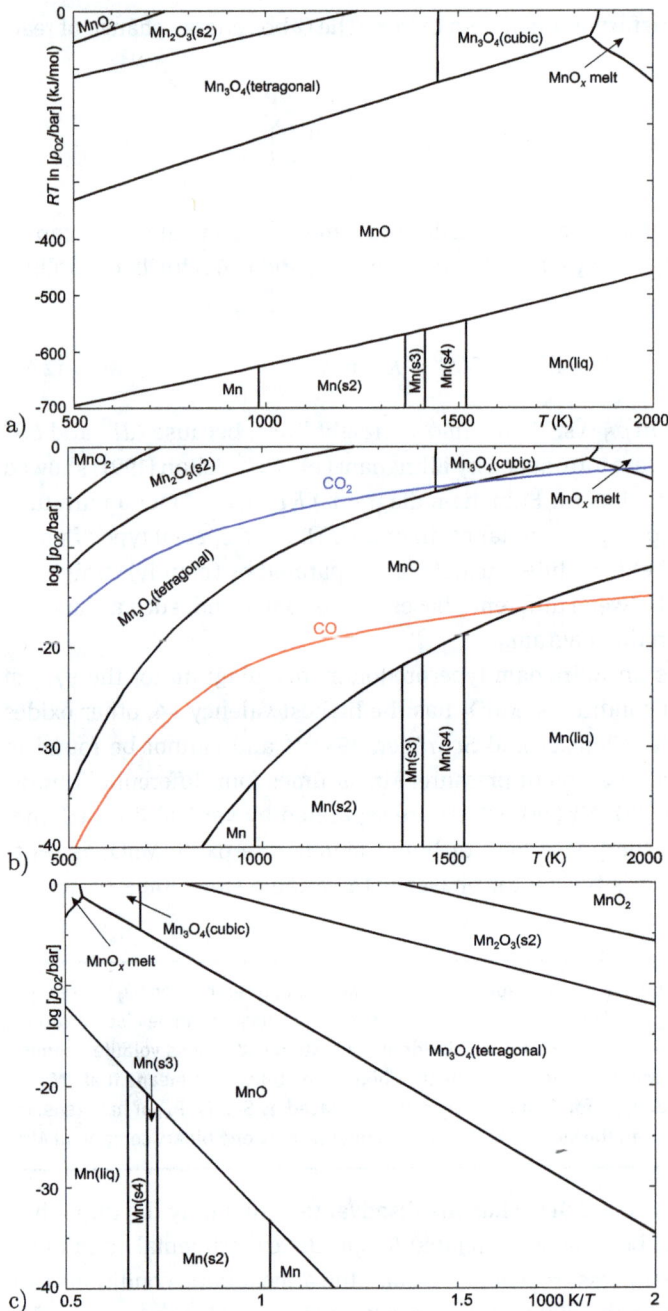

Figure 2.51: Three types of predominance diagrams for the system Mn–O$_2$. (a) Ellingham type, where ΔG for the transformation between neighboring phases is the ordinate; (b) Oxygen partial pressure as an ordinate, the $p_{O_2}(T)$ supplied by pure CO$_2$ or CO at 1 bar shown as overlay; see Fig. C.22; (c) TOMBE type with inverse temperature scale.

by T. For the practical work, it is often easier to use the (common, decadic) logarithmic oxygen partial pressure directly as an ordinate value. Such a diagram is shown in Fig. 2.51 (b) for the same T range and for a similar range of p_{O_2}. (Again, the upper border $\log[p_{O_2}/\text{bar}] = 0$ corresponds to pure oxygen.) The arrangement of phase fields, which are now distorted, is identical. It is obvious from this representation that the p_{O_2} where a specific oxidation state is stable depends significantly on T. As shown in Section 2.6.2, it is practically impossible in thermal analyzers to reach oxygen partial pressures $p_{O_2} \lesssim 10^{-6}$ bar by simple pumping and rinsing with "inert" gas. Besides, for metals with many possible valence states (here Mn^0, Mn^{2+}, Mn^{3+}, Mn^{4+}), it is often impossible to find any $p_{O_2} = $ const. that keeps the corresponding metal oxide stable over an extended T range, e. g., during a thermoanalytic measurement.

Predominance diagrams $\log[p_{O_2}/\text{bar}]$ vs. T for binary systems of many chemical elements with oxygen can be found in Appendix C. These diagrams should be always checked in preparation of "critical" measurements at elevated $T > 500 \ldots 800\,°C$ to guarantee that the sensitive parts of the thermal analyzer (carbon protective tubes, sample holder, crucibles) and that the samples remain stable. Typical critical situations are:

- p_{O_2} for the given T is so high that parts of the analyzer or a sensitive sample are oxidized.
- p_{O_2} for the given T is so low that the sample is reduced to metal, which alloys metallic crucibles.

Nair et al. (2018a,b) proposed a third type of predominance diagrams, shown in Fig. 2.51 (c) for the $Mn-O_2$ system. The name TOMBE is an abbreviation for "Thermodynamics of MBE", and it was used by these authors for a discussion of experimental conditions for molecular beam epitaxy. The inverse temperature scale for the abscissa in combination with the logarithmic pressure scale for the ordinate combines the benefits of both other diagrams: the phase boundaries are preferably straight, and the oxygen partial pressure can be read directly at the ordinate.

3 Applications

3.1 Characterization of raw materials

3.1.1 Calcination of carbonates

As described in Section 2.1.1, DTA and TG measurements are often coupled to "Simultaneous Thermal Analysis" (STA). Figure 3.1 shows the thermal decomposition (calcination) of a commercial $CaCO_3$ that was checked for purity. Depending on the CO_2 concentration in the surrounding gas and on the heating rate, calcination according equation (3.2) occurs at $\approx 800 \ldots 900$ °C. $CaCO_3$ is a convenient substitute for CaO in many processes were calcium oxide is needed as component, because this reaction runs completely to the product side. CaO ("quicklime") is highly reactive already with traces of H_2O and CO_2 in the atmosphere, and quantitative charging of this substance is complicated.

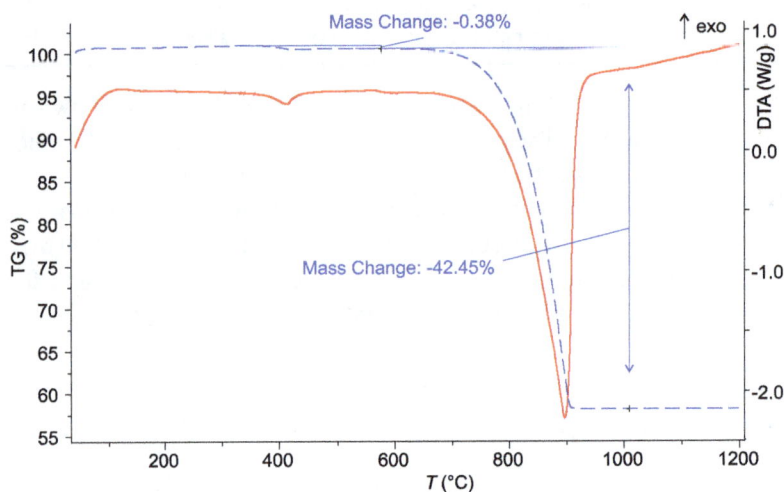

Figure 3.1: Simultaneous DTA/TG of 86.29 mg commercial $CaCO_3$ powder with nominal 99.99 % purity in flowing argon.

The TG curve of Fig. 3.1 shows a small step near 400 °C, which is accompanied by an endothermal effect. The subsequent larger TG step is obviously the calcination, but with a slightly smaller mass loss than expected. Detailed analysis shows that the following reactions subsequently appear:

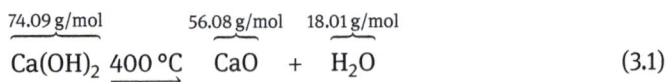

$$\overbrace{Ca(OH)_2}^{74.09\,g/mol} \xrightarrow{400\,°C} \overbrace{CaO}^{56.08\,g/mol} + \overbrace{H_2O}^{18.01\,g/mol} \tag{3.1}$$

https://doi.org/10.1515/9783110743784-003

$$\underbrace{CaCO_3}_{100.09\,g/mol} \xrightarrow{800\,°C} \underbrace{CaO}_{56.08\,g/mol} + \underbrace{CO_2}_{44.01\,g/mol} \qquad\qquad (3.2)$$

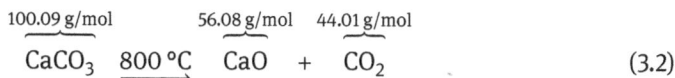

The mass losses by both reactions are 24.31 % (3.1) or 43.97 % (3.2), respectively. With the measured TG steps, from Fig. 3.1 we get as sample composition $0.38/24.31 \cdot 100\,\% = 1.6\,\%$ $Ca(OH)_2 + 42.45/43.97 \cdot 100\,\% = 96.5\,\%$ $CaCO_3$. This sums up to 98.1 %; the rest can assumed to be adsorbed humidity.

It seems surprising that a commercial chemical with 99.99 % nominal purity contains > 1 % of another substance, but this does not contradict the specification, because the purity of most chemical compounds is given "on metals basis", and hence nothing is said about anions. The presence of hydroxides in commercial carbonates is an issue that occurs occasionally. We can speculate about the origin, and it seems feasible that after wet-chemical precipitation, the raw materials are washed and dried. If the drying is performed at too high temperature, then partial decomposition (3.2) cannot be ruled out, and the highly hygroscopic CaO can subsequently react with humidity from the environment.

The relationship between $CaCO_3$, $Ca(OH)_2$, and CaO can be illustrated with Fig. 3.2, which is a predominance diagram in analogy to Section 2.8.3. For this calculation, a constant absolute humidity of 1 % was assumed for the gas phase, which is a typical value for air under ambient conditions. We can see that at low T and high $p(CO_2)$, $CaCO_3$ is stable either as calcite or aragonite. Depending on $p(CO_2)$, the decomposition to CaO takes place at several hundred degrees centigrade. Only for low T and at relatively low $p(CO_2)$, $Ca(OH)_2$ is stable. (Indeed, the chemical reaction $Ca(OH)_2 + CO_2 \rightarrow CaCO_3 + H_2O\uparrow$ is the "setting" reaction, where slated lime solidifies by carbonation under the release of water.)

Commercial $CaCO_3$ chemical is usually produced under aqueous conditions by precipitation with CO_2 or alkali carbonates, and the wet material has to be washed and dried subsequently. The question is under which conditions this drying process is performed. If large batches of the wet product are handled, then the water steam can drive out atmospheric CO_2 significantly below its natural value, which is the upper limit of the gray region in Fig. 3.2. Then the phase boundary to the CaO phase field is easily trespassed already at 350...400 °C. Once CaO is formed, it can be converted to $Ca(OH)_2$ by ambient humidity after cooling.

Bertram and Klimm (2004) report on the STA analysis of some other carbonates where the procedure described above had to be modified. The reason was that the decomposition temperature of some carbonates is so high that already the evaporation of volatile oxides starts. For Li_2CO_3, some Al_2O_3 was added to reach decomposition before the evaporation of Li_2O becomes significant. Sulaiman et al. (2016) observed by DSC of Li_2CO_3/Al_2O_3 mixtures exothermal peaks below 400 °C, which could be attributed to the formation of lithium aluminates like $LiAlO_2$ and $LiAl_5O_8$ under the release of CO_2. The measurement of $SrCO_3$ was performed by Bertram and Klimm (2004) in vacuum to shift the calcination equilibrium to the product side (Fig. 2.25).

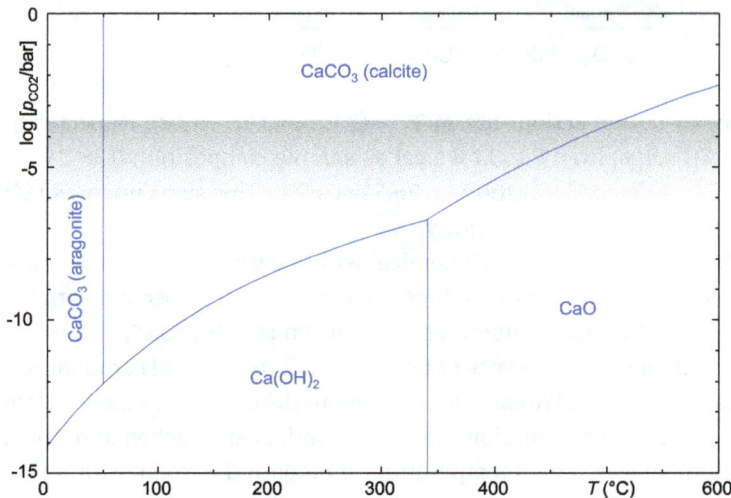

Figure 3.2: Calculated stability fields in the system $CaO–H_2O–CO_2–Ar$ (total pressure 1 bar, Ar excess, $p(H_2O) = const. = 0.01\,bar$). The ambient atmospheric CO_2 concentration (415 ppm) corresponds to the upper border of the grey bar.

3.1.2 Annealing loss of oxides

The oxides of different metals are often used for the production of constructive and functional materials, e. g., as components for ceramics or for the growth of single crystals. Accurate charging requires accurate data for the metal content (the "assay") of the corresponding material. This is often not straightforward, because manufacturers give purity data usually "on metals basis". In Section 3.1.1, it was already shown that a commercial $CaCO_3$ with nominal 99.99 % purity can contain more than 1 % hydroxide without violating its specification.

Sometimes, it is claimed that working with oxides is easier, because they can be annealed in air prior to the weighing of charges for production processes; these "dried" oxides are expected to have exactly the expected metal assay. However, this is not always true, because many oxides have a strong affinity to H_2O and CO_2, which are abundant constituents of ambient air and can lead to unexpected byproducts in the commercial substances.

Some of the oxides of the rare-earth (RE) elements, which should include here also scandium (Sc), yttrium (Y), and lanthanum (La), show a surprisingly high affinity to both H_2O and CO_2. Also, the RE elements themselves and other RE compounds such as their salts tend to bind them under humid conditions; this was shown for LaF_3 by Moritz et al. (2001). The affinity is extremely strong for La_2O_3, which is demonstrated in Fig. 3.3, where samples from two suppliers are investigated by simultaneous DTA/TG measurements. It is useful to compare these measurement to the results with $La_2(CO_3)_3 \cdot nH_2O$, which were presented in Section 2.4.2.3.

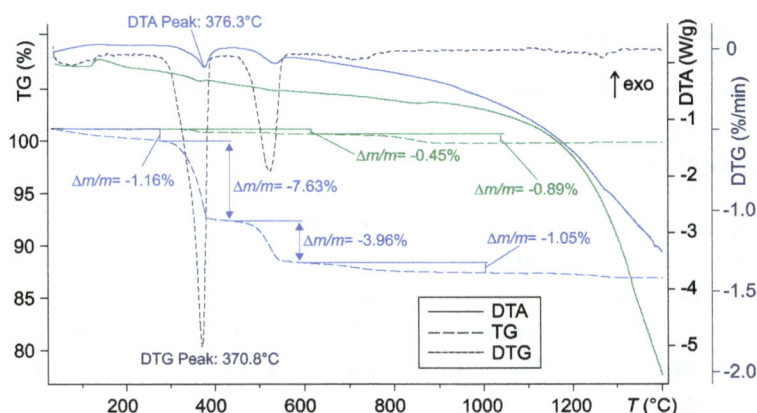

Figure 3.3: Simultaneous DTA/TG (NETZSCH STA409C, Al_2O_3 crucibles, $\dot{T}$ = 10 K/min, Ar flow) of two La_2O_3 batches (green curves: m_s = 96.77 mg, blue curves: m_s = 55.67 mg), both nominally with at least 99.99% purity. For the blue TG curves also the 1st derivative DTG is shown.

Both samples in Fig. 3.3 show several TG steps, but the mass losses $\Delta m/m$ are significantly smaller for the green TG curve. At least the position of the first TG step (peak in the DTG curve at 370.8 °C) is found for both materials. A second, strong step occurs for the blue curve near 500 °C, but for the green curve, near 850 °C. Moreover, the blue curve shows a last step >1200 °C, which is absent for the green curve.

The endothermal DTA peaks in Fig. 3.3, which are especially strong for the blue curve, are obviously related to endothermal chemical reactions that result in the TG steps, because volatiles are released from the sample. It is interesting to note that the strongest mass loss rate (DTG peak) appears slightly before the DTA peak. This is normal, because the signal of the mass loss immediately reaches the balance. In contrast, the DTA signal reaches the thermocouples with some delay Δt, because thermal conduction needs time. From the observed peak temperature difference ΔT we easily calculate the time delay with $\Delta t = \Delta T/\dot{T}$ = 0.55 min = 33 s, which is a typical value for DTA sample holders.

It is not straightforward to assign the subsequently appearing effects to single chemical reactions, because a manifold of phases exists in the system La–O–H–C: Besides La_2O_3, $La(OH)_3$, and the ordinary carbonate $La_2(CO_3)_3$, also $La_2(CO_3)_3 \cdot nH_2O$ (see Section 2.4.2.3), LaOOH, $La_2O_2CO_3$, and other compounds were reported in the literature (see, e. g., Füglein and Walter, 2012; Shirsat et al., 2003), and their interactions are complicated and partially kinetically influenced.

Not only La_2O_3, but also RE_2O_3, MgO, and many salts tend to adsorb humidity and CO_2 from the ambient atmosphere in not well-defined amounts, which influences the assay value of these materials. For accurate charging of such chemicals, prior to weighing, it is inevitable either
- to "dry" them (including the drive out of CO_2) by heating to an appropriate temperature (this temperature should be high enough to destroy the unintentional additives, but still so low that melting, evaporation, or reaction with the container material is avoided) or

– to take them as they are, but to account for the additives by a correction factor derived for the actual batch of this chemical by a TG measurement.

3.1.3 Stoichiometry of oxides

Many chemical elements, and especially the transition metals, form a variety of oxides Me_xO_y with different metal valence $2y/x$. Generally, for lower temperature T and higher oxygen partial pressure p_{O_2}, oxides with higher valence are stable, and vice versa. Ellingham-type predominance diagrams introduced in Section 2.8.3 give an instructive impression on stability fields for different Me_xO_y. Figure C.24 in Appendix shows this for the system $Fe–O_2$.

According to this diagram, metallic iron (Fe^0) can be oxidized to Fe(II)O, Fe(II,III)$_3$O$_4$, and Fe(III)$_2$O$_3$. Another valance level Fe^{6+} is not found in the diagram because it is stable only under strongly basic conditions together with other metals as ferrate(VI) ion, e. g., as $BaFeO_4$ (Gump et al., 1954). Iron(II) oxide (wüstite) is labeled in Fig. C.24 as $Fe_{1-x}O$ because it is not only a berthollide (see Fig. 1.28) phase (this is the case also for magnetite Fe_3O_4). Rather, for $Fe_{1-x}O$, we always find $x > 0$, which means that the exact composition FeO is not included in the stability field of this phase. Ganschow et al. (2016) could obtain $Fe_{1-x}O$ crystals with compositions ranging from $x = 0.017 \pm 0.005$ to $x = 0.096 \pm 0.003$ by variations of the growth rate. Like in an earlier study by Klimm and Ganschow (2005), iron(II) oxalate hydrate $Fe(COO)_2 \cdot 2H_2O$ was used as starting material in an atmosphere of 85 % Ar + 10 % CO_2 + 5 % CO, which stabilizes Fe^{2+} over a wide T range.

From Fig. C.24 it is obvious that Fe_2O_3 cannot be molten under atmospheric conditions that are easily accessible. Even under 1 bar of pure oxygen ($\log[p_{O_2}/\text{bar}] = 0$), it would lose oxygen according

$$3Fe_2O_3 \longrightarrow 2Fe_3O_4 + \frac{1}{2}O_2, \tag{3.3}$$

which corresponds to a mass loss of 3.34 %. Depending on p_{O_2}, the resulting Fe_3O_4 is expected to melt at $\approx 1500...1600\,°C$.

Figure 3.4 shows another version for parts of the system $Fe–O_2$ in more detail. This is a conventional presentation in the $x - T$ coordinates. Additionally, gray lines and labels show p_{O_2} at equilibrium in Pascal[1] units (1 Pa = 10^{-5} bar). The broad homogeneity (stability) fields of wüstite and magnetite are obvious. The "slag" phase at high T is identical to "FeO$_x$(liq)" in Fig. C.24. The highest oxygen isobar $p_{O_2} = 10^5$ Pa = 1 bar corresponds to the top border in the predominance diagram Fig. C.24.

Figure 3.5 shows simultaneous DTA-TG measurements with Fe_2O_3 powder from different suppliers. All measurements are baseline corrected. The red curves, measured

---

[1] Blaise Pascal (19 June 1623–19 August 1662).

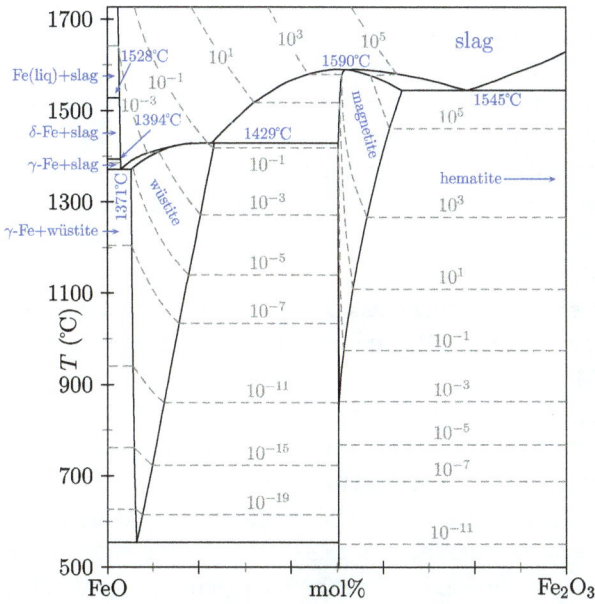

Figure 3.4: FeO–Fe_2O_3 phase diagram, redrawn with data from the ACerS-NIST (2014) database, entry 12340 based on calculations by Degterov et al. (2001). Gray lines and labels are O_2 isobars in Pa.

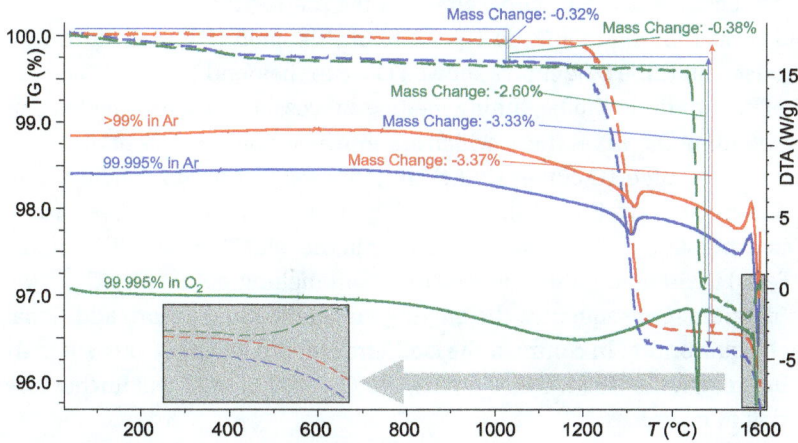

Figure 3.5: Simultaneous DTA/TG of Fe_2O_3 samples with m_s = 43...48 mg from different batches with nominal purity >99% or 99.995%. (NETZSCH STA409 "F3", Al_2O_3 crucibles, $\dot{T}$ = 10 K/min, gas flow 50 ml Ar or O_2, respectively).

with a material of nominal purity >99 % in argon with 99.999 % purity, show the expected behavior: The mass remains almost constant until ca. 1200 °C. The following TG step of −3.37 % is fairly good on the level expected from reaction (3.3). The blue and green TG curves, measured with a nominally much purer Fe_2O_3 material, show a significant initial mass loss of 0.32 or 0.38 %, up to ca. 400 °C. We should remember that already in Section 3.1.1 it was written that purity data for inorganic chemicals are usually given "on metals basis". Hence nothing is said about contamination by anionic impurities or volatiles. For Fe_2O_3, we can assume that adsorbed water leads to the partial formation of Fe(III) oxide hydroxide FeOOH. Different modifications of this substance are known to lose water upon mild heating and convert to Fe_2O_3 (see, e. g., Ishikawa et al., 1992; Musić et al., 2004).

Also, the purer Fe_2O_3 sample showed a mass loss at 1200 °C if measured in argon (blue curve), almost of the same magnitude like the red curve, again accompanied by an endothermal DTA peak. This marks the reduction of Fe_2O_3 (hematite) to Fe_3O_4 (magnetite), shown in equation (3.3). If the measurement is performed in oxygen, then the chemical equilibrium is shifted to the side of the educts, which results in a higher decomposition temperature (green curve), compared to Ar.

The mass loss during the measurement in O_2 (2.60 %) is significantly smaller, compared to the measurement of the same material in Ar (3.33 %). This difference can be explained with the phase diagram in Fig. 3.4, because the homogeneity region of the magnetite phase extends for the p_{O_2} = 10^5 Pa isobar (pure oxygen) much wider to the O-rich side, compared to the much lower p_{O_2} < 1 Pa, which can be assumed as rest impurities in Ar with 99.999 % purity. Besides, the broader homogeneity range of magnetite at high p_{O_2} leads to a less accurately fixed stoichiometry of this phase and hence to a less flat course of the TG curve after passing the TG step.

The liquidus maximum of Fe_3O_4 in Fig. 3.4 marks its melting temperature T_f = 1590 °C, and indeed all three DTA curves showed close to their end at 1600 °C a sharp endothermal effect, marking the beginning melting process. It is interesting to compare the corresponding TG curves there, which are shown at a larger scale in the insert of Fig. 3.5. The red and blue TG curves (in Ar) show a mass loss accompanying melting, whereas the green curve (in O_2) shows a rising mass. Also, for this observation, Fig. 3.4 gives an explanation: The p_{O_2} = 10^5 Pa isobar in the "slag" phase (with variable composition FeO_x) is situated right from the congruent melting point of Fe_3O_4 in the middle of the diagram. Consequently, the melting magnetite must absorb additional oxygen to reach equilibrium. In contrast, the isotherms for the lower p_{O_2} present in Ar are left from the "magnetite" field. Consequently, Fe_3O_4 melting in Ar will further lose oxygen, resulting in mass loss.

3.1.4 Hydrolysis of chalcogenides

Many metals have a strong chemical affinity to group 16 elements (chalcogens) and are forming compounds, which are named oxides (with O), sulfides (with S), selenides (with Se), or tellurides (with Te). Oxides and sulfides are the most important constituents of ores as sources of metals. Some of them tend to react under humid conditions with moisture, often under the formation of hydroxides. In the case of oxides, this hydrolysis process is often reversible, because a reaction of the kind

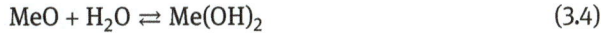

$$MeO + H_2O \rightleftharpoons Me(OH)_2 \qquad (3.4)$$

tends to proceed backward (to the educts) upon heating. In contrast, the hydrolysis of sulfides like ZnS

$$ZnS + 2H_2O \rightarrow Zn(OH)_2 + H_2S \uparrow \qquad (3.5)$$

is often not reversible; rather, hydrogen sulfide H_2S leaves the material and leads to the typical smell of rotten eggs, which is found for many sulfides. Upon heating, the hydroxide can be converted to oxide, which corresponds to the back reaction of (3.4). As a consequence, we must be aware that metal sulfides often may be contaminated by impurities of the corresponding oxide or hydroxide. In the following, we will explain why these oxidic contaminations can have a severe impact on the sulfide material.

Figure 3.6 shows a TG measurement of ZnS with simultaneous emanating gas analysis (EGA) by a quadrupole mass spectrometer (QMS, see Section 2.4.1). The TG curve shows a continuous mass loss of 2.2% from the beginning to ca. 600 °C, which continues weaker up to ca. 1000 °C. An increased signal for $m/z = 18$ (H_2O) proves that the initial mass loss can be attributed mainly to the release of water that is partially physically adsorbed and partially bound as $Zn(OH)_2$. Strong peaks appear around 600 °C for $m/z = 48$ and $m/z = 64$. Data from NIST (2020) allow us to relate both signals to sulfur dioxide SO_2. For this gas, resulting from fragmentation, S_2 ($m/z = 64$) is the strongest signal, and SO ($m/z = 48$) is the next signal, which is almost half as strong as S_2.

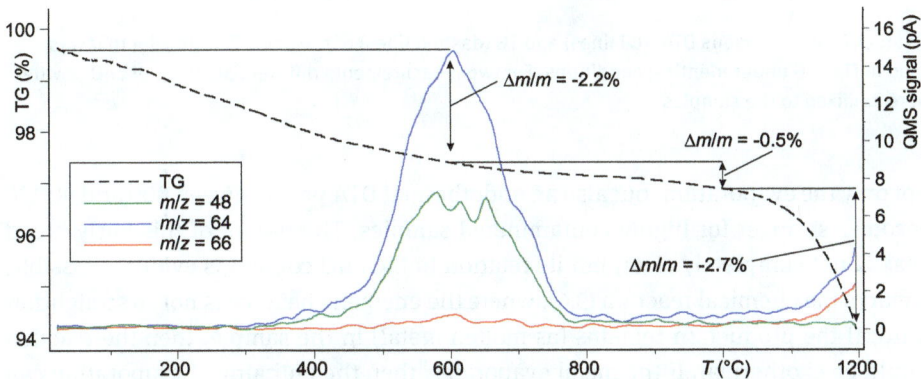

Figure 3.6: Simultaneous TG/QMS of commercial ZnS powder in a NETZSCH STA409C (Al_2O_3 crucibles, $\dot{T} = 10$ K/min). QMS signals $m/z = 48$ (SO, from fractioning of SO_2), $m/z = 64$ (S_2, fractioning of SO_2; and isotope ^{64}Zn), and $m/z = 66$ (isotope ^{66}Zn) are shown.

Sulfur dioxide is the result of a chemical reaction between the main component ZnS and its oxide contamination,

$$ZnS + 2\,ZnO \rightarrow 3\,Zn\ (\uparrow) + SO_2\ \uparrow, \tag{3.6}$$

where the gaseous product SO_2 is expected to quickly leave the sample. However, metallic zinc has a high volatility too, because its boiling point is rather low, $\gtrsim 900\,°C$.

Consequently, the significant mass loss at the end of the measurement, starting at ca. 1000 °C, results only partially from the evaporation of the main component ZnS. Figure 3.7 shows DTA/TG curves (green) for the same material that was used in Fig. 3.6 together with curves for samples with different admixtures of ZnO. It is evident that the mass loss is drastically increased for samples with high contamination by ZnO.

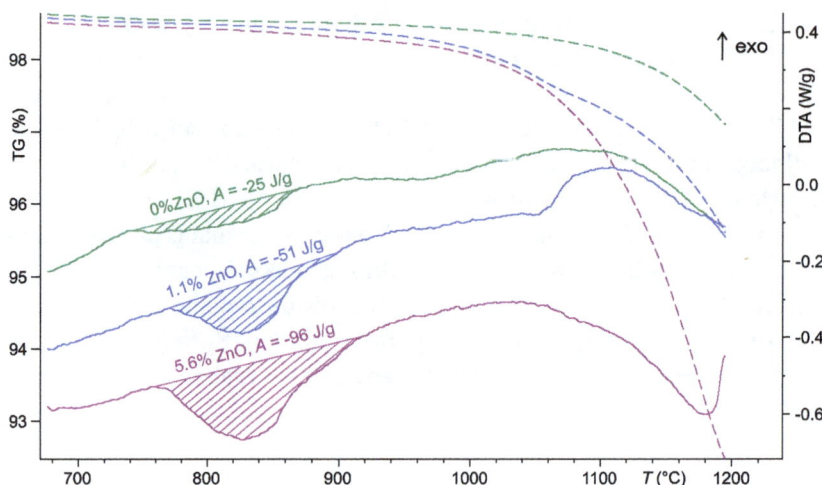

Figure 3.7: Simultaneous DTA (full lines) and TG (dashed lines) of the same ZnS powder that was used in Fig. 3.6 under identical conditions. For two measurements different amounts of ZnO powder were admixed to the samples.

Not only the evaporation, but also an endothermal DTA peak between 800 and 900 °C becomes stronger for highly contaminated samples. The nature of this fairly broad peak is not completely clear, but its relation to the ZnO content is evident. Possibly, it marks the chemical reaction (3.6), where the energetic balance is not so straightforward: If the product Zn remains (as molten metal) in the sample, then the reaction would be exothermal. If the metal evaporates, then the enthalpy of evaporation can shift the energetic balance to the endothermal side.

Figure 3.8 shows an equilibrium calculation of an ZnS excess (0.8 mol) with ZnO (0.2 mol) in 1 mol argon. In nature and under ambient conditions, ZnS occurs in both

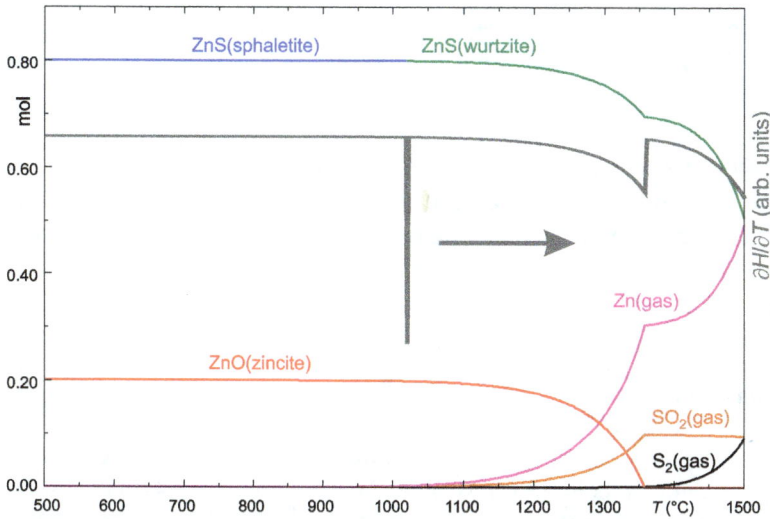

Figure 3.8: Equilibrium of 0.8 mol ZnS + 0.2 mol ZnO during heating in an Ar atmosphere of $p = 1$ bar. Calculated with FactSage 8.0 (2020). The ZnO curve starts to drop at 850...900 °C. The enthalpy balance $\partial H / \partial T$ (exo up) is shown without scaling by the gray curve.

modifications, sphalerite (= "zincblende") and wurtzite,[2] and in this calculation a transition from wurtzite (at low T) to sphalerite is shown at the equilibrium temperature $T_t \approx 1020$ °C. However, it is known that the transition occurs (depending on grain size and mechanical conditions) not always under equilibrium conditions. Besides, a partial solubility of oxygen in ZnS and the formation of other phases like $ZnSO_3$, $ZnSO_4$, and $Zn_3O(SO_4)_2$ cannot be ruled out (Skinner and Barton, 1960; Lakin et al., 1980; Schultze et al., 1995). These phases were partially discussed above in Section 2.7.3.

The gray curve in Fig. 3.8 is the calculated energetic balance $\partial H / \partial T$ of the equilibrium calculations, which is proportional to the expected DTA signal. The sharp peak resulting from the sphalerite/wurtzite transition can be misleading, because the calculation assumes thermodynamic equilibrium. However, this condition may be not fulfilled if both ZnS modifications coexist. The enthalpy change is smaller, because then only a part of such metastable sample transforms. The second endothermal effect is connected with the emanation of oxygen (as SO_2) and Zn from the contaminated ZnS sample and is also observed in the experiments (in Fig. 3.7, especially, for the sample with the strongest contamination by ZnO).

Already Hedvall (1938)[3] pointed out that the chemical reactivity of solids is often enhanced in the vicinity of phase transitions. Schultze et al. (1995) observed a higher

---

2 Charles Adolphe Wurtz, also Karl Adolph Wurtz (26 November 1817–12 May 1884).

3 Johan Arvid Hedvall (18 January 1888–24 December 1974).

reactivity of sphalerite type ZnS with oxygen, in comparison to wurtzite, and claimed that the Hedvall effect might be responsible for this difference.

Taking into account the uncertainties when the sphalerite/wurtzite transition really takes place and often the parallel existence of both ZnS modifications (Skinner and Barton, 1960), the onset of reaction (3.6) seems to be related to the transition between both modifications. This reaction can affect the anion/cation balance of the solid, because the sulfide anion S^{2-} and the oxide anion O^{2-} combine to the very volatile species SO_2, leaving behind (at least in the first step, before Zn evaporates too) the pure metal Zn.

3.1.5 Purification of fluorides

CaF_2 single crystals are often used as optical components such as lenses, because the material is transparent down below $\lambda = 140$ nm and dispersion $\partial n / \partial \lambda$ is small (n – refractive index). However, these beneficial properties are severely degraded by oxygen impurities. Adding "scavengers" (typically, PbF_2, Yonezawa et al., 2003; sometimes, ZnF_2, Ko et al., 2001) to the CaF_2 melt can significantly reduce oxygen concentration because the reaction

$$CaO + MeF_2 \longrightarrow CaF_2 + MeO \uparrow \tag{3.7}$$

removes it. (PbO and ZnO have high volatility, especially, as the growth process is often performed under reduced pressure and under reducing conditions.) Figure 3.9 shows results from a simultaneous DTA/TG/QMS measurement with $CaF_2 + ZnF_2$ in graphite crucibles. CaF_2 melts near 1420 °C, but for clarity, here neither the DTA nor the TG signal are shown. A significant mass loss due to $ZnF_2/ZnO/Zn$ evaporation started around 1400 °C, close to the melting point. Natural zinc is a mixture of five stable isotopes, and the most abundant are ^{64}Zn (48.6 %), ^{66}Zn (27.9 %), and ^{68}Zn (18.8 %). These signals were registered by the QMS, obviously, in the charge state +1. Parallel to the Zn signals, also O (or CH_4; both have $m = 16$) show weak peaks. However, the major portion of the oxygen reacts with carbon from the graphite crucible or from the vitreous carbon skimmer and is found as CO^+ ($m/z = 28$). We can conclude that ZnF_2 is an efficient scavenger for the removal of oxygen from CaF_2, but some overheating of the melt beyond the melting point is required to completely remove the scavenger.

CaF_2 crystals are usually grown in a Bridgman process inside graphite crucibles, as described, e. g., by Molchanov et al. (2005). The presence of graphite results in strongly reductive conditions, and from Fig. C.32 under these conditions, we read at the red curve (CO, in equilibrium with C) $p_{O_2} \approx 10^{-16}$ bar. These conditions are similar to the interior of the DTA/TG/QMS apparatus that was used in Fig. 3.9 (vitreous carbon protective tube).

If the purification of CaF_2 melts according to equation (3.7) with ZnF_2 is successful, then ZnO is expected to leave the melt quantitatively by evaporation (Fig. C.32), but the

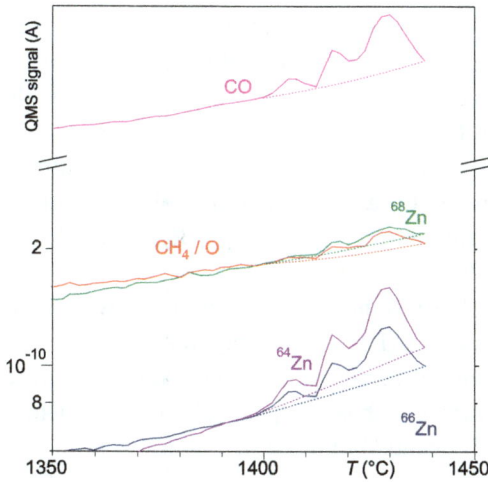

Figure 3.9: Simultaneous DTA/TG/QMS of 67.07 mg CaF_2 with 3.83 mg ZnF_2 additive as scavenger (NETZSCH STA 409 with graphite furnace, vitreous carbon protective tube with skimmer and BALZERS QMS, Klimm, 2010a). For clarity, only mass spectrometer signals for five m/z ratios are shown.

abundant carbon environment reduces it according to

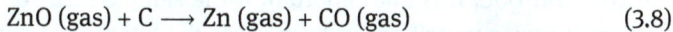

$$ZnO\ (gas) + C \longrightarrow Zn\ (gas) + CO\ (gas) \tag{3.8}$$

under the formation of carbon monoxide. Indeed, besides the simultaneous signals of the three most abundant Zn isotopes, the signal $m/z = 28$ also shows a strong simultaneous 3-peak pattern. The area under the CO peaks can be evaluated and is different for different CaF_2 batches. It turns out that this area quantitatively correlates with the optical quality of crystals that were grown from the corresponding batch: a higher concentration of oxide impurities in the material results in larger QMS peaks for CO and in worse optical quality of grown crystals. (The smaller QMS signal for $m/z = 16$, shown as a red curve in Fig. 3.9, is probably due to oxygen atoms resulting from partial fragmentation of CO in the QMS, NIST, 2020.)

3.2 Measurement of thermodynamic properties

3.2.1 Melting points

If a crystalline phase is heated, then it will typically transform to the liquid phase at some temperature T_f, which is called the melting point and denotes the fusion or melting temperature. Melting is a first-order phase transition and hence accompanied by the exchange of a certain latent "heat of fusion" ΔH_f. The following exceptions may occur:

1. The substance can undergo chemical decomposition. This is the case for many organics where the decomposition process is often irreversible. Also, peritectic melting (see Section 1.5.2.2) is a kind of decomposition but is principally reversible and appears at a well-defined temperature T_{per}.
2. If the triple point where solid, liquid, and gaseous phases are in equilibrium is at some low $p_{tri} < p_0 \approx 1$ bar (p_0 – ambient pressure). Then the solid evaporates directly, without forming a liquid. Examples are arsenic (p_{tri} = 36.3 bar, Gokcen, 1989) and CO_2 (p_{tri} = 5.185 bar, NIST, 2020).
3. Some solid phases Φ can become instable at some transition temperature T_t without melting; rather, another solid phase Φ' is formed, accompanied by the latent "heat of transition" ΔH_t. Three predominance diagrams in Fig. 2.51 demonstrate that for the metal manganese, even four solid phases appear subsequently before Mn(s4) melts: Mn $\rightleftarrows$ Mn(s2) $\rightleftarrows$ Mn(s3) $\rightleftarrows$ Mn(s4) $\rightleftarrows$ Mn(liq).

---

i The exotic phenomenon of "inverse melting" where a disordered (amorphous or liquid) phase converts into a crystalline phase on raising the temperature will be disregarded in this book. It was seen by Rastogi et al. (1993) in a polymeric system and is also known for a few other substances like ^{3}He below $T = 0.3$ K around $p = 30$ bar. Already Tammann (1903) speculated that inverse melting can occur if the entropy of the crystal is higher than the entropy of the liquid phase.

For DTA and DSC, it is not important if the sample melts or if it undergoes another first-order phase transition. In both cases the latent heat results in an endothermal deviation of the DTA/DSC curve, which lasts until the corresponding transition of the sample is finished. Subsequently, the curve returns nearly exponentially back to the baseline, as described in Section 2.5.3.

The temperature where the first transition (between solid phases, or melting) occurs can be obtained from heating curves with the "extrapolated onset" construction. In Sections 2.6.3 and 2.6.4, it was shown that onset temperatures give a fairly objective and reproducible value for T_f. The determination of extrapolated onsets can be difficult if a second transition appears soon after the first one. The upper curve in Fig. 3.10 demonstrates this for $BaCl_2$. (The hydrate water has left the sample already at much lower temperatures in the first heating run and is not shown there.) The first peak relates to the low-T/high-T phase transition (see Hull et al., 2011), and its onset can be well determined. However, the second peak due to melting follows soon. Consequently, no flat baseline appears between both effects, which makes the onset construction somewhat vague. The lower (blue) curve in Fig. 3.10 was measured immediately after the red curve with a significantly lower heating rate $\dot{T}$ = 2 K/min. As shown in Section 2.6.4, then DTA peaks become much smaller, which sufficiently separates both peaks. Now the determination of the extrapolated onset is possible also for the melting peak.

It should be noted that also the heating run with 2 K/min was performed with the same T correction file that was used for the 10 K/min heating, and this calibration

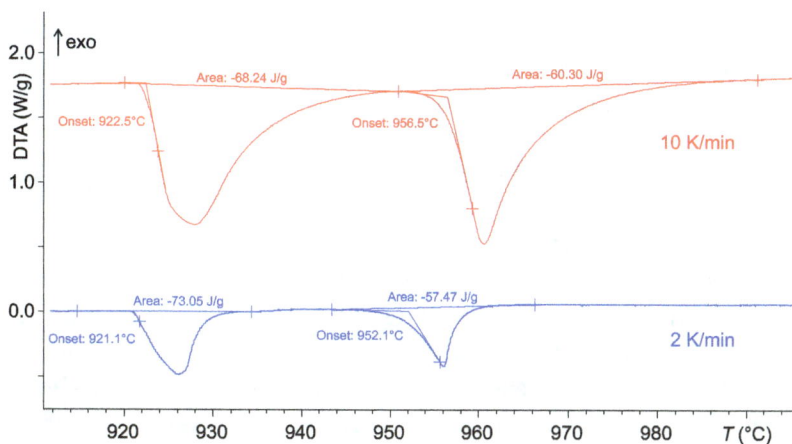

Figure 3.10: Second and third DTA heating curves of 31.02 mg $BaCl_2 \cdot 2H_2O$ (corresponds to 26.44 mg anhydrous $BaCl_2$) with a NETZSCH STA 449"F3" in Al_2O_3 crucibles and with flowing Ar atmosphere. The second heating was performed with 10 K/min, and the third heating with 2 K/min. Compare also with Fig. 2.24.

was performed with the typical DTA rate of 10 K/min, as described in Section 2.5.1. This could be the origin of differences between the onset temperatures from both measurements. Identical or at least similar rates in calibration and sample measurements are generally better. Already Fig. 2.24 showed that also other changes of the technical conditions, like crucibles or atmosphere, can lead to a better separation of proximate effects.

The extrapolated onset temperature is found at the intersection of a linear baseline prolongation, starting at the low T side of the peak with a tangent to the first inflection point of the peak (before it reaches its maximum). The onset temperature often depends only weakly on experimental parameters like heating rate, sample mass, or even crucible material. **!**

Solid phases with congruent melting behavior (see Section 1.5.2.1) have a sharp melting point, and the width of the melting peak is only a result of device parameters, like described in Section 2.5.3. This is the reason why solidus and liquidus lines in mixed crystal systems meet at the end members at the melting point $T_f = T_{sol} = T_{liq}$. For intermediate compositions of most phase diagrams (except at eutectic or azeotrope points), the melting process starts at T_{sol} and terminates at $T_{liq} > T_{sol}$.

The thermodynamic simulation of the whole melting process in a manner described in Section 2.8.2.1, and the optimization of the thermodynamic parameters to meet the experimental results is one way to obtain reliable T_{liq}. However, this procedure is elaborate, and not always reliable data are available.

Another, often sufficient way is to measure first the melting peak widths ΔT^0_{width} of the pure end members, which are often identical or at least only slightly different. We can take these ΔT_{width} as a reasonable approximation for the broadening of every melting peak by the thermoanalytic device. $T_{onset}(x) = T_{sol}(x)$ and $\Delta T_{width}(x)$ are measured for all intermediate compositions x, too. Then

$$T_{liq}(x) \approx T_{sol}(x) + T_{width}(x) - \Delta T_{width}^0 \tag{3.9}$$

can be an approximation for the liquidus temperature.

Nevertheless, the accurate determination of the liquidus temperature from DTA measurements is usually more challenging and less accurate than the determination of the solidus: If a solid is heated, then it is usually guaranteed that the melting process starts at T_{sol}, which is found from the extrapolated onset. In contrast, T_{liq} is almost there where the melting peak "fades out", and the DTA curve returns to the baseline. We can think that T_{liq} can be found from the onset of the crystallization peak in cooling curves. Figure 2.21, as an example, shows that this assumption is not true: During cooling, crystallization is often delayed and appears very often significantly below T_f (for congruently melting substances, like the chemical element gold in Fig. 2.21) or below T_{liq} for crystallization from mixtures.

Davenport and Bain (1970)[4,5] introduced time-temperature-transformation (TTT) diagrams where an isothermal annealing temperature is plotted versus the logarithm of the time, which is required to perform a fixed degree of transition. In this original publication, equilibria between γ-Fe (austenite) and α-Fe (ferrite, not to be mixed up with iron oxide-based compounds) + Fe_3C (cementite) were described.

Meanwhile, TTT diagrams are used for the quantitative description of crystallization kinetics in many systems, including "metallic glasses" discovered by Klement jun et al. (1960).[6] Löffler et al. (2000)[7] performed isothermal DTA studies with a glass forming Pd–Cu–Ni–P alloy by heating samples 350 K above T_{liq} = 823 K and by subsequent quick cooling with –25 K/s = –1500 K/min to a selected temperature T. At that temperature the samples were held isothermally until crystallization was detected by a temperature rise owing to the release of heat of fusion. In a second series of experiments, DTA cooling runs were performed with $\dot{T}$ = –1.35 ... – 0.20 K/s = –812 ... – 12 K/min, and in these experiments, no crystallization was observed for rates ≥ 0.35 K/s. For lower rates, crystallization was observed between 690 K ($\dot{T}$ = –0.20 K/s) and 670 K ($\dot{T}$ = –0.30 K/s).

Figure 3.11 shows a TTT diagram obtained by Krüger and Deubener (2016) from DSC measurements of a single lithium disilicate ($Li_2Si_2O_5$) sample in several hundred

---

4 Edmund Sharington Davenport (born 1897).

5 Edgar Collins Bain (14 September 1891–27 November 1971).

6 William Klement (PhD thesis CalTec, Pasadena 1962, in the P. E. Duwez Lab.)

7 Jörg F. Löffler (born 1969).

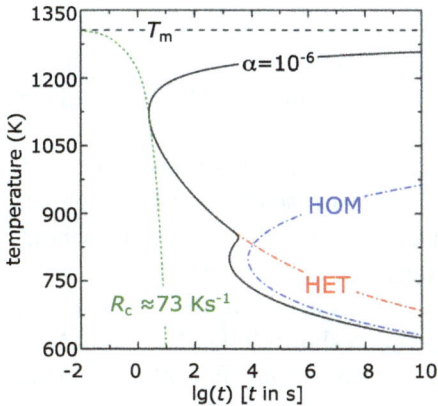

Figure 3.11: TTT diagram of lithium disilicate for a crystallized fraction of $\alpha = 10^{-6}$ (black solid line), including homogeneous crystallization in the volume (blue dashed-dotted line, HOM) and heterogeneous crystallization at the surface (red dashed-dotted line, HET). The critical cooling rate is $R_c \approx 73\,\mathrm{K\,s^{-1}}$ (green dotted line). $T_m = T_f = 1306$ K is the melting point. Graph copied from Krüger and Deubener (2016) under the Creative Commons Attribution License (CC BY).

undercooling runs. If the material is cooled from T_f at a constant rate, then the function $T(\lg[t])$ follows a path similar to the green dotted line; for higher cooling rates, the function is steeper, and for lower cooling rates, it is flatter. Crystallization occurs only inside the "double-nose" (almost c-shaped) area, and this area is trespassed only for cooling rates below the "critical rate" $R_c = \dot{T}_c \approx 73\,\mathrm{K\,s^{-1}}$. The "double-nose" shape results from the competition of heterogeneous crystal seed formation HET at the surface of the DSC crucibles (which were made of Pt-Rh alloy) and homogeneous crystal seed formation HOM in the volume of the melt. Of course, we have to expect a significant impact of the crucibles used on the HET curve. Schawe and Löffler (2019)[8] revealed different branches in TTT diagrams for nucleation and growth in Au-based metallic glasses. Therefore they have used "fast differential scanning calorimetry" (FDSC) with cooling rates up to a value of $-40,000\,\mathrm{K\,s^{-1}}$ with a Mettler-Toledo Flash DSC 2+.

The crystallization behavior shown in Fig. 3.11 is typical also for other crystallization processes. The crystallization will usually be observed at lower $T < T_m$, because the upper branch of the crystallization curve approaches T_m only asymptotically, a phenomenon called supercooling. **!**

It turns out that the reliable determination of the liquidus temperature solely from one DTA/DSC measurement is often not possible, especially for samples exhibiting a strong tendency to supercooling and glass formation (e. g., silicates, phosphates, borates). From Fig. 3.11 we see that the supercooling tends to become smaller for lower

---

8 Jürgen E. K. Schawe (born 1959).

cooling rates. This was exploited by Ferreira et al. (2010) for DSC measurements, e. g., in the system $Li_2O–B_2O_3$. These authors extrapolated the end points of DSC melting peaks obtained with different heating rates $\dot{T} \approx 5\ldots20$ K/min to $\dot{T} = 0$, which gave consistent values for T_{liq}.

Repeated heating/cooling cycles between a fixed lower temperature $T_0 < T_{sol}$ and an upper temperature T', which is increased stepwise by small increments ΔT, like indicated in Fig. 3.12, can be an alternative way to reveal T_{liq}. The idea behind is that supercooling usually is significantly suppressed or even eliminated if seeds of the crystallizing phase are present. However, this is the case in the two-phase field between T_{sol} and T_{liq}. For the temperature program shown in Fig. 3.12, the analysis should be performed for the cooling segments, which present an exothermal crystallization peak if they start from some $T' > T_{sol}$. The area A of this crystallization peak increases with increasing T' until T' trespasses T_{liq}.

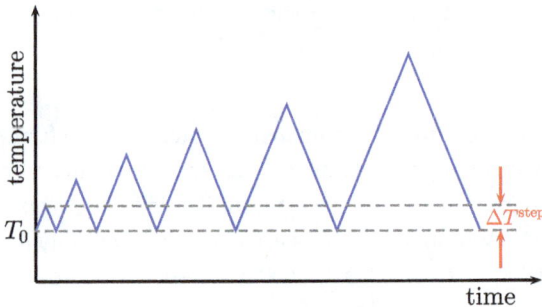

Figure 3.12: Schematic temperature program for the determination of liquidus temperatures with $T(t)$ cycles between T_0 and a stepwise increased $T' = T_0 + n\Delta T^{step}$, $n = 1, 2, 3, \ldots$.

Figure 3.13 analyzes cooling curves obtained from a zinc sample. Of course, for this congruently melting material, we have $T_{sol} = T_{liq} = T_f \approx 419\,°C$. The upper two cooling curves, starting from $T' = 412$ or $413\,°C$, respectively, show no sign of a crystallization peak. For the last three curves, which start from $T' = 417, 418$, or $419\,°C$, the crystallization peak is fully developed and has the same peak area. This peak area A can be defined with a baseline starting from the left side, because this gives a well-defined end, also on the right side of the peak.

The insert in Fig. 3.13 shows as dots the measured peak areas as functions of the (upper) starting temperature T'. It turns out that Verhulst's[9] logistic function

$$A(T') = \frac{a}{1 + \exp[-\frac{T'-T'_{mid}}{b}]} \tag{3.10}$$

---

9 Pierre François Verhulst (28 October 1804–15 February 1849).

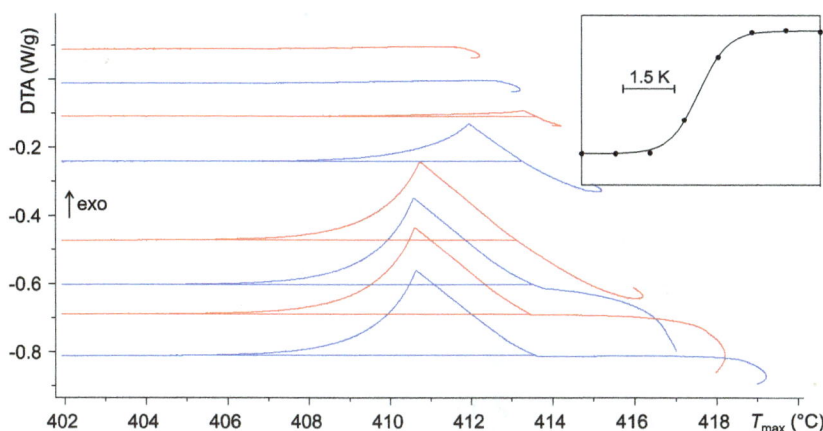

Figure 3.13: DTA cooling curves obtained from 18.77 mg Zn in Al_2O_3 crucibles, measured with a NET-ZSCH STA 449 "F3" in flowing Ar, T' increased stepwise by $\Delta T^{step} = 1\,K$, $\dot{T} = -1\,K/min$. The peak area A as a function of the temperature T' where cooling starts is shown in the insert, together with the transient width ΔT_{trans}.

can well fit the rise $A(T')$. The curve starts from 0, reaches $a/2$ at the midpoint T'_{mid}, and goes to its maximum a. The width ΔT_{trans} of the transient region scales with e^b. A finite width $\Delta T_{trans} = 1.5\,K$ is found for the example in Fig. 3.13, because the thermal resolution of the device is limited.

The cycling method described here gives fairly reliable data only on the width of the melting range, $T_{liq} - T_{sol}$. However, as shown in Fig. 3.13, the temperatures themselves are often not well reproduced. These differences result mainly from two circumstances: (1) Thermal calibration of the device is performed with heating runs, but here cooling runs are analyzed. (2) The calibration is usually performed with a standard heating rate of, e. g., 10 K/min, but due to the typically small ΔT^{step}, the rates of the analyzed cooling segments are much smaller.

Figures 3.14 and 3.15 are examples from the binary system CsCl–KCl, which shows, according to Sangster and Pelton (1987), unlimited mutual solid miscibility with an azeotrope point $x_{aze}(KCl) = 0.36$, $T_{aze} = 815\,°C$. The composition measured in Fig. 3.14, $x(KCl) = 0.4039$, is not far from x_{aze}. It shows a transient width $\Delta T_{trans} = 5.0\,K$. ($\Delta T_{trans}$ tends to become larger for higher cooling rates.) A repetition of the Zn measurement from Fig. 3.13 with $\dot{T} = 2\,K/min$ and $\Delta T^{step} = 2\,K$ resulted for this material in $\Delta T_{trans} = 2.9\,K$. Comparing both values, we can conclude that the difference between liquidus and solidus for the sample in Fig. 3.14 can be estimated to a very small value $5.0\,K - 2.9\,K \approx 2\,K$.

If the sample does not evaporate or is otherwise altered, then during the extended temperature programs necessary for the cycling measurements, the method can be used also at high temperatures, where the accuracy and temperature resolution of thermoanalytic devices generally become worse. An example was given by

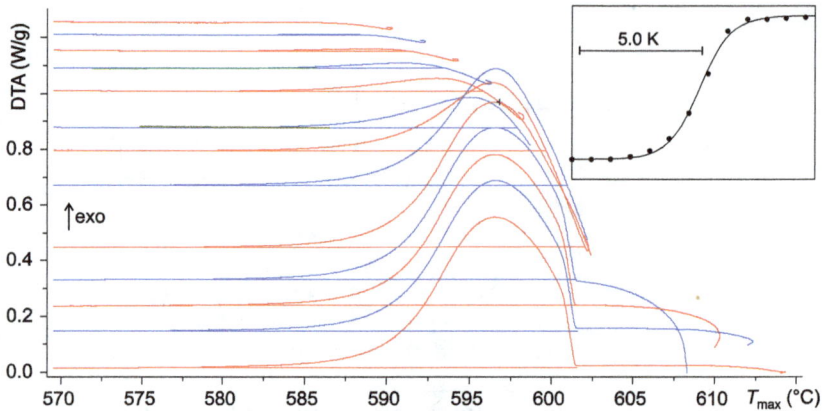

Figure 3.14: DTA cooling curves obtained from a mixture of 83.64 mg CsCl + 25.10 mg KCl (m_s = 108.74 mg, x(KCl) = 0.4039), ΔT^{step} = 2 K, $\dot{T}$ = −2 K/min.

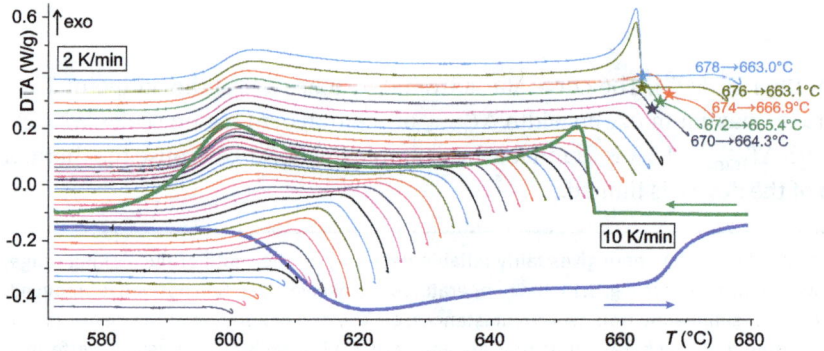

Figure 3.15: DTA cooling curves obtained from 49.76 mg CsCl + 41.26 mg KCl (m_s = 91.02 mg, x(KCl) = 0.6518), ΔT^{step} = 2 K, $\dot{T}$ = −2 K/min. The asterisks mark crystallization peak onsets T_{cryst} for some curves, and the corresponding labels stand for $T' \rightarrow T_{cryst}$. Thick lines show preliminary heating/cooling curves performed with a standard rate of ±10 K/min.

Szysiak et al. (2011) for the pseudo-binary system NdAlO$_3$–YAlO$_3$ with $T_{liq} \approx 1880\ldots$ 2100 °C. Both components show large but limited mutual solubility ($\approx$ 30 %) in the solid phase, and all compositions Y$_{1-x}$Nd$_x$AlO$_3$ with $0 \leq x \leq 0.35$ have narrow melting peaks with onsets in the range 1880...1900 °C. Szysiak et al. (2011) performed thermal cycling with samples in this concentration range and found the narrowest ΔT_{trans} = 4 K for a sample x = 0.20, which was interpreted as an azeotrope point on the YAlO$_3$ side of the system.

For very broad crystallization peaks with large difference $T_{liq} - T_{sol}$, it is sometimes difficult to determine a reliable baseline, and measurements of the peak area A are vague. Nevertheless, then also thermal cycling can be helpful. Figure 3.15 shows another example from the CsCl–KCl system with KCl concentration higher than that used in Fig. 3.14. From Sangster and Pelton (1987) we read for this composition T_{sol} = 628 °C and T_{liq} = 667 °C (difference 39 K). Figure 3.15 shows cooling curves with $\dot{T}$ = −2 K/min, starting from T' = 600 °C up to T' = 678 °C with step ΔT^{step} = 2 K. Additionally, a heating and a cooling curve of the same sample measured with $\dot{T}$ = ±10 K/min are presented.

The direct comparison makes obvious that crystallization peaks for $\dot{T}$ = −2 K/min start to occur for upper temperatures T' almost there, where the blue heating curve (for $\dot{T}$ = +10 K/min) has its extrapolated onset. However, in the green cooling curve measured with $\dot{T}$ = −10 K/min, the crystallization peaks starts too late: In the uppermost cooling curves from the cycling measurements crystallization starts earlier, with a maximum value 666.9 °C for the red curve that began at T' = 674 °C. A remarkable crystallization peak area appears first for the olive cooling curve starting at T' = 606 °C (4th curve from the bottom), and the difference $T_{liq} - T_{sol}$ ≈ (666.9 − 606) °C ≈ 61 K can be estimated at this composition. This are ca. 20 K more than estimated from the publication by Sangster and Pelton (1987).

3.2.2 Specific heat capacity c_p

As already introduced in Section 2.3.2.2, a $c_p(T)$ measurement by DSC relies basically in the comparison of three subsequent DSC curves, which are measured under identical conditions and with identical crucibles:

1. The baseline is measured with two empty (reference and sample) crucibles.
2. The standard measurement where a reference sample of known $c_p(T)$ resides inside the sample crucible. Often, crystalline α-Al_2O_3 ("sapphire") is used as reference if (small) bulk samples have to be measured. For powdered samples, Al_2O_3 powder as standard is often better.
3. The sample measurement where the sample under investigation replaces the standard.

Different modifications of such measurements were developed, which can be used if appropriated thermal analyzers are available and if the sample type makes this useful.

T-step method. Mraw and Naas (1979) compared in their study of the mineral pyrite (FeS_2) the "scanning method" with continuous heating for $c_p(T)$ measurements mentioned above, with an alternative they named "enthalpy method". In this procedure, which is now often described as "step heating" or simply step method, T is raised stepwise between isothermal segments, as depicted in Fig. 3.16. It turns out that the area of the DSC signal over the dashed interpolated baseline, in the transient between subsequent isothermal segments $T_{init.}$ and T_{final}, is proportional to the enthalpy increment $\Delta H_{incr.} = H_{final} - H_{init.}$. $\Delta H_{incr.}$ can be quantified by comparison with a standard sample, such as sapphire. Then $\overline{c_p} = \Delta H_{incr.}/\Delta T$ (ΔT =

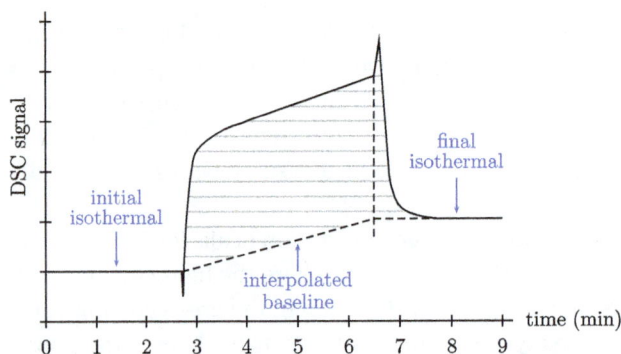

Figure 3.16: Schematic heat flow (= DSC signal) versus time during the step method where T is raised from an initial isothermal segment to a final isothermal. The shaded area is proportional to the enthalpy increment between both equilibrium states. Drawing adopted from Mraw and Naas (1979).

$T_{final} - T_{init.}$) is the average specific heat capacity in the interval. Claudy et al. (1988) investigated glass transitions in T-step mode and separated their thermodynamic and kinetic features. Recently, Silva et al. (2021) measured the functions $c_p(T)$ of Fe-Nb intermetallic compounds with a Setaram Tian–Calvet calorimeter (see Section 2.3.1). They used large samples of approximately 1 g and temperature steps $\Delta T = 35$ K. The heating rate between steps was 3 K/min, and the isothermal holding time at each step was 90 min.

Temperature-modulated DSC (TM-DSC). Here T fluctuates for several minutes with an amplitude of typically a few 0.1 K around some mean temperature $T_1, T_2, \ldots$ (Fig. 3.17). Correspondingly, the DSC signal fluctuates with the same frequency. Like in the T-step method, the strength of the DSC signal depends on the heat capacity of the sample. A reference sample with known $c_p(T)$ (e. g., sapphire) is measured separately, and by comparison $c_p(T)$ for the sample under investigation is obtained. In contrast to the T-step method, where an average $\overline{c_p}$ over an T range is measured, TM-DSC gives one accurate c_p value for each temperature step T_i where fluctuations are performed.

Fast Differential Scanning Calorimetry (FDSC). "Flash DSC" devices from different manufacturers allow the performance of DSC measurements with extremely high heating rates, from $<$ 10 K/min over several 1,000 K/min up to 3,000,000 K/min. Such rates are possible only with very small sample masses in the order from μg down to nanograms, which allows time constants $\leq$ 1 ms (Mettler Toledo, 2021). The sample is placed for such measurements directly on a chip that bears resistance heater elements with sets of thermocouples surrounding the sample. FDSC allows the investigation of fast kinetic processes (see, e. g., Grassia et al., 2018). Quick et al. (2019) discuss the application of FDSC for c_p measurements.

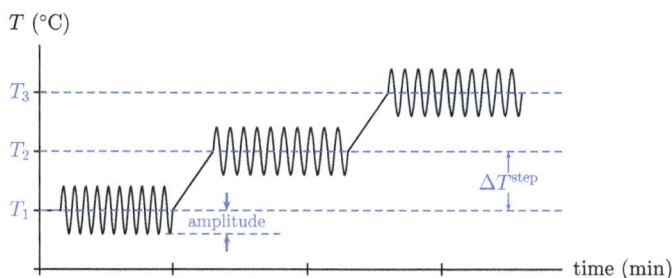

Figure 3.17: For c_p measurements by TM-DSC, T is raised stepwise by ΔT^{step} from T_1 to $T_2, T_3, \ldots$. At every step, T fluctuates (almost) harmonically by an amplitude of typically several 0.1 K.

3.2.3 A thermodynamic dataset

Lanthanum indate $LaInO_3$ is a substance that crystallizes in an orthorhombically distorted perovskite structure. The structural and chemical properties make this material a potentially interesting substrate for the epitaxial deposition of $BaSnO_3$ layers, which possess very high electron mobility and electrical conductivity (Prakash et al., 2017). Galazka et al. (2021) demonstrated that the growth of $LaInO_3$ single crystal in substrate dimensions is possible.

Almost no thermodynamic data for $LaInO_3$ are available from the literature; just the melting temperature $T_f = (1880 \pm 15)$ °C was measured pyrometrically during crystal growth from the melt by Galazka et al. (2021). This section describes a basic thermodynamic characterization of the substance by thermal analysis.

As a first point, we have to accept that, unfortunately, DTA up to the melting point – and this way the more accurate measurement of T_f and, moreover, ΔH_f – is technically not possible so far: The predominance diagram Fig. C.37 shows that In_2O_3 and hence also $LaInO_3$ are stable as condensed phases only in atmospheres containing $\geq 1\%$ O_2. To reduce evaporation, Galazka et al. (2021) used an atmosphere containing 9...14 % O_2 with iridium crucibles as melt containers. Unfortunately, this is not possible for DTA, because all commercial devices have carbon and/or tungsten in the hot zone.

As a first step, the high-temperature specific heat capacity was measured by DSC. This was done as described in Section 2.3.2.2, by comparison of heating curves (each 4 runs with 20 K/min from 40 °C to 1000 °C in lidded Pt crucibles) of α-Al_2O_3 powder as standard with a ground part of single crystalline $LaInO_3$. Typically, the first heating curves deviated remarkably from the second to fourth heating. The data points in Fig. 3.18 show the average of runs 2, 3, and 4. Some points in the beginning deviate from smooth behavior because some transition time was required to settle on a constant heating rate. Also, above 1100 K, experimental points deviate from the expected smoothly rising trend. Only the black points were fitted to a polynomial of the kind

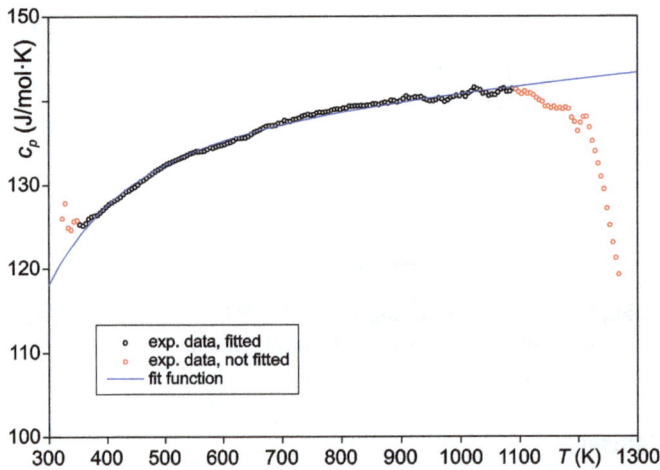

Figure 3.18: $c_p(T)$ measurement of a powdered LaInO$_3$ crystal sample with Al$_2$O$_3$ powder as standard; see Section 2.3.2.2 (DIN 51007/ASTM E1269/ISO 11357). Only the black data points were fitted by a polynomial $c_p = a + bT + d/T^2$ with $a = 136.8228$, $b = 5.8324 \times 10^{-3}$, $d = -1.847436 \times 10^6$ (in J/(mol K)).

(2.6), just the term $c \cdot T^2$ was not needed in this case to satisfactorily describe the experimental data.

The formation enthalpy ΔH_f from the component oxides

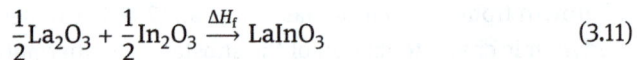

$$\frac{1}{2}\text{La}_2\text{O}_3 + \frac{1}{2}\text{In}_2\text{O}_3 \xrightarrow{\Delta H_f} \text{LaInO}_3 \tag{3.11}$$

was obtained from the DTA measurements in Fig. 3.19. However, the preparation of a useful sample is not straightforward in this case. Only In$_2$O$_3$ can be easily charged by weighing the substance on a laboratory balance. La$_2$O$_3$, in contrast, is hygroscopic (see Fig. 2.7), which makes accurate weighing of small quantities in the laboratory almost impossible.

To circumvent this problem, the sample powder for the measurements in Fig. 3.19 was prepared with a small excess of La$_2$O$_3$, and scaling of the DTA signal is given there in milliwatt, rather than the typical W/g (or mW/mg, like used by some manufacturers). Under such conditions, we can assume that the In$_2$O$_3$ component reacts completely, and from the known mass share of In$_2$O$_3$ and the corresponding m_s it is possible to calculate ΔH_f^{1628} at the reaction temperature. ($T_{\text{peak}}^{20\,\text{K/min}} = 1355.1\,°\text{C} = 1628.25\,\text{K}$ was used as the reaction temperature.) The exothermal reaction peaks in Fig. 3.19 are rather small and can be determined only at sufficiently high rates $\dot{T} \geq 20\,\text{K/min}$. The shift to higher T for higher $\dot{T}$ is typical for thermally activated kinetic processes; see Section 2.7.

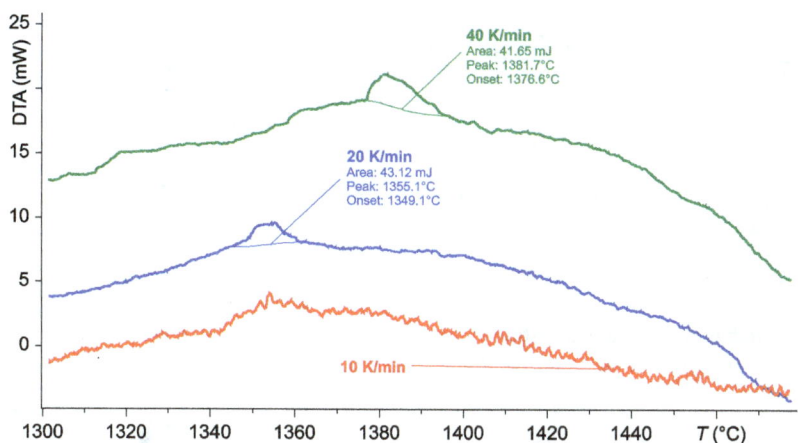

Figure 3.19: DTA heating runs of La_2O_3/In_2O_3 powder mixtures with La_2O_3 excess. The sample masses were: for 10 K/min: 85.84 mg; for 20 K/min: 73.29 mg; for 40 K/min: 93.45 mg. (NETZSCH STA 449 "F3", DTA/TG sample holder, Ar/O_2 flow, Pt crucibles).

It is usual to give formation data for standard conditions and from the chemical elements; in this case,

$$La + In + \frac{3}{2}O_2 \xrightarrow{\Delta H_f^0} LaInO_3. \tag{3.12}$$

Fortunately, ΔH_f^0 for La_2O_3 and In_2O_3 are found in the FactSage 8.0 (2020) databases. Thus the Born–Haber cycle as in Fig. 3.20 can be constructed. The left enthalpy data for the heating of the component oxides from standard condition to $T_{peak}^{20\,K/min}$ result from a simple and reliable FactSage calculation. The exothermal reaction enthalpy 192.73 J/mol was calculated from the peak area of the blue curve in Fig. 3.19. The right enthalpy is released upon cooling $LaInO_3$ back to standard conditions and can be calculated by equation (1.19) from the experimental $c_p(T)$ data. For the simple polynomial expression given in the caption of Fig. 3.18, the integral is easily calculated by hand. More complicated expressions can be calculated, e. g., using WolframAlpha (2021). The formation enthalpy of $LaInO_3$ from the component oxides is the difference from these three values, and hence $\Delta H_f \approx -13.9$ kJ/mol.

A Born–Haber cycle as shown in Fig. 3.20 will give ΔH_f from the components that were used in the measurement, which here are the component oxides. If ΔH_f from the chemical elements is requested, then the standard heats of formation of these component compounds from the elements have to be added. They are typically negative and correspond to the H values given, e. g., in Table 1.3. NIST (2020) is a valuable resource for such data.

A thorough thermodynamic characterization of a phase would require additional data for its entropy. If c_p can be measured down to low $T \geq 0$ K, then S is easily calculated by

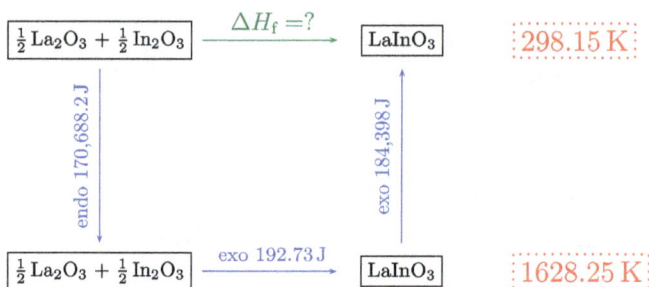

Figure 3.20: Born–Haber cycle for the formation enthalpy of $LaInO_3$ from the component oxides: $\Delta H_f = 170,688.2\,J/mol - 192.7\,J/mol - 184,398\,J/mol \approx -13.9\,kJ/mol$.

equation (1.24). Except for a very few "frustrated systems" (e. g., "spin ice", Ramirez et al., 1999), $c_p(T) \approx c_V(T)$ vanishes as $T \to 0$ quickly with T^3. Consequently, the contributions to S at the lowest T are usually small and can be reasonably estimated. Different authors described c_p measurements at low-T by adiabatic calorimetry, e. g., Venero and Westrum (1975)[10] and Konings et al. (1998), or more recently with a "Physical Property Measurement System" (Shevchenko et al., 2017).

If experimental data for the calculation of S are not available, then the values for a complex compound can be estimated simply as the sum of data from binary compounds. This is the way used, e. g., by FactSage 8.0 (2020) in the "Compound" module if a new phase is constructed numerically, and often such estimation is not bad. Latimer (1951)[11] published tables with contribution of many simple and complex ions and shows that the entropy of a complex compound can often be well approximated by the sum of its constituents. Estimations for high melting carbides were presented by Worrell (1964),[12] for oxides by Wu et al. (2017), and for organic compounds by Benson and Buss (1958)[13,14] and Domalski and Hearing (1993).

3.3 Determination of phase diagrams

For the experimental determination of binary phase diagrams without intermediate compound, it is often sufficient to measure ≈ 10 DTA or DSC curves, and usually heating curves are preferred, because then the observed thermal effects occur with high accuracy and reproducibility at their equilibrium temperatures. Certainly, we should

---

10 Edgar Francis Westrum, Jr. (16 March 1919–7 May 2014).

11 Wendell Mitchell Latimer (22 April 1893–6 July 1955).

12 Wayne L. Worrell (25 October 1937–18 February 2012).

13 Sidney William Benson (26 September 1918–30 December 2011).

14 Gerald Hatten Buss (27 January 1933–18 February 2013).

always inspect cooling curves too, but with some care: Supercooling is a quite common phenomenon and can lead to crystallization events that occur significantly (often several 10 K) below their equilibrium temperature.

Heating curves are preferred for the determination of phase diagrams, because then phase transitions including melting occur usually at their equilibrium temperatures. Often, a heating rate $\dot{T} = 10$ K/min is a good compromise for $T_{max} \leq 1500$ °C. As outlined in Section 2.6.4, for lower rates, DTA peaks are narrower, which allows better separation of nearby effects, but unfortunately then the peaks are also smaller, and weak effects can be overseen. The sensitivity of thermal analyzers drops significantly for very high T (see Section 2.5.2), and thus higher $\dot{T} = 15 \ldots 20$ K/min can be useful.

!

Often the thermal effect that occurs first in the corresponding heating/cooling program can be seen best. For heating, in a phase diagram, this is often the solidus or a phase transition below it. Vice versa, in cooling curves, usually, the liquidus is the first effect. Indeed, liquidus temperatures T_{liq} can be recognized sometimes better in cooling curves than in heating curves. But as written above, we should consider T_{liq} values from cooling curves as a downward estimate: For sure, the observed value is not too high, but possibly too low as a result of supercooling. Sometimes, even almost perpendicular exothermal "jumps" can be observed on cooling curves after some supercooling and the following "delayed" crystallization. In such cases the measured thermal effect is with high probability below T_{liq}, as shown for the crystallization of gold in Fig. 2.21. On the other side, supercooling leads to temporary nonequilibrium and may enhance the strength of the first crystallization – and may make it at least visible, even if it appears below T_{liq}.

If the phase diagram must be measured up to the formation of a melt, then an elaborate sample preparation is often not necessary. Instead, we can start the measurements with one pure component and perform a DTA measurement up to its melting point. This is demonstrated in Fig. 3.21 (a) for pure KCl in the top curve. Then for the determination of the LiCl–KCl system, step by step LiCl was added to the crucible of the first measurement, and the molar fraction of LiCl was added as a label to all curves. All these subsequent measurements (and, if possible, also the first one with the pure component) must contain at least two heating segments, because not before the first heating above T_{liq} the sample is homogenized in the molten state, and useful data are obtained in the next heating. It may be useful to add even a third heating segment to check if homogenization was successful; in this case the curves from heating segments 2 and 3 should be almost identical.

If this way ca. 4...5 measurements are performed, then we can start a second series in the same way. This is shown in Fig. 3.21 (b) for pure LiCl (top curve) with subsequent additions of KCl. At least for the pure components, a TG curve should be measured: Some substances lose mass due to evaporation volatiles like water (crystal water or just humidity) or carbon dioxide (from the decomposition of carbonates at high T). Then a reduced mass has to be used for the calculation of the sample composition.

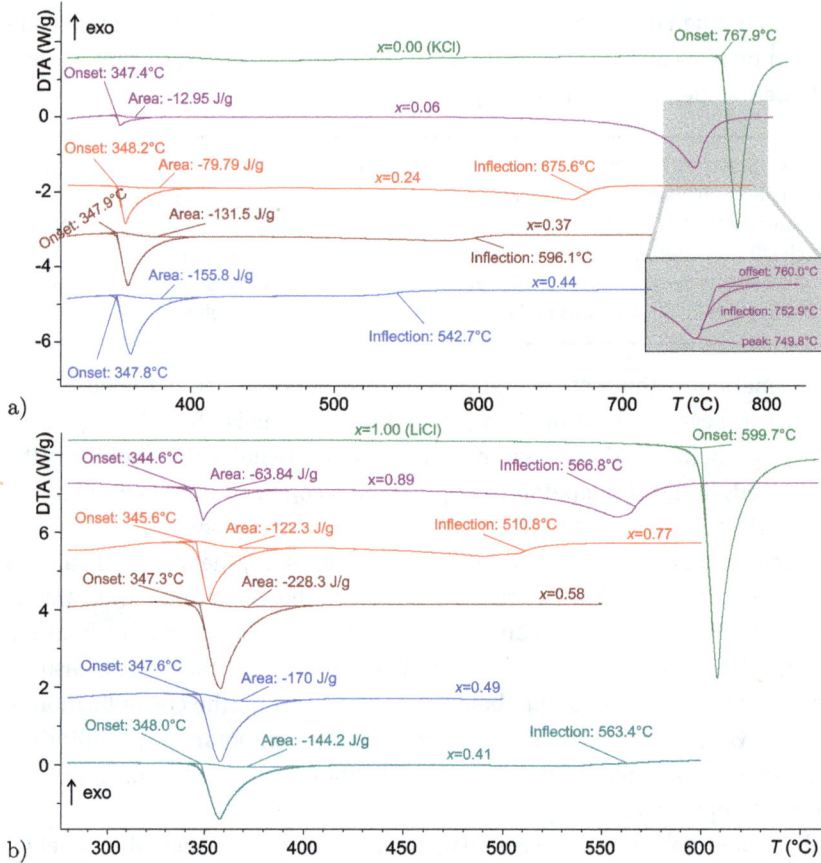

Figure 3.21: 11 DTA heating curves ($\dot{T}$ = 10 K/min, second heating) of x LiCl + (1 − x) KCl samples. The molar fraction of LiCl is given as parameter. (a) Started with pure KCl, LiCl added subsequently; (b) started with pure LiCl, KCl added subsequently.

3.3.1 Eutectic system: LiCl–KCl

The two panels in Fig. 3.21 show the measurements that were performed to investigate the system (1 − x) KCl + x LiCl. Both pure components show just one sharp melting peak at 767.9 °C (KCl) or 599.7 °C (LiCl). Immediately after adding small amounts of the other component (6 % LiCl in Fig. 3.21 (a); 11 % KCl in Fig. 3.21 (b)), an endothermal peak appears at an almost constant T_{eut} ≈ 347 ± 2 °C. (The average value is given here, and the maximum deviation holds for the experimental error.) The value for the x = 0.89 curve should be handled with care, because there the deviation from the average is significantly larger than for all other samples.

It was explained in Section 1.5.1.3 that in a eutectic system, melting starts at T_{eut} = const., and continues up to $T_{liq}(x)$. However, the rate of this melting process is not constant: From the lever rule (1.33) we can derive that the onset of melting is always strong at T_{eut}, especially for compositions close to x_{eut}. Then the melting rate (per minute or per Kelvin, which is the same for $\dot{T}$ = const.) becomes smaller, and grows again as $T \to T_{liq}$, especially for compositions close to the pure components. This enhanced melting process can be seen for the samples x = 0.06 and x = 0.24 in Fig. 3.21 (a), and on the other side of the phase diagram for x = 0.89 and (significantly less pronounced) for x = 0.77 in Fig. 3.21 (b). Sometimes, besides the "eutectic peak" at T_{eut}, this second effect is called the "liquidus peak". This is, however, not really correct. Instead, we have a continuous melting process that starts strong at T_{eut} and prevails up to T_{liq}, where the whole sample is liquid.

Often, only the determination of T_{eut} is straightforward and accurate, because it is the extrapolated onset of the eutectic peak (see Section 2.6.3). Without performing a detailed numerical simulation of heat flows inside the thermal analyzer, it is not so easy to determine T_{liq}, except for the pure components where T_{liq} is just the onset of the melting peak (curves x = 0 and x = 1 in Fig. 3.21). The gray insert in Fig. 3.21 (a) for x = 0.06 demonstrates two extrema, which can be reasonable estimations for different systems, and one compromise:
- The "extrapolated offset" is an upward estimate for T_{liq}. This is the intersection of the prolongated baseline (coming from high T) with the tangent at the inflection point of the "liquidus peak" that returns to the baseline.
- The "peak temperature" of this "liquidus peak" is a downward estimate for T_{liq}. Until there the melting process becomes stronger, and what follows is basically the return of the thermal signal to the baseline, which is significantly determined by the thermal resistance of the measurement system; see Section 2.5.3.
- Often, the inflection point itself is a reasonable compromise between the extrema given before. If such an inflection point can be found be the analysis, then it could be chosen for T_{liq}.

Nevertheless, it should be taken into account that the determination of liquidus temperatures by thermal analysis is usually less accurate than the determination of solidus temperatures. (This holds also for most phase diagrams found in the literature.) Especially for samples near x_{eut}, T_{liq} is so close to T_{eut} that both cannot be separated. In Fig. 3.21 (b), this is the case for x = 0.59 and x = 0.49. (Indeed, no experimental T_{liq} points are shown in the phase diagram in Fig. 3.23.) For an approximate description of a eutectic system, the following questions have to be answered:
1. Where is T_{eut}? $\longrightarrow$ There the horizontal eutectic line must be drawn.
2. Where are the melting points of the component? $\longrightarrow$ From these points at x = 0 and x = 1 the liquidus lines start downward.
3. Where is x_{eut}? $\longrightarrow$ This is the point where both liquidus curves must meet on the level of T_{eut}.

The first questions are easily answered with measurements of the pure components (question 2) or with the onsets of the eutectic peaks that appear at T_{eut} in the measurements of nearly all mixtures (question 1). The answer for question 3 can often be found with Tammann plots introduced in Section 1.5.1.1. For the current system, such a plot is shown in Fig. 3.22. The areas A of the eutectic peaks were obtained with an asymptotic baseline. This means that the peak area is limited to the top by a line that starts parallel to the DTA (DSC) curve from both sides of the peak. At intermediate points, this limiting line is on a position that changes proportionally to the partial area of the peak that is already passed. Such a construction has two benefits over a straight line between both ends of the peak:

1. The absolute position of the DTA (DSC) signal is a function of $c_p(T)$ for the sample (see Section 2.3.2.2). However, during passing the eutectic peak, a major portion of the sample changes it aggregation state from solid to liquid, which may significantly change c_p. This change from $c_p^{(sol)}$ to $c_p^{(liq)}$ is smooth and proceeds almost in the same manner like the consumption of heat of fusion, and hence to the partial peak area.
2. Moreover, there is a practical aspect: If the limit of the peak area starts asymptotical to the thermal signal, then the position where the area measurements starts and ends is not so important, because the main contribution to A comes from the positions where the peak is strong. This is not so for straight limits: There the (often arbitrary) choice of start and end positions contributes significantly more to A.

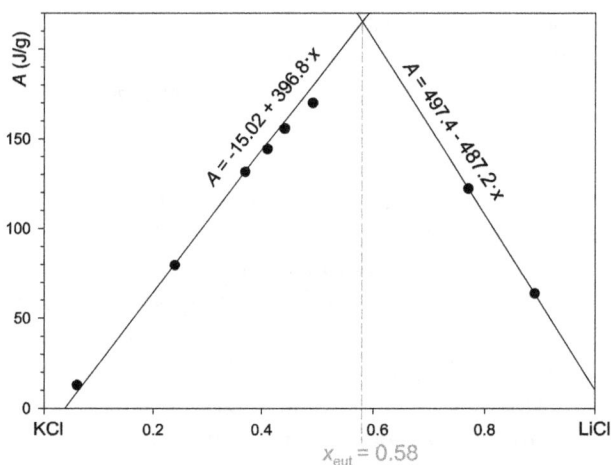

Figure 3.22: Tammann plot (Tammann, 1905) with the eutectic peak areas A from Fig. 3.21. The functions $A(x)$ are linear with maximum at the eutectic composition x_{eut}.

Figure 3.22 shows that the functions $A(x)$ rise linearly from both sides, approaching a maximum at their intersection at x_{eut}, as expected from the lever rule (1.33). The value of x_{eut} can be easily calculated from the linear fit functions given in the figure. For an ideal eutectic system, both linear fit functions should pass the level $A(x) = 0$ for $x = 0$ and $x = 1$, respectively. It is unphysical that in Fig. 3.22, we have $A > 0$ for $x = 1$ instead, but this discrepancy can be understood from the weak experimental basis on this side: There the fit function is defined by just two experimental points. The experimental basis seems better on the KCl side, where the fit could be performed through six experimental points. Nevertheless, also here the point $A = 0$, $x = 0$ is not met. It is not possible to decide on the basis of the current data if this deviation is also the result of experimental scatter or if, indeed, a small solubility of LiCl in solid KCl exists. In this case, it would be natural that we find $A(x) \to 0$ as $x \to x_{\text{max}}^{\text{Li:KCl}}$ if $x_{\text{max}}^{\text{Li:KCl}}$ is the maximum solubility of LiCl in solid KCl.

If a mutual solubility of the components exists for the solid phases in eutectic systems, then this solubility has often its maximum near the eutectic temperature. In any case, it must vanish at the melting point and must become smaller for low T, because the entropic contribution to the Gibbs energy (1.29) shrinks.

Figure 3.23 compares the experimental points that were read directly from the DTA curves in Fig. 3.21 and x_{eut} that was obtained from the Tammann plot in Fig. 3.22 with a theoretical phase diagram, which was calculated with assessed thermodynamic data from commercial databases that come with the FactSage 8.0 (2020) package. It should be noted that data on a significant solubility of LiCl in solid KCl are not available in these databases.

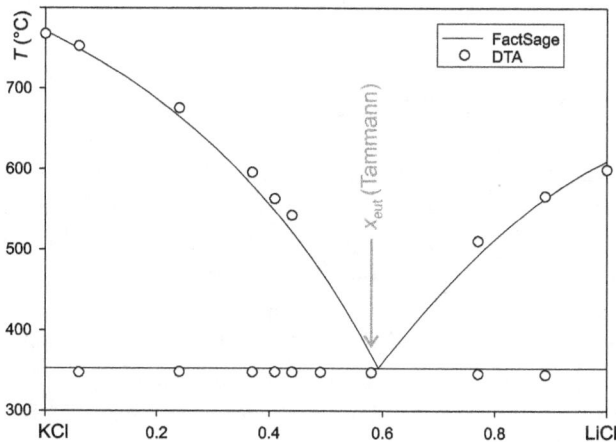

Figure 3.23: With FactSage 8.0 (2020) calculated phase diagram LiCl–KCl (lines) and experimental points derived from Fig. 3.21. The eutectic composition $x_{\text{eut}} = 0.58$ obtained from the Tammann plot in Fig. 3.22 is indicated by an arrow.

However, it should be taken into account that at least a minor mutual solubility of the components of a eutectic system will always exist. For drawing a phase diagram, it is just a question of whether minor solubilities can be shown on the scale of the concentration axis. If a minor solubility can be represented graphically, then this is sometimes called a "rim solubility". If the mutual solubility fields are large and in the order of the two-phase region, then this two-phase region is called a "miscibility gap", an example of which will be given in the next section.

3.3.2 Eutectic system with rim solubility: KCl–KI

Thermal analysis of the system KCl–KI was first reported by Le Chatelier (1894), and the results of a FactSage 8.0 (2020) calculation were shown above in Fig. 1.23. The partially contradictory results in the literature were reviewed by Sangster and Pelton (1987).

Figure 3.24 shows a reevaluation in two series of DTA measurements. For each series, first, a pure component (KCl or KI, respectively, ca. 30 mg) was measured in Ar flow. Subsequently, small amounts (starting with $\approx$1 mg) of the other component were added, and two heating/cooling runs were performed with this sample. In the first heating/cooling cycle the sample was molten and homogenized, and the second heating runs are shown in Fig. 3.24. In Fig. 3.24 (a), we see that starting from pure KCl, the melting peak drops in temperature and becomes broader down to $x = 0.60\ldots0.71$. Below $x = 0.60$, all peaks have an onset near 595.5 °C. Fig. 3.24 (b) shows initially a similar behavior for the KI side: The melting peak drops and becomes broader. However, already at $x = 0.08$ the same effect with onset at 595...596 °C appears. (We can "guess" it already for $x = 0.04$.)

There are only a few thermal effects that can occur in binary systems always at the same temperature, independent of composition (see Section 1.5.1.1). One of them is a eutectic (see Section 1.5.1.3). An explanation can be given also with the upper red phase boundary in Fig. 1.23: The shape of the two-phase field "liquid + solid" in a mixed crystal system is determined by the equilibrium of G^{liq} and G^{sol}. If for both phases, the term G^{ex} in (1.26) is negligible, then the system is ideal, and the two-phase field has the typical lens shape (Fig. 1.22). Interaction forces between the species in a phase are the origin of Gibbs excess contributions G^{ex}. In liquids, such interactions are usually weaker, because atoms or molecules move quite freely. It is not so in solids, where binding forces try to fix the atoms in their given positions. Principally, such binding forces can be attractive (and thus reduce G) or repulsive (and increase G), but in the case of strong attractive interactions, it is rather realistic to assume that true bonds and a new intermediate compound are formed. Hence repulsive interactions with $G^{\mathrm{ex}} > 0$ are more realistic than the opposite case.

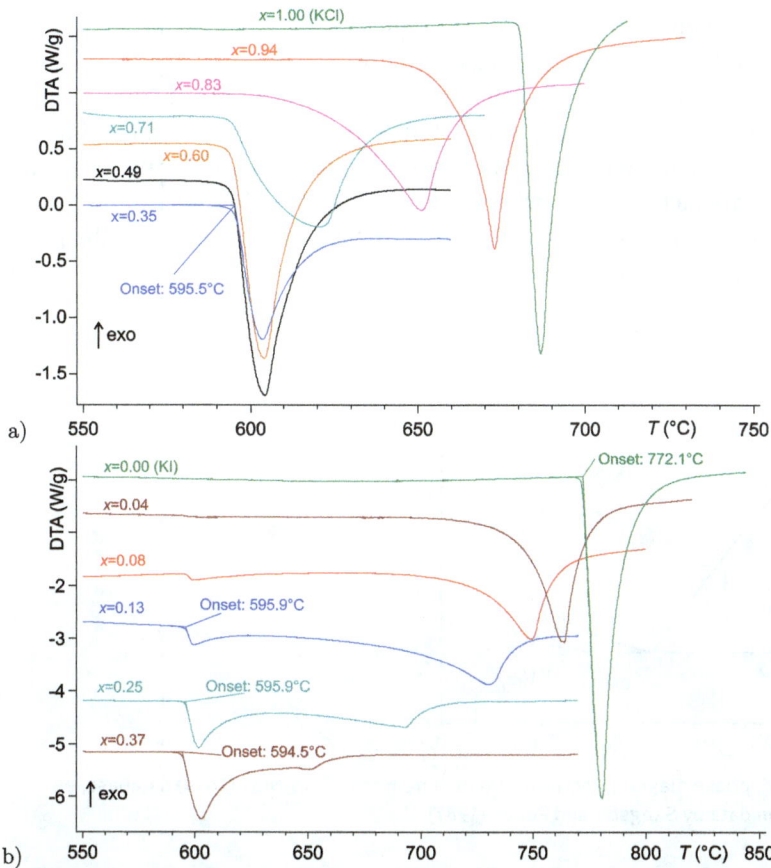

Figure 3.24: DTA heating curves ($\dot{T} = 10$ K/min) of $13\,x$ KCl $+ (1-x)$ KI samples, the molar fraction of KCl is given as a parameter. (a) Started with pure KCl, KI added subsequently; (b) started with pure KI, KCl added subsequently. (Note that the T scales are slightly different for both pictures, because KI melts higher than KCl.)

If an azeotrope occurs in a mixed crystal system, then it can principally be a common maximum or a common minimum of liquidus and solidus. However, it is more typical that the solid phase has positive excess Gibbs energy, which destabilizes this phase. Then the two-phase field is bend downward, and the azeotrope point is a minimum.

!

A crossover from a mixed crystal system to a eutectic system with "rim" solubility can occur if G^{ex} is very large (positive). Then the miscibility gap phase boundary may be shifted so high that it achieves the solidus. In Fig. 1.23, this corresponds to the upper dotted red line. Between the intersections of the solubility gap and the solidus a horizontal tie line can be drawn, which is the eutectic line. Below this line, we have the phase field "K(Cl,I)$_{\mathrm{ss}}$(I) + K(Cl,I)$_{\mathrm{ss}}$(II)", and above it, we have two-phase fields "liquid + K(Cl,I)$_{ss}$". This K(Cl,I)$_{ss}$ is Cl-rich on the right side and I-rich on the left side. We can conclude that the KCl–KI system is such a eutectic system with broad mutual miscibility, especially on the KCl side, because there the eutectic peak appears not before $x = 0.71$. This means for the diagram in Fig. 1.23 that the calculated blue miscibility gap is surely too low; rather, the upper red dotted curve approaches the truth. (However, it expresses not satisfactorily the obvious asymmetry of the solubilities.)

The experimental results in Fig. 3.24 are in excellent agreement with the phase diagram in Fig. 3.25.

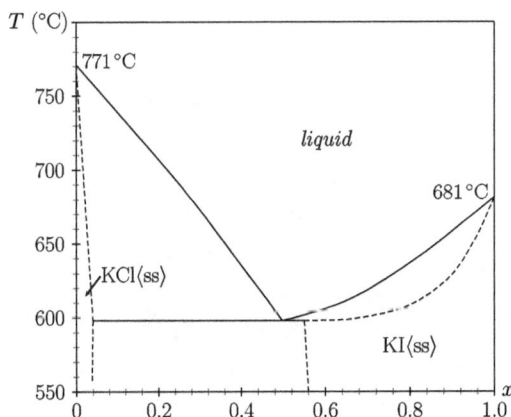

Figure 3.25: KCl–KI phase diagram, redrawn with data from the ACerS-NIST (2014) database, entry 7804, based on data by Sangster and Pelton (1987).

It is not always possible to count simply the minima or maxima in DTA or DSC curves and to attribute them to singular phenomena in the phase diagram. For instance, the bottom three curves in Fig. 3.24 (b) have one peak with onset 595...596 °C and a second "peak" between 650 and 735 °C. This second "peak" just marks (almost) the end of the melting process, the liquidus T_{liq}. Nevertheless, melting continued over the whole T range from T_{eut} to T_{liq}. Also, for the bottom curves in Fig. 3.24 (a) with always only one peak, continuous melting occurs between T_{eut} and T_{liq}; just the determination of T_{liq} is not so straightforward.

3.3.3 System with intermediate compound: LiF–YF$_3$

Thoma et al. (1970) reported that for the heavier rare-earth elements RE = Eu...Lu, and Y, the systems LiF–REF$_3$ contain an intermediate phase LiREF$_4$, which crystallizes in the tetragonal scheelite[15] structure. The stability of the LiREF$_4$ phases becomes

---

15 Carl Wilhelm Scheele (9 December 1742–21 May 1786).

larger for heavier RE metals, and for RE = Tm...Lu, the scheelite phases melt congruently. In contrast, for RE = Eu...Er, peritectic decomposition under the release of solid REF_3 occurs. Crystals can be grown from the melt, because the compounds undergo no solid state phase transitions. Isomorphism of all $LiREF_4$ allows doping with another RE′, and their thermophysical properties (high optical transmittance, low thermal lensing) make doped $LiREF_4$ crystals interesting for laser applications. Ranieri et al. (1996) demonstrated homogeneous RE dopant concentrations in $LiYF_4$. More recently, Fedorov et al. (2002) mentioned that also for RE = Sm, the intermediate scheelite is formed but undergoes decomposition already below the liquidus into LiF(sol) and SmF_3(sol).

Thermal analysis of fluorides is difficult, especially if the corresponding metal has a high valence state ≥ 3. The situation is very complicated for the REF_3. Such fluorides tend to undergo hydrolysis already with traces of water, e. g., from the gas flow in the thermal analyzer under the incorporation of oxygen and the release of HF. With a few ($\approx 1\ldots 6$) per cent oxygen, solid solutions of the kind $REF_{3-2x}O_x$ are formed. Some of these berthollide oxide fluorides are eutectoids, and their formation can be confused with solid state $\alpha - \beta$-transitions of the REF_3 itself. Others melt congruently or peritectically, and their melting can be misinterpreted as the melting of the REF_3. For details, the reader is referred, e. g., to Sobolev et al. (1976).[16]

Estimate the mass of gaseous water that flows during a 6 hours measurement through a thermal analyzer if the device is rinsed by Ar (50 ml/min) with 99.999 % purity having a typical humidity of 3 ppmv? (ppmv = "parts per million in volume"). Which amount of terbium fluoride can be hydrolyzed by this water to oxyfluoride if the reaction equation $TbF_3 + H_2O \rightarrow TbOF + 2HF \uparrow$ is assumed?

Sobolev et al. (1976) gave some experimental results for their DTA measurements: The REF_3 compounds were prepared from commercial RE_2O_3 by chemical reaction with the gaseous decomposition products of teflon, which contain significant amounts of HF (see Fedorov et al., 2002). Furthermore, for thermal analysis, there was used a noncommercial setup, with Ni or Mo crucibles, a heater system without ceramic parts (to reduce the introduction of humidity), and tungsten/rhenium thermocouples, which were calibrated with the melting points of different fluorides, such as CaF_2, SrF_2, BaF_2. The sample mass was larger than in typical contemporary devices, in the range 1...2.5 g. The large sample mass certainly affects the resolution of nearby DTA peaks, but it has the advantage that minor water traces have a relatively smaller impact on the sample.

As another method to overcome hydrolysis related problems, Sobolev et al. (1976) performed DTA in a static helium atmosphere, which avoids the permanent introduction of oxygen and/or humidity by a gas flow (see the question above this paragraph).

---

16 Boris Pavlovich Sobolev (born 26 March 1936).

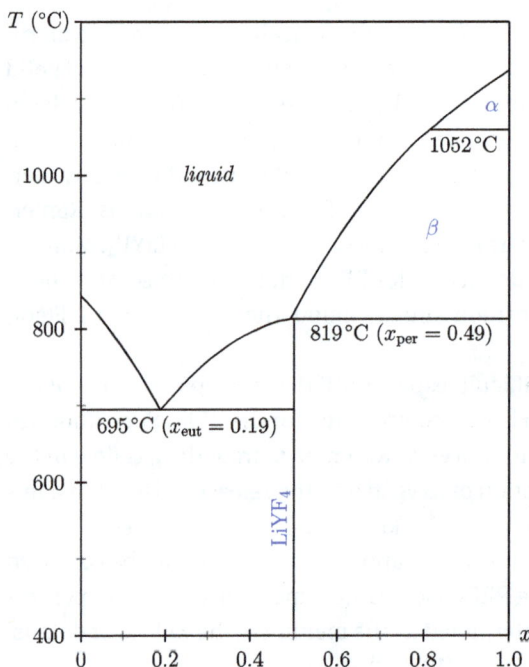

Figure 3.26: LiF–YF$_3$ phase diagram, redrawn with data from the ACerS-NIST (2014) database, entry 1477, based on data by Thoma et al. (1961). $x = 0$ is LiF.

Additionally, remaining impurities were gettered by heated Ti sponge inside the thermal analyzer. The authors reached this way low oxygen contamination levels around 0.04 wt% in the samples.

The LiF–YF$_3$ phase diagram shown in Fig. 3.26 is in good agreement with the experimental results of most authors (cf. Thoma et al., 1961, 1970; Fedorov et al., 2002) and also with a thermodynamic assessment of the system by Dos Santos et al. (2012). Figure 3.27 shows a series of DTA curves with compositions spanning the whole system from pure LiF (top) to pure YF$_3$ (bottom). Already on a first glimpse, we recognize from top to bottom three types of peaks that appear at almost constant T for a certain range of compositions:

1. Peak 1 occurs at $T \leq 700\,°C$; it is visible for LiF-rich compositions $0 < x \leq 0.4798$, and is strongest around $x = 0.2$. It is not related to LiF itself (otherwise, it would be present also for $x = 0$). Obviously, peak 1 is related to the LiF/LiYF$_4$ eutectic and is strongest at $x_{\mathrm{eut}} \approx 0.2$. The exact value of x_{eut} can be determined with a Tammann plot, like in Fig. 3.22.
2. Peak 2 at $T \approx 820\,°C$ seems to start already around $x = 0.3829$, but at lower T there. Then it moves upward and reaches its final value $820\,°C$ near $x = 0.5$. It remains there for all other samples (except pure YF$_3$, $x = 1$) but becomes continuously smaller. Its maximum strength at $x \approx 0.5$ indicates that it is related to this

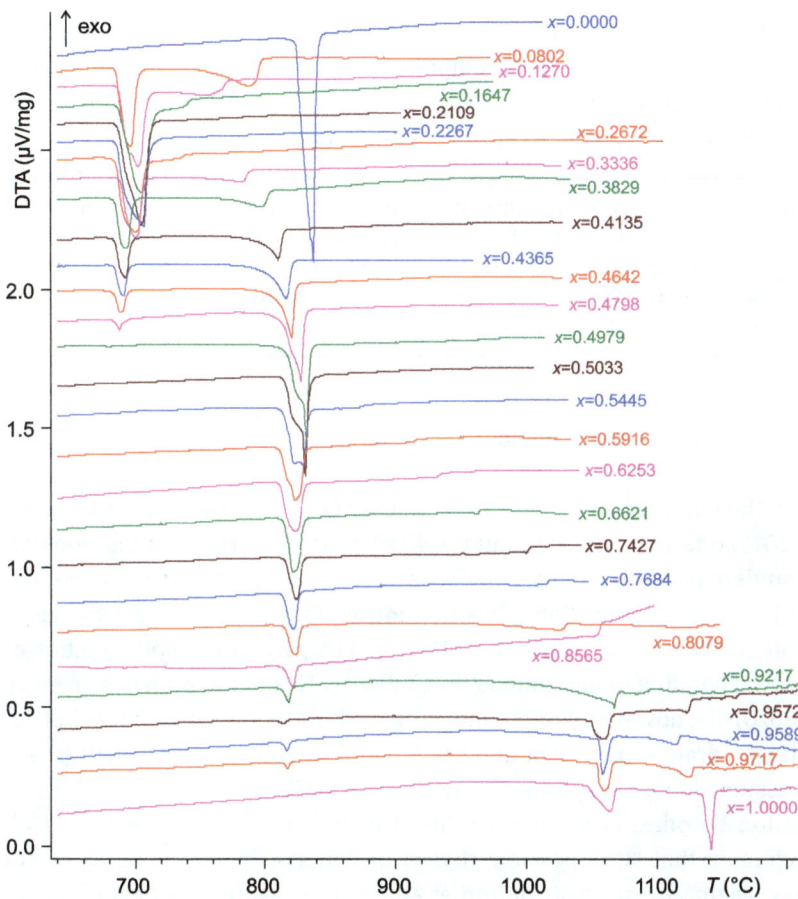

Figure 3.27: DTA heating curves (second heating, $\dot{T}$ = 10 K/min) of 28 binary mixtures ranging from pure LiF (x = 0.000) to pure YF$_3$ (x = 1.000). NETZSCH STA449C "Jupiter", Pt95/Au5 crucibles, atmosphere flowing Ar 99.999%.

composition and hence to the intermediate phase LiYF$_4$; because the onset temperature is constant for all x > 0.5, this phase melts peritectically at this onset temperature T_{per} ≈ 820 °C. The continuous rise of the onset for x < 0.5, which starts already from a shoulder for x = 0.2672, marks the LiYF$_4$ liquidus between x_{eut} and x_{per} ≈ 0.5.

3.   Peak 3 at T ≈ 1050 °C appears only for very YF$_3$-rich compositions $x \gtrsim 0.92$, including pure YF$_3$ (x = 1). Indeed, this peak is related to a solid state transition between an orthorhombic β-YF$_3$ structure, which is stable at room temperature and a not well-defined trigonal α-YF$_3$ structure at high T (Ranieri et al., 2008). The peak appears only in the narrow concentration range where solid YF$_3$ is present, because it relies on the solid phases of YF$_3$.

We see from the original curves in Fig. 3.27 that it is typically considerable easier to identify effects that occur for different compositions at the same temperature. Here these are a eutectic, a peritectic, and a solid state phase transition. A monotectic is another effect, which appears over a certain concentration range at constant T.

🕴 The temperature of a solid state phase transition is only required to be constant if the composition of this phase is fixed (daltonide). If a berthollide phase with variable composition undergoes a phase transition, then the temperature of this transition can be a function of concentration. Carruthers et al. (1971) demonstrated this for $LiNbO_3$, where the ferroelectric transition temperature lies between 1020 °C and 1080 °C, depending on the Li:Nb ratio, which is not exactly 1.0.

3.3.4 System with intermediate compound: KCl–BaCl$_2$

Since decades, the book edition "Phase Diagrams for Ceramist" (now, electronically, ACerS-NIST, 2014) offers a convenient and well-referenced source for a first look on several thousands of phase diagrams, with focus on oxide and halide/salt systems. For the system KCl–BaCl$_2$, the collection offers two entries shown in Fig. 3.28. Obviously, an intermediate compound exists, which is placed in Fig. 3.28 (b) at a molar fraction of $BaCl_2$ $x = \frac{1}{3}$, corresponding to the composition K_2BaCl_4. This phase even has an entry in the ICDD (2020) file, however, without indexing the X-ray reflections. The authors of Fig. 3.28 (a) did not draw in the compound itself, but obviously an intermediate phase K_2BaCl_4 was assumed too, with peritectic melting at 656 °C.

The situation becomes even more confused with a paper by Krogh-Moe et al. (1967), who claimed that "it is possible that even non-stoichiometric ratios of KCl and BaCl$_2$ may occur". In contrast, Hattori et al. (1981) measured the heat of fusion of K_2BaCl_4 by DSC without mentioning deviations from the K:Ba = 2:1 stoichiometry. Also, Gasanaliev et al. (1982) reported congruent melting of K_2BaCl_4 with eutectics to both sides at 648 °C and 638 °C, respectively. More recently, Vinogradova et al. (2005) and Moiseeva (2019) grew crystals of "K_2BaCl_4" by the Bridgman method, where the melt inside a silica ampoule was moved out off the hot zone with a rate of 1.5...2 mm/h. Both authors observed that only the first few centimeters of the grown crystal appeared optically clear, but the parts that solidified later were opaque and polycrystalline. This observation is a hint that
1. either the compound melts peritectically, as in Fig. 3.28 (a), or that
2. some congruently melting composition deviates from the composition of the starting material for the crystal growth experiments, $x = \frac{1}{3}$.

Thermal analysis of the KCl-BaCl$_2$ system helps us to decide which option is true. In a first step the true assay of the pure component chlorides was checked by TG. Prior to the measurements shown in Fig. 3.29, a "correction measurement" with empty crucibles was performed, which results in flat TG curves if no mass loss occurs (compare

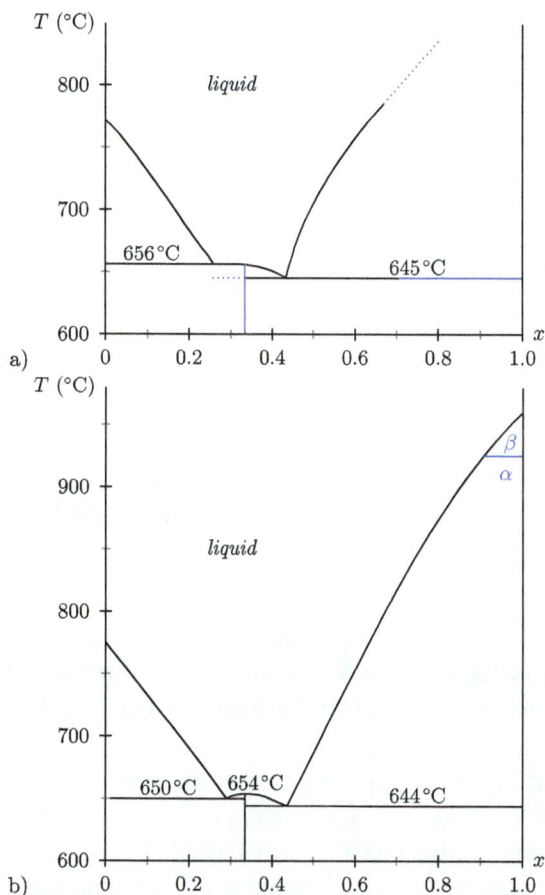

Figure 3.28: Two variants of the KCl–BaCl$_2$ phase diagram, redrawn with data from the ACerS-NIST (2014) database. Blue lines were added for completeness; $x = 0$ is pure KCl. (a) Entry 1262, based on data from 1932: the compound K$_2$BaCl$_4$ ($x = 0.3333$) was not explicitly drawn, but the peritectic melting at 656 °C is obvious. (b) Entry 3053, based on data from 1955: K$_2$BaCl$_4$ shown with congruent melting at 654 °C.

to Fig. 2.6, which was measured without such a correction). KCl shows only a minor mass loss of 0.11 %, obviously resulting from the release of humidity traces at the sample surface. As pointed out by Haase and Brauer (1978),[17] barium chloride, in contrast, forms at least two hydrates BaCl$_2 \cdot$ H$_2$O and BaCl$_2 \cdot$ 2H$_2$O. The latter is the conventional commercial product, which was used for the measurements.

The red curve in Fig. 3.29 shows well the two TG steps of almost identical height, which result from the stepwise release of the two water molecules per formula unit.

---

17 Georg Brauer (11 April 1908–26 February 2001).

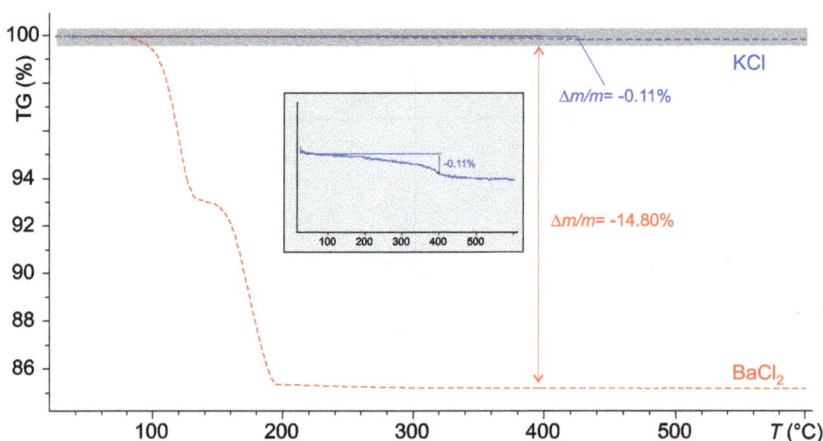

Figure 3.29: Thermogravimetry of the pure starting materials KCl and $BaCl_2 \cdot 2H_2O$, which were used for the DTA measurements of Fig. 3.30. Prior to these TG measurements, a "correction" measurement with empty reference and sample crucibles was performed to eliminate buoyancy effects.

The observed $\Delta m/m = -14.80\,\%$ is slightly higher than the theoretical value $-14.75\,\%$, and again, the small difference is probably due to adsorbed humidity. The measured $\Delta m/m$ were used to calculate exact concentrations of DTA measurements for the KCl-$BaCl_2$ system.

Figure 3.30 (a) gives an overview (not all measurements are displayed there) for the whole concentration range from $x = 0$ (KCl) to $x = 1$ ($BaCl_2$). The experimental melting point 771.9 °C for KCl agrees very well with the FactSage 8.0 (2020) value 771 °C, and from there the "melting peak" moves down toward some peak $T_2 = \text{const.}$ $BaCl_2$ shows a double peak resulting from a first-order phase transition between α-$BaCl_2$ (low T, space group $Pnam$) and β-$BaCl_2$ (high T, space group $Fm\bar{3}m$, Hull et al., 2011) at 925.3 °C (in agreement with FactSage 8.0, 2020). It is difficult to reveal the exact melting point of $BaCl_2$, because the transition peak and the melting peak are partially overlapping. Nevertheless, the value 958.5 °C from Fig 3.30 (a) is in fairly reasonable agreement with 962 °C from FactSage 8.0 (2020). Also, from there the liquidus drops with KCl addition, however, to a different peak at $T_1 = \text{const.}$

The curves in Fig 3.30 (a) are compatible with both phase diagrams in Fig. 3.28. Hence a series of additional measurements was performed, where the transition between the peaks at T_1 and T_2 is expected to occur, and where a potential intermediate compound is expected to exist. The curves in Fig 3.30 (b) show that $T_1 \approx 647$ °C appears for $x \geq 0.33684$ and $T_2 \approx 657$ °C appears for $x \leq 0.28633$. The curve $x = 0.30523$ is interesting: There one peak appears at slightly (but significantly) higher $T_f = 660.5$ °C. Such behavior, with a local maximum of the liquidus temperature, is compatible only with a congruently melting intermediate compound, similar to Fig. 3.28 (b). Then the curve $x = 0.31965$ marks the drop from the liquidus maximum to T_1.

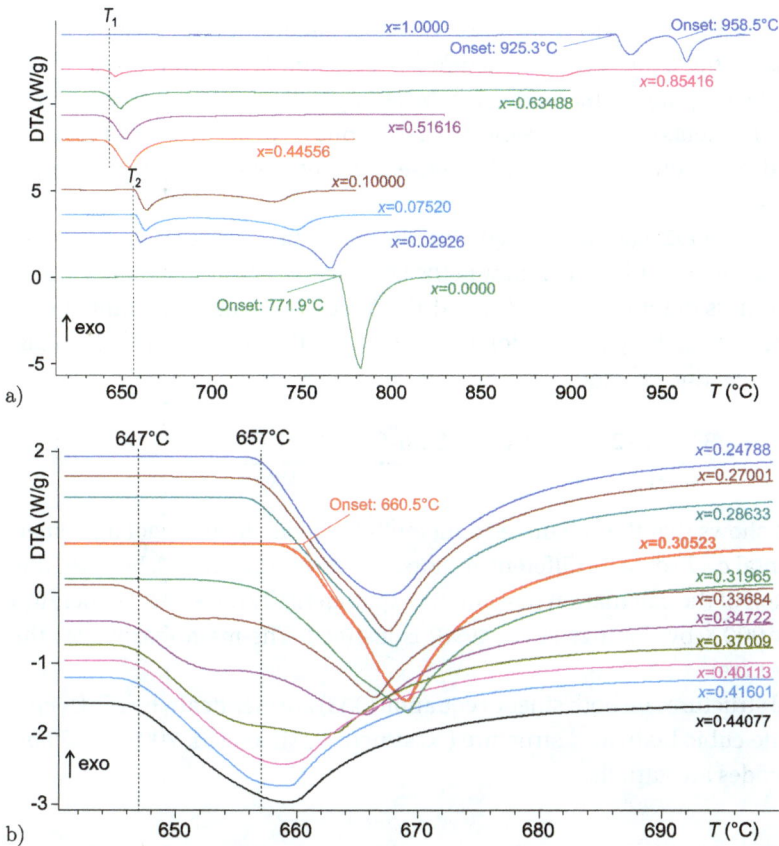

Figure 3.30: DTA heating curves ($\dot{T}$ = 10 K/min) of $(1-x)$ KCl + x BaCl$_2$ samples, the molar fraction of BaCl$_2$ is given as parameter. (a) Top: 5 curves started with BaCl$_2$, KCl added subsequently; bottom: 4 curves started with pure KCl, BaCl$_2$ added subsequently. (b) 11 samples around the expected composition of "K$_2$BaCl$_4$", $x = \frac{1}{3}$, of the same system.

It seems surprising that the congruent melting point was found not at the expected $x = \frac{1}{3} \approx 0.33333$, but at $x' \approx 0.3$, which corresponds to a composition of approximately K$_7$Ba$_3$Cl$_{13}$. Such a phase is not yet described in the literature, but certainly a sound description of it would require an accurate X-ray investigation of the crystal structure, which is still lacking. However, it should be emphasized that the slightly different stoichiometry of the intermediate phase in the KCl-BaCl$_2$ system explains very well the recent experimental observations by Vinogradova et al. (2005) and Moiseeva (2019).

3.3.5 Missing data for a complex compound: Y$_3$Al$_5$O$_{12}$

Klimm et al. (2007) described the isopleth section Y$_3$Al$_5$O$_{12}$–Nd$_3$Al$_5$O$_{12}$ of the concentration triangle Y$_2$O$_3$–Nd$_2$O$_3$–Al$_2$O$_3$, which has technical relevance because solid solutions $(Nd_xY_{1-x})_3Al_5O_{12}$ ("Nd:YAG") are one of the most important laser host materials. They crystallize in the cubic garnet structure (space group $Ia3d$), where the cations are coordinated by [4], [6], or [8] oxide ions. Most RE_2O_3–Al_2O_3 (RE = rare-earth element or yttrium) systems contain, besides the garnet phase, phases $REAlO_3$ (distorted perovskite) and/or a monoclinic $RE_4Al_2O_9$ phase, some of them also the "β-alumina" phase $REAl_{11}O_{18}$.

Wu and Pelton (1992) gave a thorough description of these systems, and the thermodynamic parameters are included now in the FactSage 8.0 (2020) databases. Unfortunately, the authors did not include Y_2O_3–Al_2O_3 in their investigations. Klimm et al. (2007) estimated, in analogy to equation (2.38), data for the missing $Y_3Al_5O_{12}$ phase from the formal reaction scheme

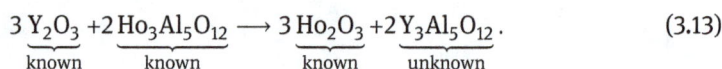

$$3\underbrace{Y_2O_3}_{known} + 2\underbrace{Ho_3Al_5O_{12}}_{known} \longrightarrow 3\underbrace{Ho_2O_3}_{known} + 2\underbrace{Y_3Al_5O_{12}}_{unknown}. \tag{3.13}$$

Figure 3.31 shows that the calculated data for $Y_3Al_5O_{12}$ are in excellent agreement with experimental $c_p(T)$ data by different authors. This is very beneficial, because thus the extrapolation of the calculation beyond the experimentally observed temperature range (T_{max} = 900 K by Konings et al., 1998) is justified. The main reasons for this good agreement are:

1. The crystal structures on both sides of equation (3.13) are identical: the RE_2O_3 crystallize in the cubic bixbyite[18] structure ("C structure, Zinkevich, 2007), and both complex oxides are garnets.

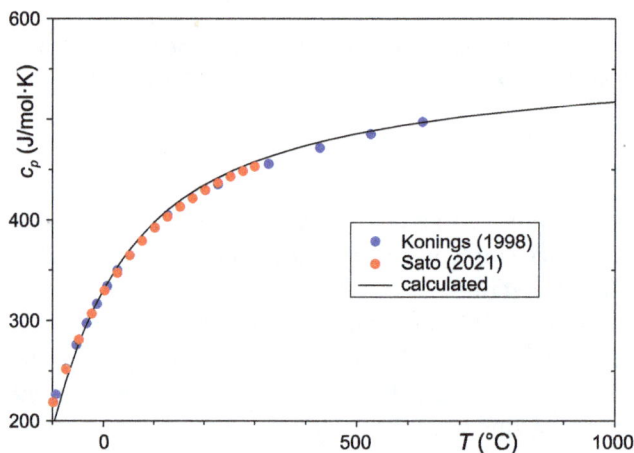

Figure 3.31: Specific heat capacity c_p of $Y_3Al_5O_{12}$ calculated from FactSage 8.0 (2020) data with eq. (3.13) compared to experimental data from Konings et al. (1998) and Sato and Taira (2021).

---

18 Maynard Bixby (28 June 1853–18 February 1935).

2. The ionic radii of Y^{3+} (104.0 pm) and Ho^{3+} (104.1 pm) in octahedral coordination after Shannon (1976) are almost identical.
3. Holmium and yttrium are chemically very similar.

As a consequent of these points, the exchange of Ho^{3+} by Y^{3+} does not influence the phonon spectra of the solids much, and hence not this most significant contribution to c_p; see Section 1.1.3.1. (It should be noted, however, that the difference of the relative atomic masses between Y (M = 89) and Ho (M = 165) is large, but, obviously, the influence of this difference on c_p is not very strong.)

Thermodynamic properties for substances that are experimentally not available can be calculated by combining its components data using Hess' law (Fig. 1.9, and eq. (2.37)). However, if possible, the calculation of unknown parameters by a substitution reaction of the kind (2.38) gives more accurate results.

A Symbols

Quantity	Symbol	Unit
atomic mass unit	$u = 1.661 \times 10^{-27}$	kg
Avogadro constant	$N_A = 6.022 \times 10^{23}$	mol^{-1}
Boltzmann constant	$k_B = 1.38 \times 10^{-23}$	J/K
crystal lattice parameters	$a, b, c, \alpha, \beta, \gamma$	Å, nm or °, resp.
degrees of freedom	F	–
enthalpy	H	J/mol
of fusion	ΔH_f	J/mol
of vaporization	ΔH_v	J/mol
entropy	S	J/(mol K)
formula units per unit cell	Z	–
gas constant	$R = 8.314$	J/(mol K)
Gibbs free energy	G	J/mol
heat energy	Q	J
inner energy	U	J/mol
molar fraction	x	–
at solidus line	x^{sol}	–
at liquidus line	x^{liq}	–
molar mass	M	g/mol
molar volume	V_m	m^3/mol
mass	m	kg
sample mass	m_s	mg
reference mass	m_r	mg
mass density	ϱ	kg/m^3
number of components	C	–
number of moles	n	–
number of phases	P	–
power (electrical, thermal)	W	J/s
pressure	p	bar
sensitivity (of thermal analyzer)	K_s	V/W
specific heat capacity	c	J/(mol K)
at constant p	c_p	J/(mol K)
at constant V	c_V	J/(mol K)
temperature	T	K (or °C)
melting temperature	T_f	K (or °C)
phase transition temperature	T_t	K (or °C)
eutectic temperature	T_{eut}	K (or °C)
peritectic temperature	T_{per}	K (or °C)
monotectic temperature	T_m	K (or °C)
azeotrope temperature	T_{aze}	K (or °C)
solidus temperature	T_{sol}	K (or °C)
liquidus temperature	T_{liq}	K (or °C)
time	t	s, min, or h
volume	V	m^3

https://doi.org/10.1515/9783110743784-004

B Answers to problems

Page 3: Time $t = Q/W = 369.45\,\text{s} \approx 6\,\text{min}\,9\,\text{s}$.

Page 10: c_V^e depends linearly on T, and the main contribution c_V according to the Debye theory (1.15) depends on T^3 for small T, which means that then c_V grows much weaker than c_V^e.

Page 14: $\Delta H_f^{(-10)} = \Delta H_f^{(0)} - (c_p^{\text{liq}} - c_p^{\text{ice}})\Delta T$; $\Delta H_f^{(-10)} = 6001 - (76 - 36) \cdot 10$; $\Delta H_f^{(-10)} = 5601\,\text{J}/(\text{mol K})$ which means that the heat of fusion of water drops with T (see figure below).

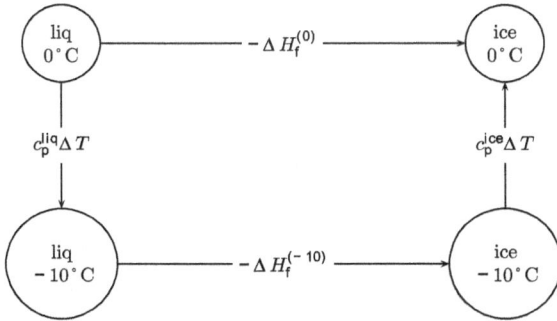

Page 19: For identical concentrations, we have $x_i = x = \frac{1}{C}$. Hence $G^{\text{id}} = R \cdot T \cdot C \cdot x \cdot \ln x = R \cdot T \cdot \ln x = -R \cdot T \cdot \ln C$; $C = 6$, $T = 1000\,\text{K}$: $G^{\text{id}} \approx -14.9\,\text{kJ/mol}$.

Page 31: The functions $G(T)$ for both phases can be regarded linear at the melting point, and both G values are equal there. Hence $H^{\text{sol}} - T_f S^{\text{sol}} = H^{\text{liq}} - T_f S^{\text{liq}}$, $T_f \Delta S_f = \Delta H_f \approx 28.158\,\text{kJ/mol}$.

Page 32: By the Gibbs phase rule $P + F = C + 2$ the sum of phases P and degrees of freedom F is restricted to the number of components C plus 2. Under isobar conditions, one degree of freedom is "consumed", and one for the binary system has $P + F = 3$. Hence $P = 3$ means that no degree of freedom is left, which applies only to singular points (e. g., eutectics, cf. Section 1.5.1.3), but not to an area. For $C = 1$, we have $F = 2$, and for $C = 2$, $F = 1$. This means that with every movement along one axis (x or T), all other conditions of the systems are fixed.

Page 46: The Gibbs phase rule $P + F = C + 2$ for a system with $C = 2$ components says that the number of phases P and degrees of freedom F must sum up to 4, and for $p = \text{const.}$, they must sum up to 3. Congruent melting: Solid (fixed composition, $F = 0$) and melt (variable composition, $F = 1$) are in equilibrium, but changing x will also change T of the equilibrium ($\Rightarrow P = 2, F = 1$). Peritectic melting: B(sol), AB(sol), and melt are in equilibrium, but there is no degree of freedom because T is fixed to T_{per} and the melt composition to x_{per} ($\Rightarrow P = 3, F = 0$).

https://doi.org/10.1515/9783110743784-005

Page 49: The sections are depicted below. 450 °C is below the ternary eutectic point E in Fig. 1.33, and only solids exist. The line LiCsF$_2$–NaF separates the two-phase fields. At 800 °C, nearly the whole system is molten, and only in the vicinity of the highest melting components, two-phase fields "LiF+melt" or "NaF+melt", respectively, appear.

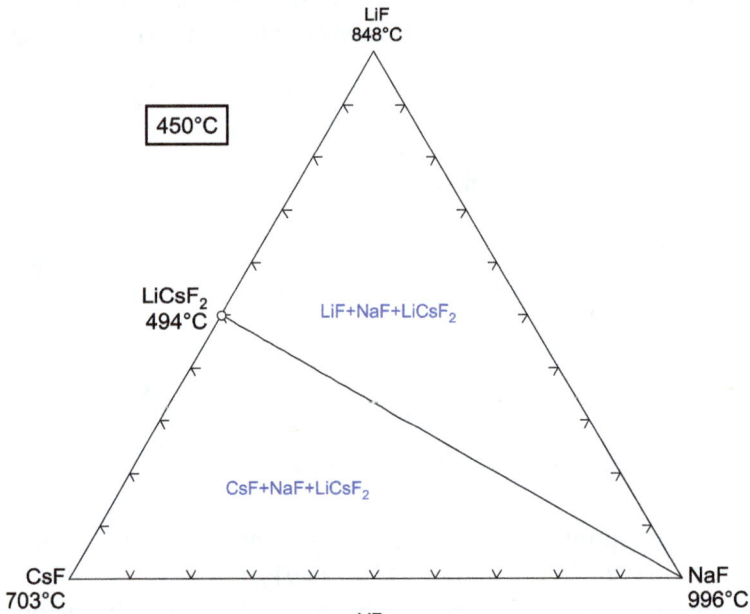

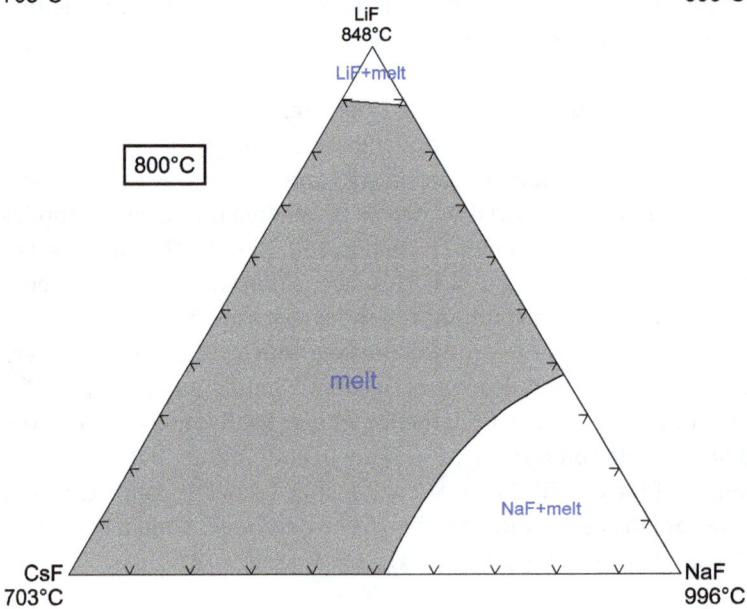

Page 53: The initial composition (80 % LiF, 10 % NaF, 10 % CsF) is marked by a red dot, which is located in the primary crystallization field of LiF. (1) From the isotherms we can estimate that crystallization of pure LiF starts near 740 °C and the crystallization path goes in the opposite direction downward. (2) At ca. 570 °C the next primary crystallization field is reached, and NaF crystallizes together with LiF down to the peritectic point P at 462 °C. There a peritectic reaction LiF + *melt* → LiCsF$_2$ takes place, which is the back reaction of (1.40), and LiF disappears under equilibrium conditions. (3) From P to the ternary eutectic point E, LiCsF$_2$ and NaF are crystallizing together, and at point E (455 °C) the rest of the melt crystallizes as a ternary eutectic mixture with approximate composition 35 % LiF, 10 % NaF, 55 % CsF.

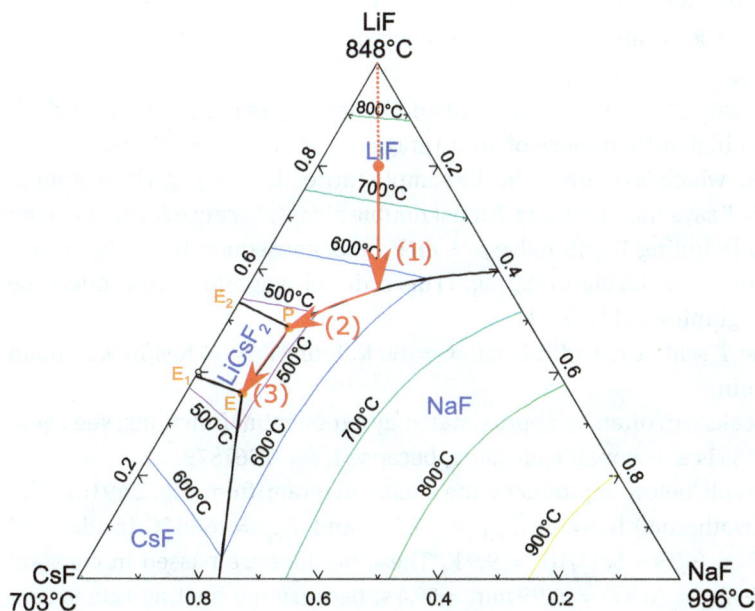

Page 58: The reason is simply mechanical: It is easier to put the "too small" Li$^+$ ion on the Na$^+$ site than to put the "too large" Na$^+$ ion on the small Li$^+$ site.

Page 67: The molar masses are 274.2 g/mol for gehlenite Ca$_2$Al$_2$SiO$_7$ and 272.6 g/mol for åkermanite Ca$_2$MgSi$_2$O$_7$. The molar fraction of åkermanite is

$$x = \frac{73.2}{272.6} \bigg/ \left(\frac{73.2}{272.6} + \frac{26.8}{274.2} \right) = 0.733 \, (= 73.3 \, \%).$$

(The difference between mol-% and mass-% is negligible, because both molar masses are so similar.)

Page 74: The last *emf* values in Table 1.4 are 38.289 mV (2200 °C) and 40.223 mV (2400 °C). (40,223 − 38,289) μV/200 K = 9.67 μV/K. The DTA signal never exceeds ≈ 45 μV. Hence 45 μV/9.67 μV/K = 4.7 K is the maximum difference, an acceptable error at such high temperatures.

Page 80: DTA and the more DSC require constant heating rates. This can only be obtained at the (empty or with an inert reference material filled) reference crucible. The sample crucible, in contrast, is heated slower, e. g., during a melting event.

Page 92: Zn melts between 400 and 500 °C, and the thermovoltage changes between both T by ca. 1 mV and hence ca. 10 μV/K. The magnitude of the DTA signal ist ≈ 0.4 μV/mg, which means, with m_s ≈ 10 mg, a signal strength of 4 μV. The T change that results from melting is $\frac{4\,\mu V}{10\,\mu V}$ K = 0.4 K.

Page 84: Energy below the peak = peak area $\times m_s$. Ar: −181 J/g × 0.06447 mg = −11.67 J; O_2: 428 J/g × 0.06245 mg = 26.73 J. Efficiency of heat transport = $\frac{26.73+11.67}{121}$ = 0.317 ≅ 31.7 %.

Page 86: The density of Ar (molar mass 40) and hence its gas density (≈ 1.8 g/dm³) are somewhat higher than those of air (average molar mass ≈ 29). It follows that the mass of Ar, which is occupied by the sample carrier, is ≈ 1.8 mg. The gas equation $p \cdot V = R \cdot T$ says that doubling T must double also V, because R and p = 1 bar are constant. Doubling V will halve $\rho = m/V$, and, consequently, the buoyancy will be lowered from 1.8 mg to 0.9 mg. (This order of magnitude can indeed be seen in the beginning of Fig. 2.6.)

Page 105: On the T scale the width is ca. w = 60 K. From $\dot{T}$ = 10 K/min we obtain $t = w/\dot{T}$ ≈ 6 min.

Page 117: DTA peaks can often be approximated by exponential functions; see equation (2.14). 37 % is a reasonable measure, because $1/e$ = 0.367879 ....

Page 151: The graph below reproduces the phase diagram from Fig. 2.49 (b) with 11 tie lines (isotherms) between T_{sol} = 587 °C and T_{liq} = 686 °C in identical distances ΔT = (686 − 587)/10 = 9.9 K. These tie lines are passed in constant time intervals $\Delta t = \Delta T/\dot{T}$ = 0.99 min = 59.4 s, because the heating rate is constant. The green numbers represent the fraction of material, which is solid at each tie line. (This fraction is the right "lever" of the tie line divided by its total length.) The blue numbers are the differences between subsequent numbers of the green column, hence the share of material molten between passing the corresponding tie lines. The composition of "just molten" material and T does not change drastically. Hence its heat of fusion will remain approximately constant. As a result, the thermal power needed for melting will be almost proportional to the numbers in the blue column. This column, however, starts at low T with a relative "melting rate" of 10 %, then slows down to 8 %, and finally rises to 15 %.

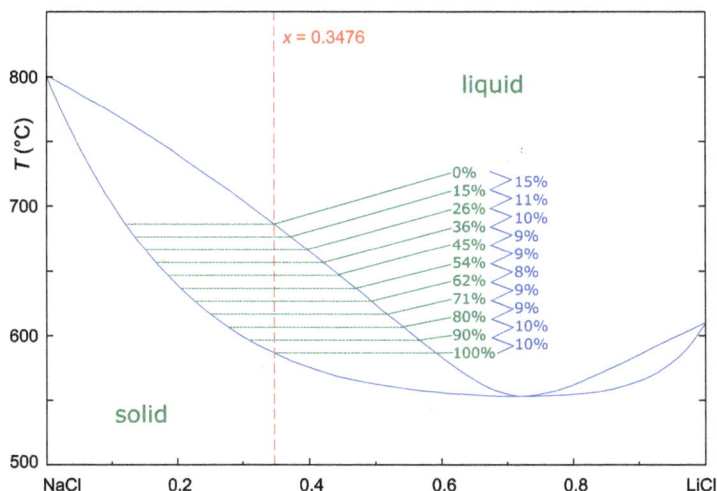

Page 191: 50 ml/min gas flow $\cong$ 0.054 ml H_2O(gas) during the measurement. Under standard conditions, this is ca. 0.043 mg H_2O. With molar masses 216 g/mol for TbF_3 and 18 g/mol for H_2O, we obtain 0.52 mg TbF_3, which can be hydrolyzed according the given reaction. (It should be noted that also other $TbF_{3-2x}O_x$ phases with significantly smaller $x \approx 0.04 \ll 1$ exist; see Sobolev et al., 1976. $x = 1$ for TbOF.) Then even more TbF_3 can be destroyed by hydrolysis with this amount of water.

C Stability of oxides

This chapter contains a collection of predominance diagrams (for the total pressure of $p = 1$ bar, calculated with FactSage 8.0, 2020, black lines) for > 40 binary systems El–O_2 in coordinates $\log[p_{O_2}/\text{bar}]$ vs. T; $0 \leq T \leq 2500$ (in °C), $-20 \leq \log[p_{O_2}/\text{bar}] \leq 0$; see Section 2.8.3. It should be noted that the total pressure $p = 10^{-20}$ bar is extremely low and is found, e. g., in space in the altitude of geosynchronous orbits (35,786 km). "El" stands for an arbitrary chemical element, preferably for a metal.

- The predominance diagrams are overlayed by a red curve, which shows the $p_{O_2}(T)$ produced by pure carbon monoxide. Below this line, free carbon is formed ("carburization"). (The optional formation of carbides with some metals is neglected.) The blue curve additionally shows the $p_{O_2}(T)$ produced by pure carbon dioxide.

- The upper boundary of the diagram corresponds to $p_{O_2} = 1$ bar and hence to pure oxygen of ambient pressure. Dry air (20.946 % O_2) corresponds to $\log[p_{O_2}/\text{bar}] = -0.6789$, a horizontal line very close to this upper boundary. Even very pure "inert gases" like commercial nitrogen or argon with 99.999 % purity usually contain a few ppm (parts per million) oxygen and water, which results in typical $\log[p_{O_2}/\text{bar}] \geq -5$. This value defines also the practical lower limit of p_{O_2}, which can be obtained by rinsing a thermoanalytic device by any gas containing no "reductive" components like CO or H_2.

- "Oxygen traps" with getter materials like zirconium (Fig. C.17, Section 2.6.2) are an alternative way to reduce p_{O_2}. Such traps can be installed either in the gas supply line, or getter parts can be placed inside the thermal analyzer, close to the sample.

- Dashed green and purple lines in the diagrams mark the limits where the sum $\sum p_i$ of the partial pressures p_i of all species carrying the corresponding element El exceeds 10^{-6} bar or 10^{-3} bar, respectively. For $\sum p_i < 10^{-6}$ bar, evaporation of the corresponding element practically does not occur, and as a rule of thumb we can assume that even long-term experiments under such conditions (p_{O_2}, T) do not result in significant evaporation of the element El itself or of its oxides ElO_x. For $\sum p_i > 10^{-3}$ bar, however, evaporation starts to become large, which in open systems can lead to concentration shifts of the sample and to pollution of the device. The position of these lines on the T axis can drastically change for different p_{O_2}, e. g., by >1500 K in the case of W–O_2 (Fig. C.21). Such strong dependency is observed if the different oxidation states possess very different vapor pressures. For W–O_2, this means that the metal itself almost does not evaporate, but already with small concentrations of oxygen, it forms WO_3, which is very volatile.

- The phase field limits for the different oxidation states are calculated for binary El–O_2 systems and can be slightly different for ternary El–El′–O_2 or for even

https://doi.org/10.1515/9783110743784-006

more complicated systems. Fortunately, the formation enthalpy of most binary compounds from the elements is much higher than the formation enthalpy of a ternary compound from its binary components. The data below demonstrate this for the system Lu–Al–O_2, which contains the binary oxides Al_2O_3 (stable modification is α-Al_2O_3 = corundum) and Lu_2O_3 (stable modification bixbyite, space group $Ia\bar{3}$, Guzik et al., 2014). These end members form the ternary (intermediate) phases $Lu_4Al_2O_9$ (monoclinic, space group $P2_1/c$, Simura and Yamane, 2020), $LuAlO_3$ (distorted perovskite, space group $Pnma$), and $Lu_3Al_5O_{12}$ (garnet, space group $Ia\bar{3}d$). The monoclinic and garnet phases melt congruently, the influence of p_{O_2} on the peritectically melting $LuAlO_3$ was discussed by Klimm (2010b) and controversially by Petrosyan et al. (2013). From the FactSage 8.0 (2020) data below we see that the formation enthalpy ΔH of these ternary oxides from the simple oxides is much smaller, compared to the ΔH of these simple oxides from the chemical elements. This holds even more if the ΔH were scaled by the number of atoms, which is larger for the monoclinic and garnet phases.

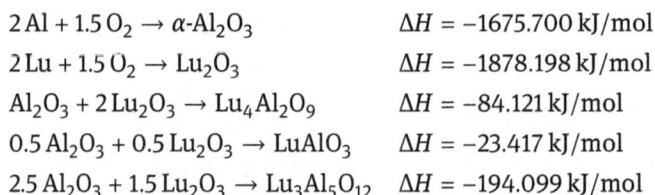

$$2\,Al + 1.5\,O_2 \rightarrow \alpha\text{-}Al_2O_3 \qquad \Delta H = -1675.700\,\text{kJ/mol}$$
$$2\,Lu + 1.5\,O_2 \rightarrow Lu_2O_3 \qquad \Delta H = -1878.198\,\text{kJ/mol}$$
$$Al_2O_3 + 2\,Lu_2O_3 \rightarrow Lu_4Al_2O_9 \qquad \Delta H = -84.121\,\text{kJ/mol}$$
$$0.5\,Al_2O_3 + 0.5\,Lu_2O_3 \rightarrow LuAlO_3 \qquad \Delta H = -23.417\,\text{kJ/mol}$$
$$2.5\,Al_2O_3 + 1.5\,Lu_2O_3 \rightarrow Lu_3Al_5O_{12} \qquad \Delta H = -194.099\,\text{kJ/mol}$$

As a result, also the stability ranges of ternary (or more complicated) oxides with unknown thermodynamic data can usually be estimated from the data of the binary oxides of the elements, which are often known. Figure C.1 demonstrates this for the Lu–Al–O_2 system; the fixed metal ratio Lu:Al = 3:5 corresponds to the garnet phase. For this system, the thermodynamic data for all binary and ternary oxides are contained in the FactSage 8.0 (2020) databases and were used to calculate the blue lines of the correct predominance diagram. We see that for $\log[p_{O2}/\text{bar}] \geq -15$, $Lu_3Al_5O_{12}$ melts at 2043 °C to a "liq. ox." liquid oxide melt. For lower oxygen partial pressure, Al_2O_3 is partially or completely lost from the condensed phases, and subsequently $Lu_4Al_2O_9$ or Lu_2O_3 are formed. Lu_2O_3 is only reduced under extremely low oxygen partial pressure, and then all components are gaseous.

The red dashed lines result from a calculation where the existence of all ternary oxides is suppressed; hence α-Al_2O_3+Lu_2O_3 coexist at low T. The two vertical lines at 1850...1900 °C represent the onset (T_{eut}) and the end (T_{liq}) of melting for the Al_2O_3/Lu_2O_3 mixture. We cannot expect here realistic data for the melting

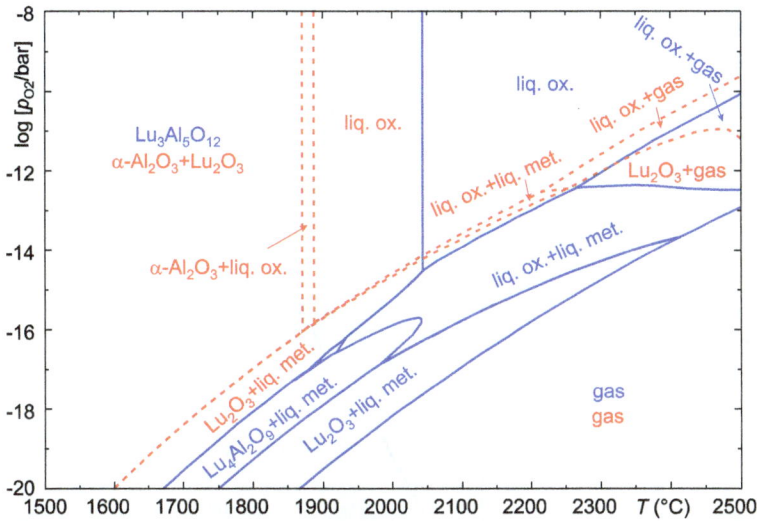

Figure C.1: Predominance diagram of the system Lu–Al–O$_2$ (Lu:Al = 3:5 = garnet composition). Blue lines and labels: the existence of the ternary phases (garnet Lu$_3$Al$_5$O$_{12}$, perovskite LuAlO$_3$, monoclinic Lu$_4$Al$_2$O$_9$) is taken into account, LuAlO$_3$ and "liq. met." (liquid metals) coexist in the narrow field without label around 1900 °C. Red lines were calculated neglecting these ternary phases: Reduction to "liq. met." starts at slightly higher p_{O_2} (ca. 0.5...1 order of magnitude).

temperature of Lu$_3$Al$_5$O$_{12}$, because the garnet phase is lacking in the calculation. However, also the red curves show reduction of Al$_2$O$_3$ for too low oxygen partial pressure, and the slope is similar to that for the correct blue phase boundaries. It is interesting to note that the oxide mixture is calculated to be slightly less stable with respect to reduction than the compound Lu$_3$Al$_5$O$_{12}$. This difference has its origin in the small but remarkable ΔH of the garnet, which is given above.

The stability limits of oxidation states in multinary systems are often only weakly influenced by compound formation and are almost an overlay of the binary El–O$_2$ diagrams for its constituents.

Solid carbon (e. g., as graphite or vitreous carbon) is stable only below the red curve shown in all stability diagrams in this chapter. Thus this line sets an upper limit of $p_{O_2}(T)$, which can be maintained inside a thermoanalytic device containing parts made of carbon (crucibles, protective tube, skimmer). Under such conditions, however, many metal oxides are not stable and are reduced to metals. A less stringent $p_{O_2}(T)$ limit is set in the system W–O$_2$; see Fig. C.21. This makes tungsten parts for high-T measurements with oxides sometimes superior, because then the conditions are not so harshly reductive.

Group 1: Alkali metals

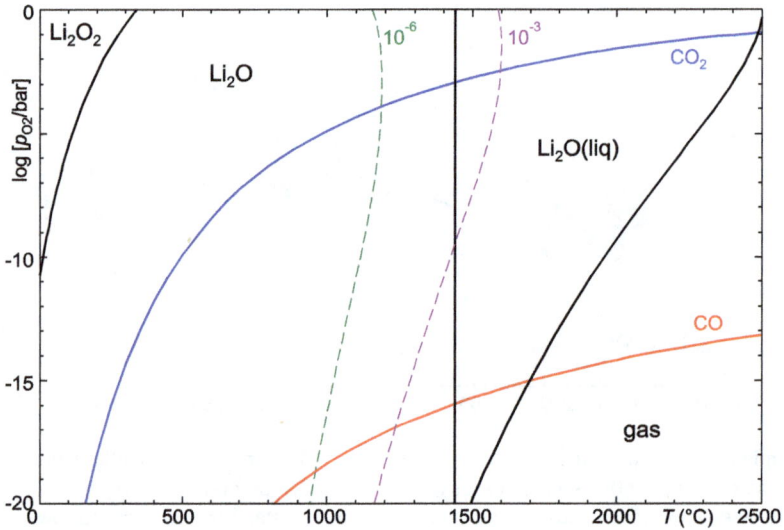

Figure C.2: The system Li–O_2.

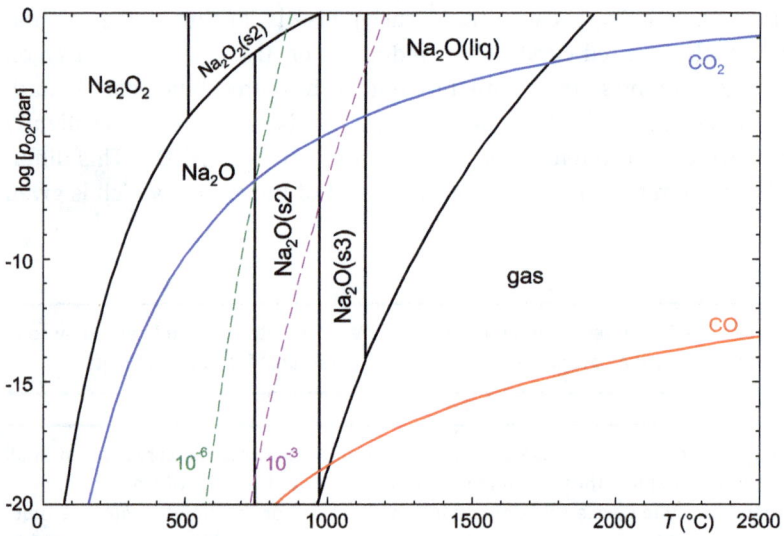

Figure C.3: The system Na–O_2.

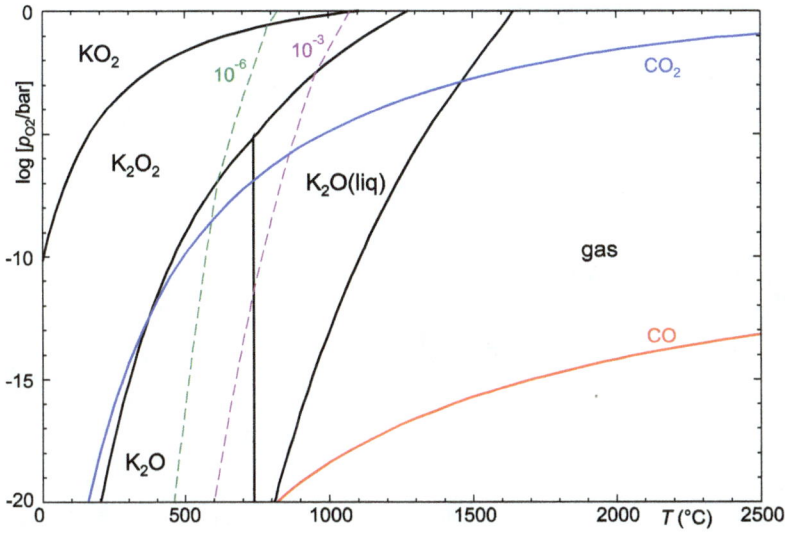

Figure C.4: The system K–O_2.

Group 2: Alkaline earth metals

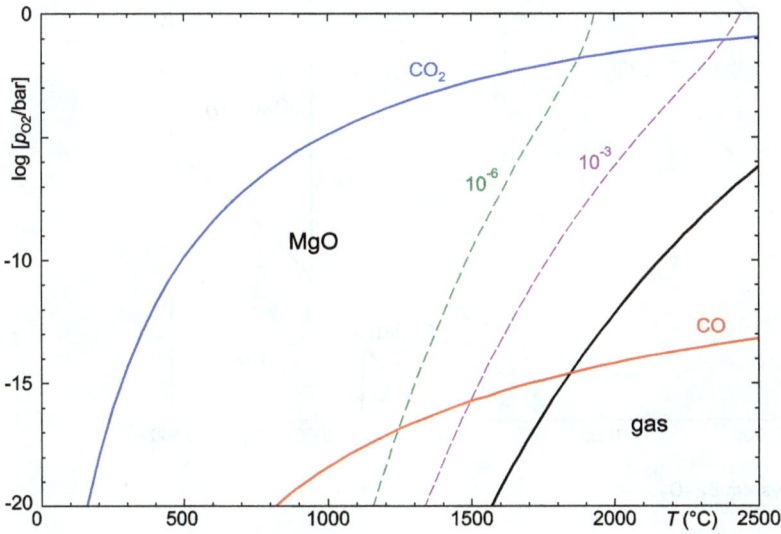

Figure C.5: The system Mg–O_2. MgO: $T_f = 2825\,°C$.

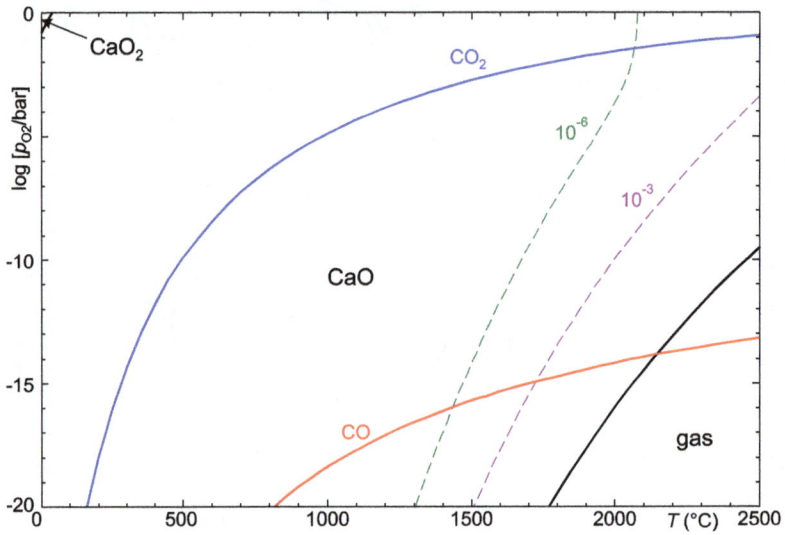

Figure C.6: The system Ca–O_2 CaO: T_f = 2572 °C.

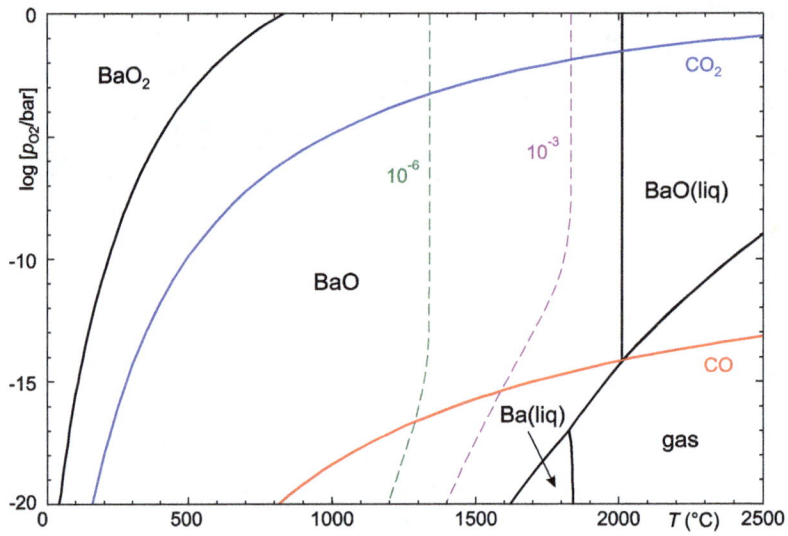

Figure C.7: The system Ba–O_2.

Group 3: Scandium, yttrium, rare-earth metals, actinides

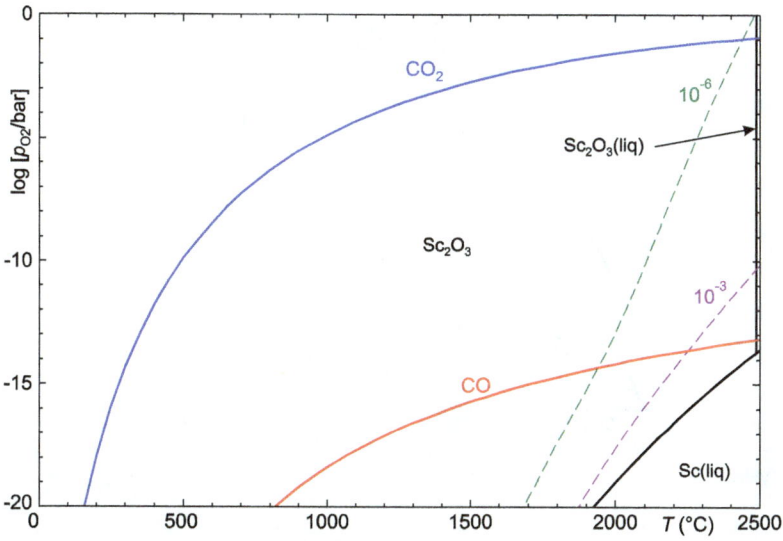

Figure C.8: The system $Sc-O_2$.

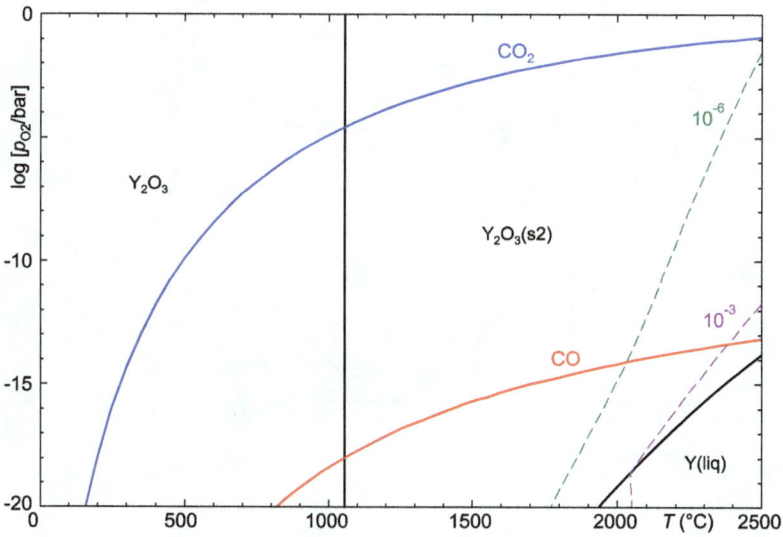

Figure C.9: The system $Y-O_2$. Navrotsky et al. (2005) report $T_f = 2382\,°C$ for Y_2O_3.

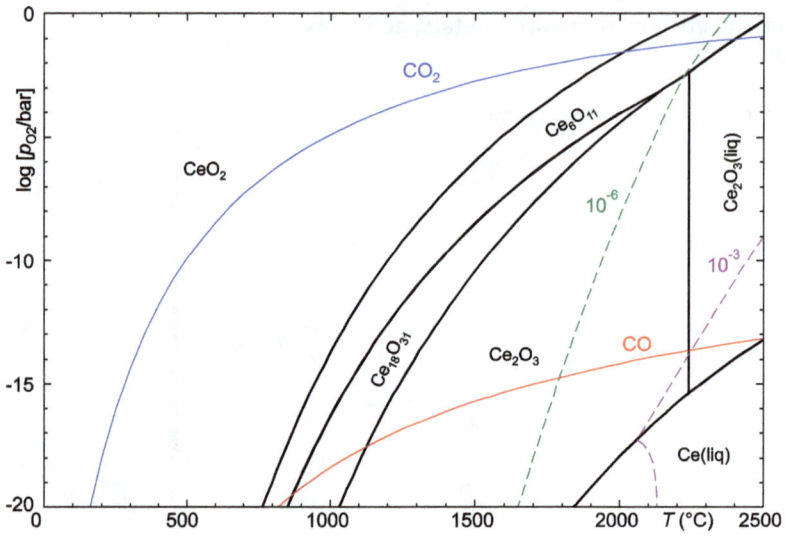

Figure C.10: The system Ce–O_2.

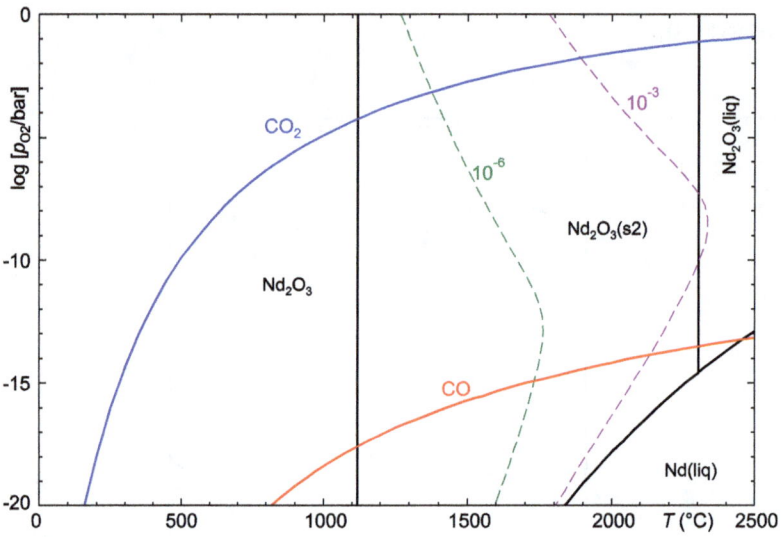

Figure C.11: The system Nd–O_2.

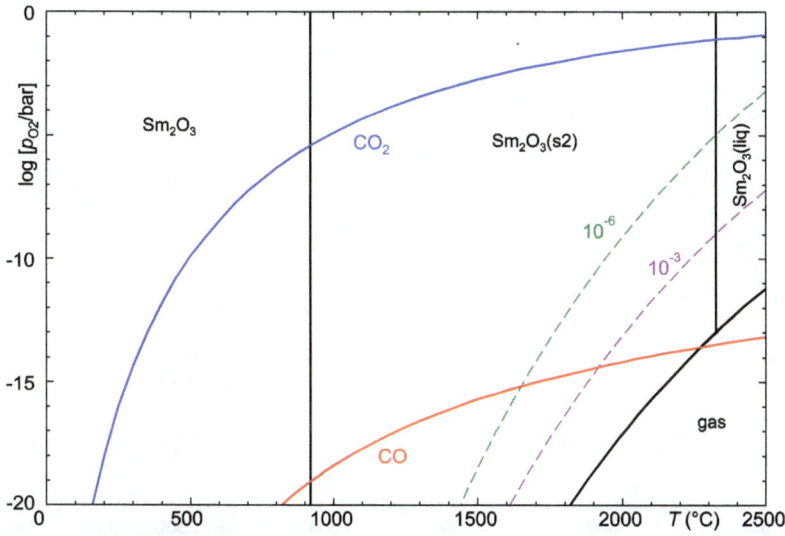

Figure C.12: The system Sm–O_2.

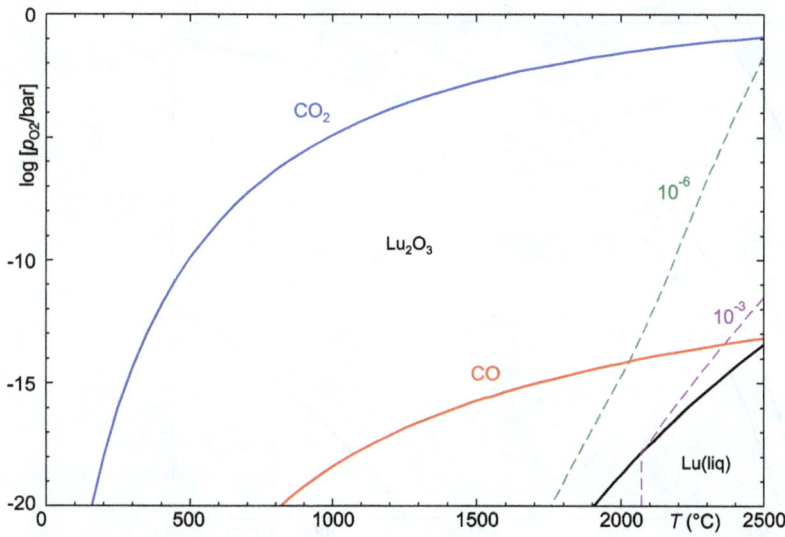

Figure C.13: The system Lu–O_2. Lu_2O_3: $T_f = 2510\ °C$.

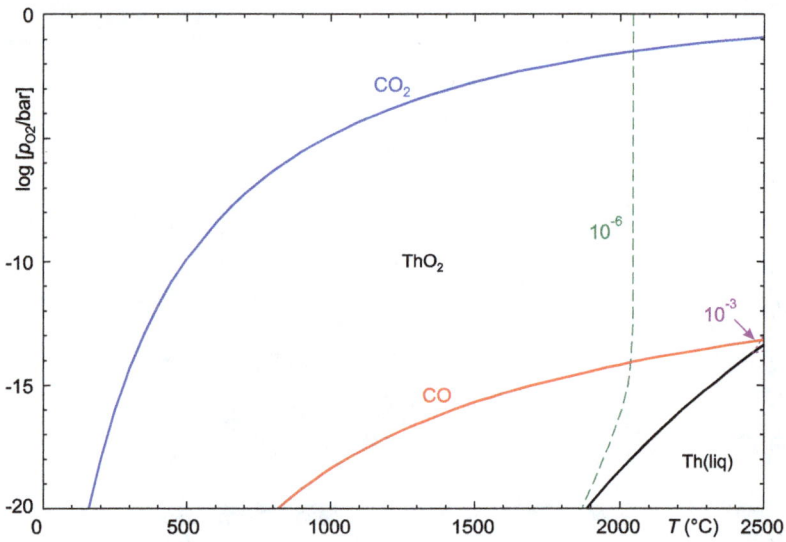

Figure C.14: The system Th–O_2. ThO_2: $T_f = 3220\,°C$.

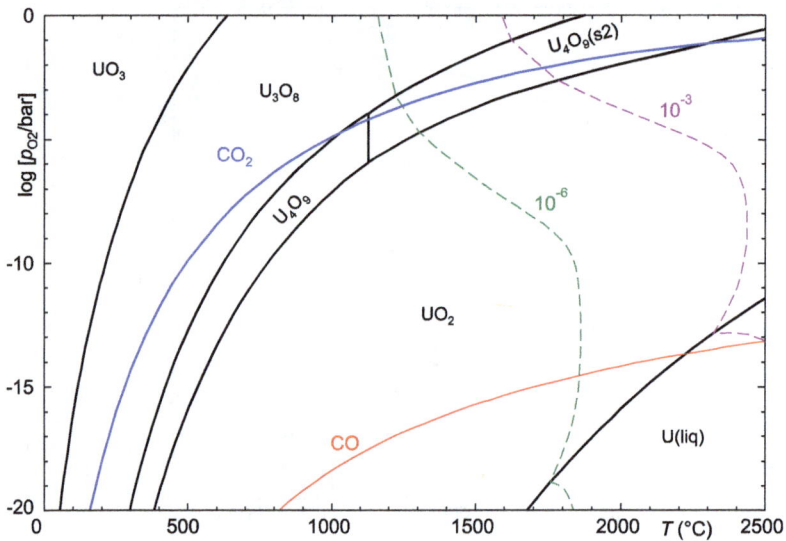

Figure C.15: The system U–O_2. UO_2: $T_f = 2842\,°C$.

Group 4: Titanium group

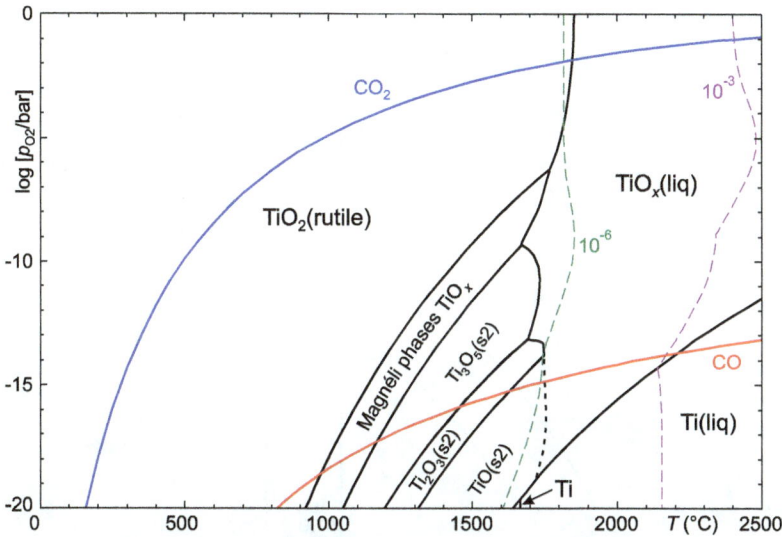

Figure C.16: The system Ti–O_2. Magnéli phases Ti_nO_{2n-1} have mixed $Ti^{3+/4+}$ valency; see Magnéli and Oughton (1951). [Arne Magnéli (6 December 1914–22 July 1996)].

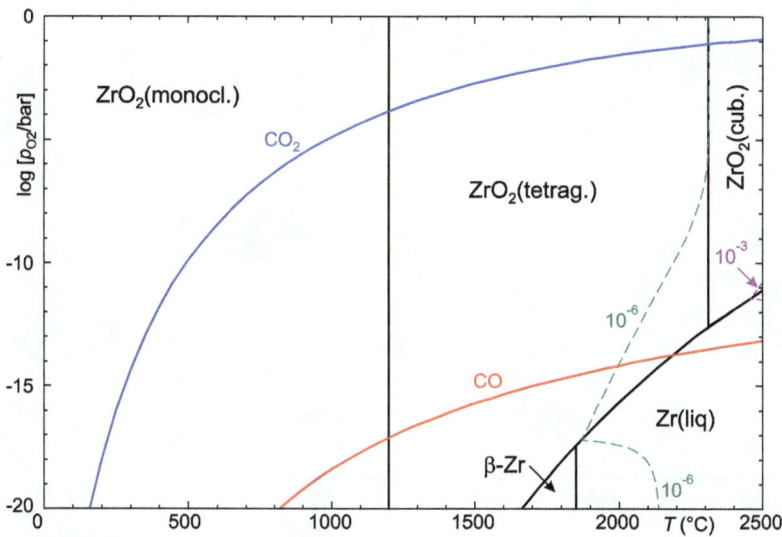

Figure C.17: The system Zr–O_2; cf. Fig. 2.27. ZrO_2: $T_f = 2678\,°C$.

Group 5: Vanadium group

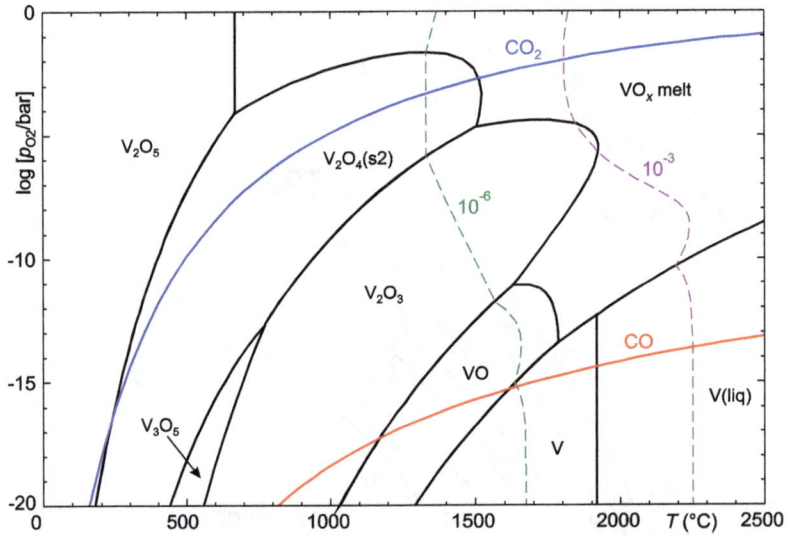

Figure C.18: The system $V–O_2$.

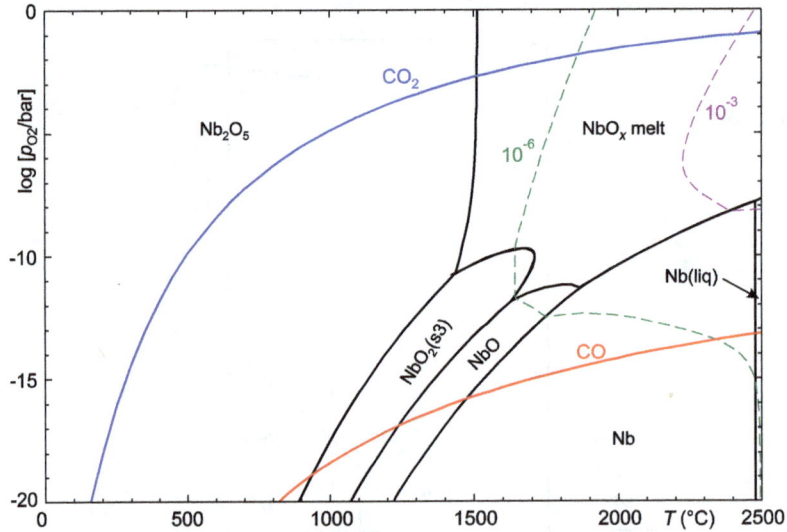

Figure C.19: The system $Nb–O_2$. Sakata (1969) reports for NbO_2 a phase transition near 800 °C.

Group 6: Chromium group

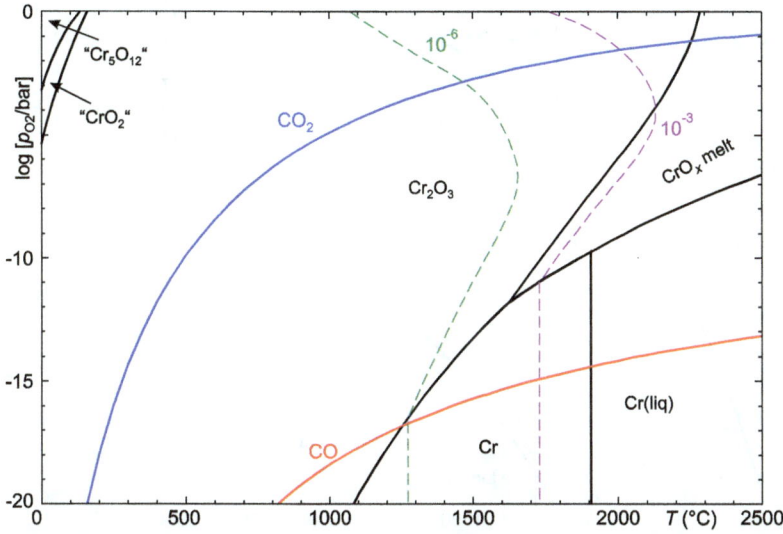

Figure C.20: System $Cr-O_2$. For CrO_2, see Bate (1978). Cr_5O_{12} is Cr(III) chromate(VI), also Cr(VI) per-oxide $CrO(O_2)_2$ (Buchera et al., 2005) and Cr(II,III) oxide Cr_3O_4 (Lin et al., 2009) are found.

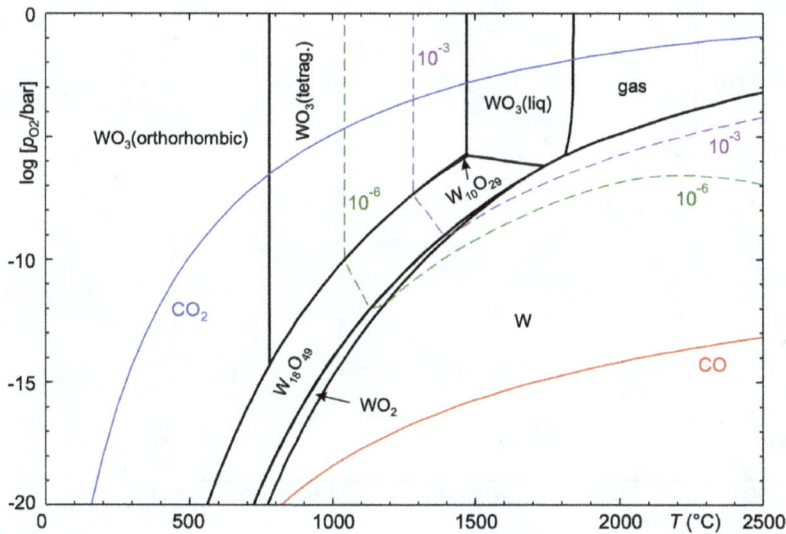

Figure C.21: The system $W-O_2$. W: $T_f = 3407\,°C$.

Group 7: Manganese group

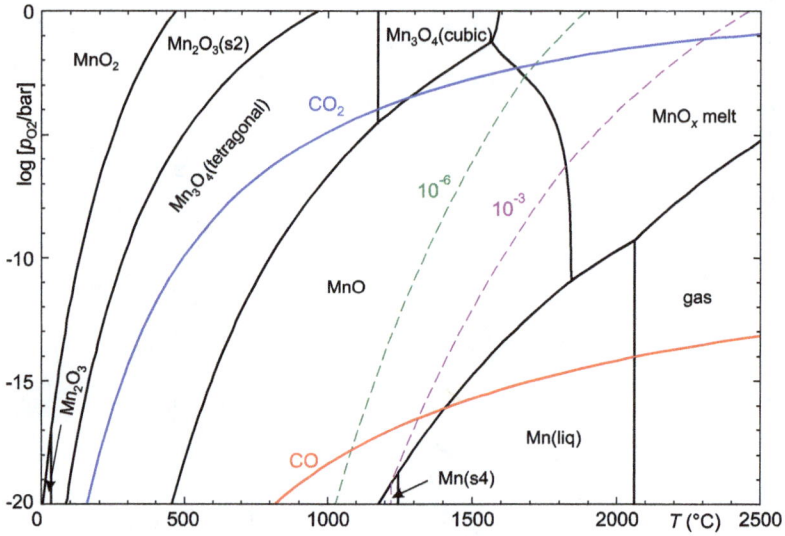

Figure C.22: The system $Mn–O_2$, cf. Fig. 2.51.

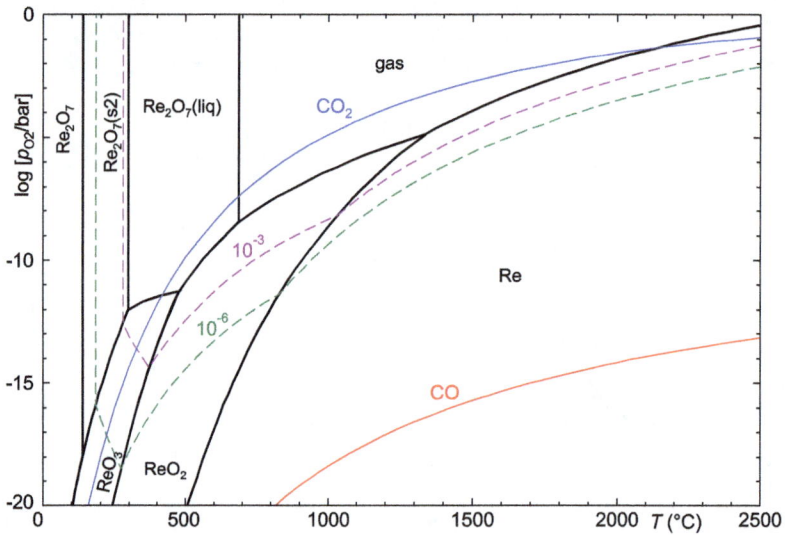

Figure C.23: The system $Re–O_2$. Re: $T_f = 3180\,°C$.

Groups 8–10: Iron, cobalt, nickel, platinum metals

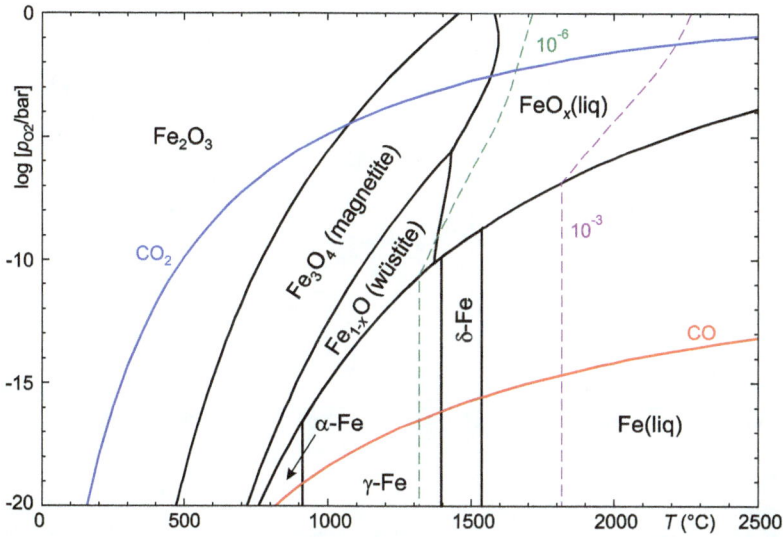

Figure C.24: The system Fe–O$_2$.

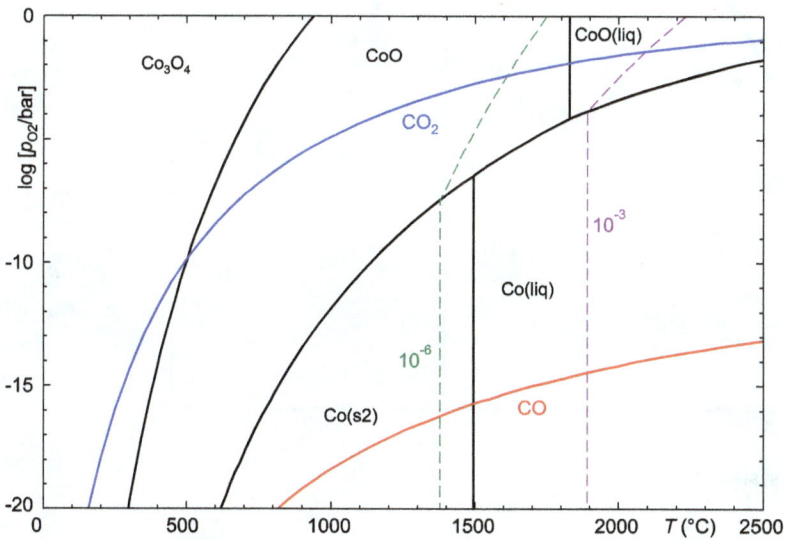

Figure C.25: The system Co–O$_2$.

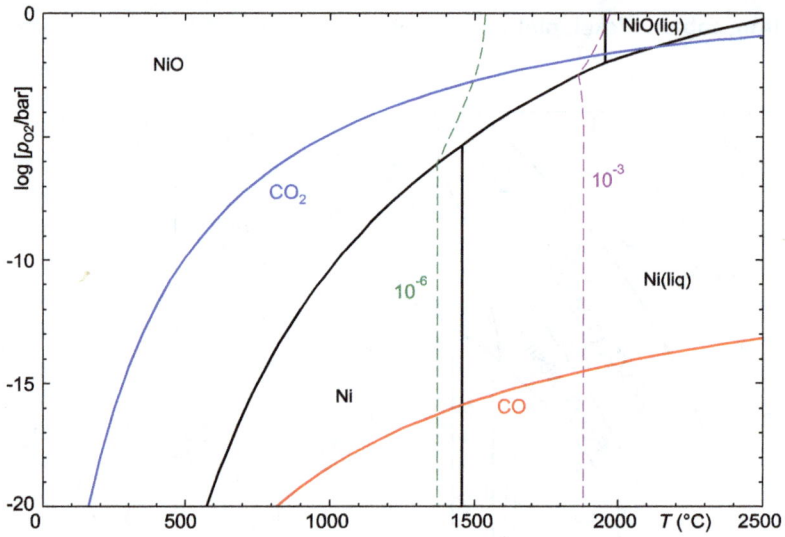

Figure C.26: The system Ni–O$_2$.

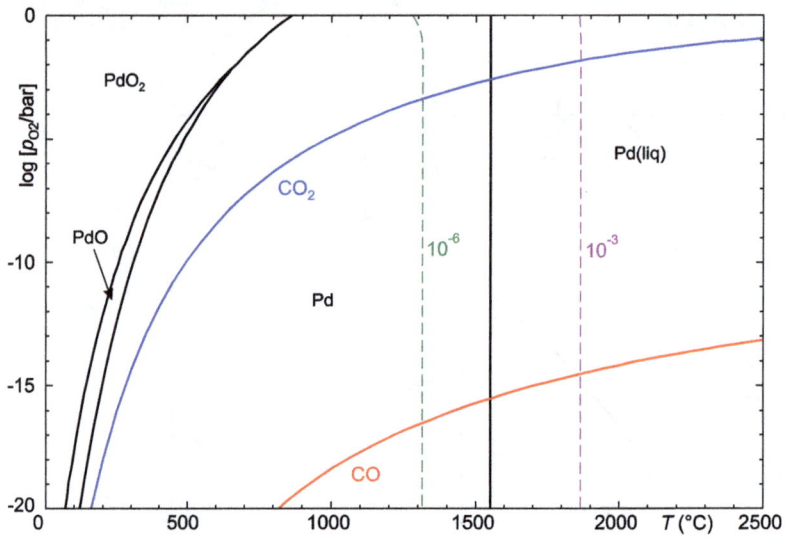

Figure C.27: The system Pd–O$_2$.

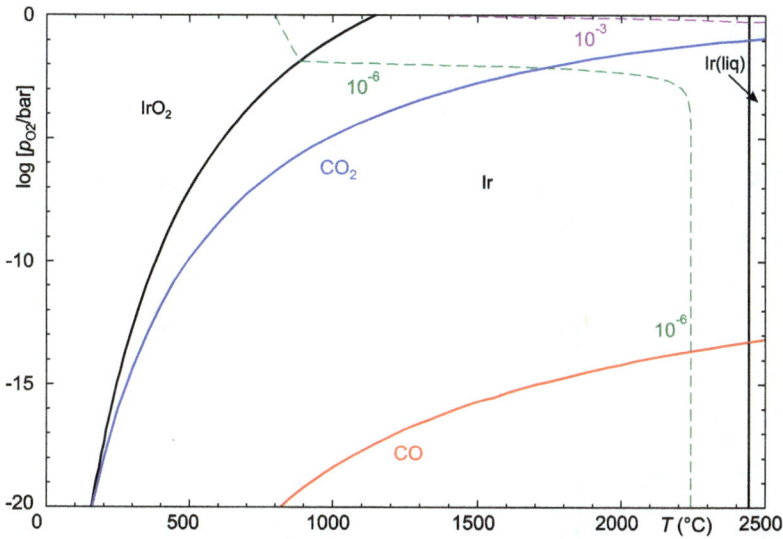

Figure C.28: The system Ir–O_2.

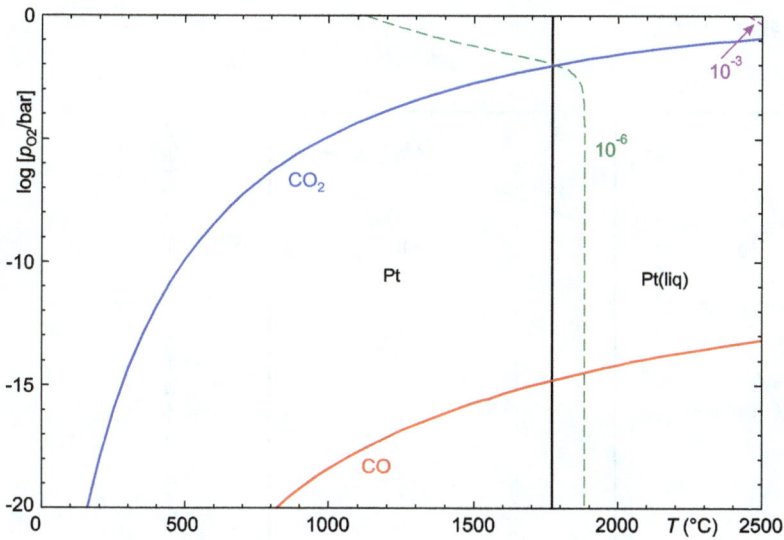

Figure C.29: The system Pt–O_2.

Group 11: Copper group

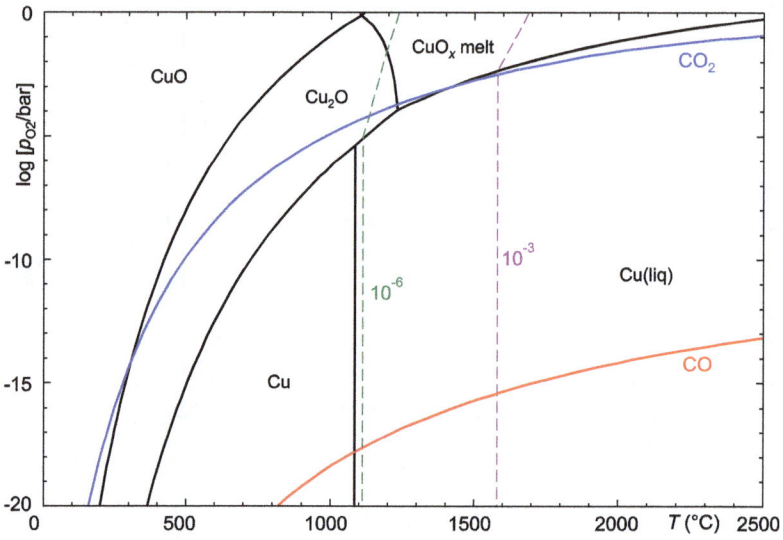

Figure C.30: The system $Cu-O_2$.

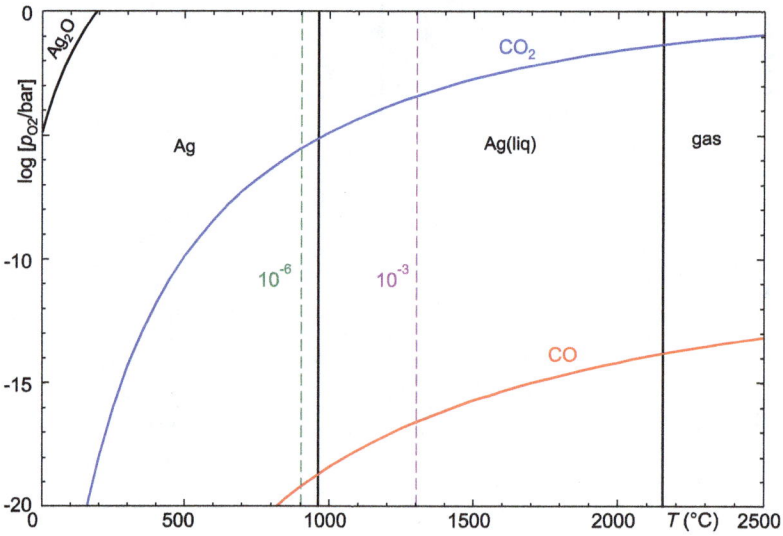

Figure C.31: The system $Ag-O_2$.

Group 12: Zinc group

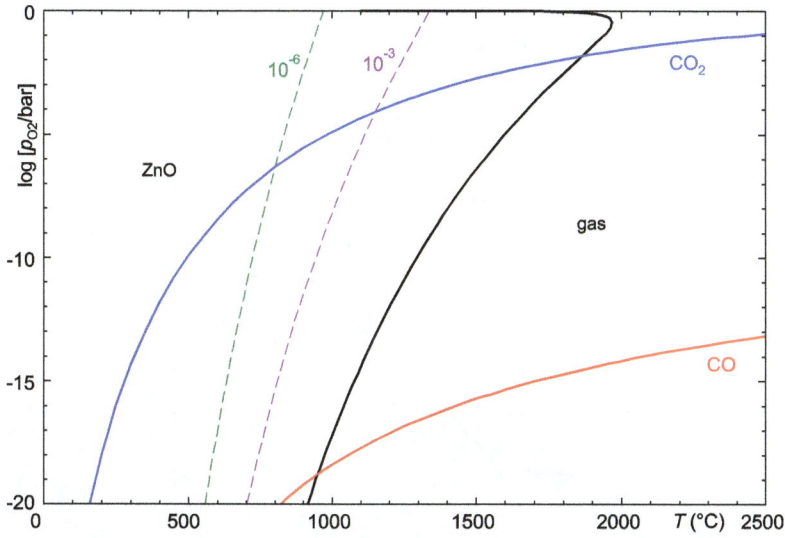

Figure C.32: The system $Zn-O_2$. ZnO: $T_f = 1975\,°C$ under pressure.

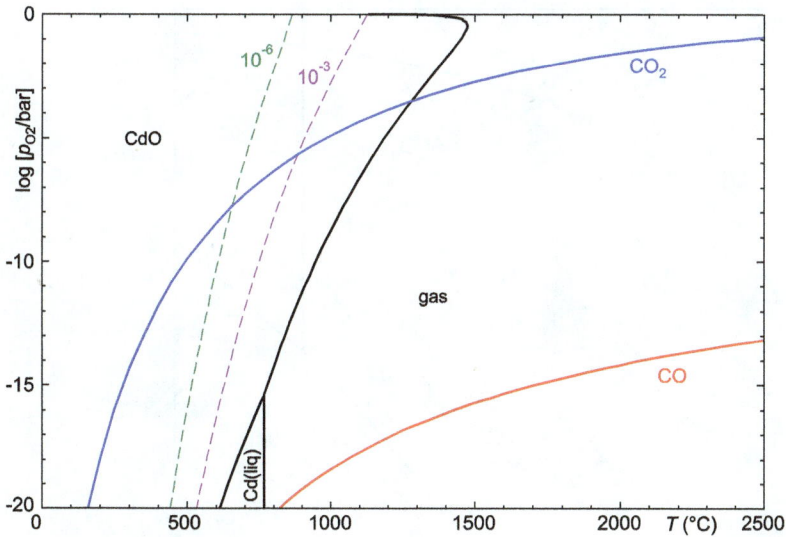

Figure C.33: The system $Cd-O_2$.

Group 13: Boron group

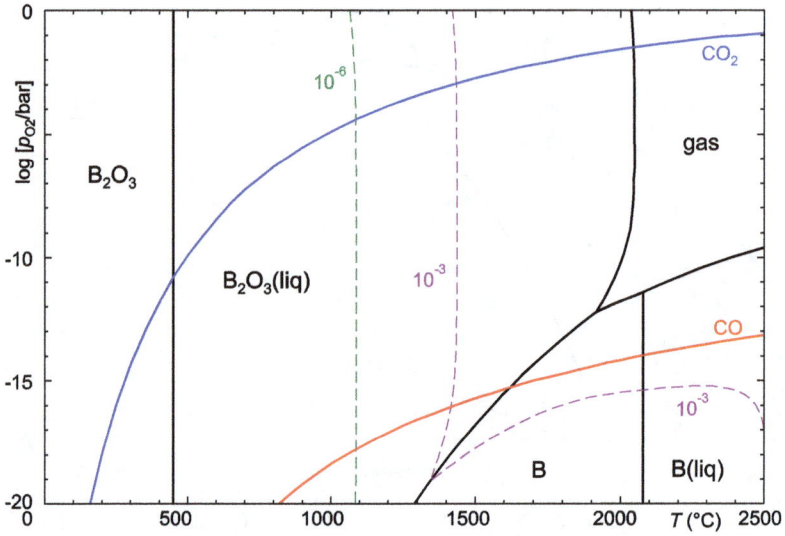

Figure C.34: The system $B–O_2$.

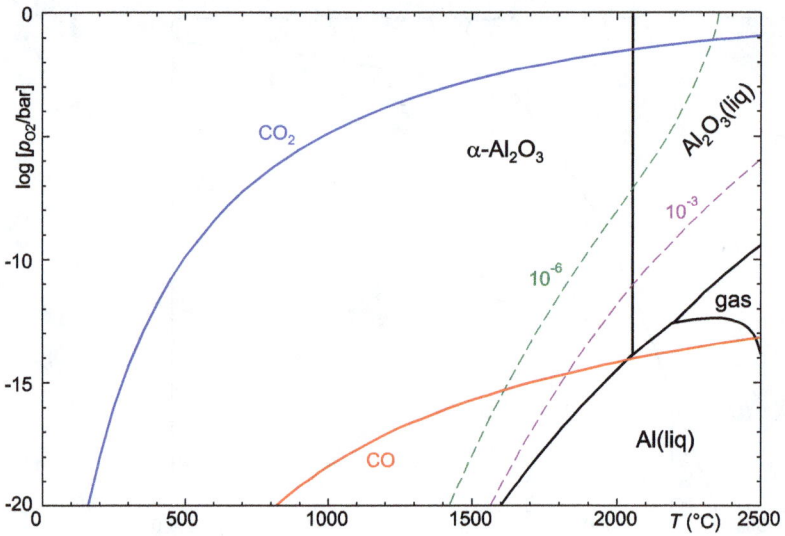

Figure C.35: The system $Al–O_2$.

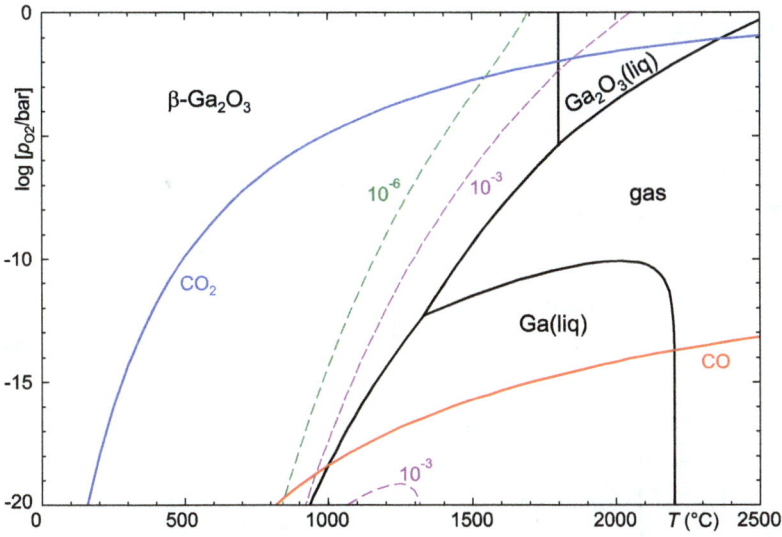

Figure C.36: The system Ga–O_2.

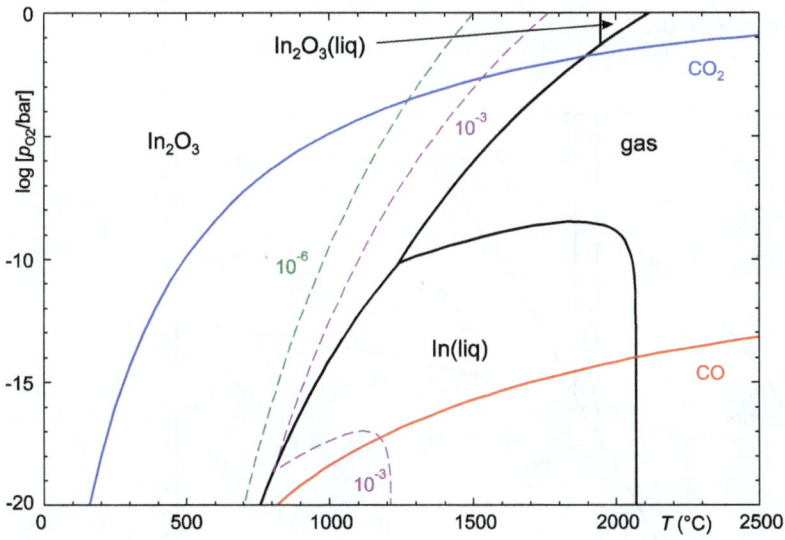

Figure C.37: The system In–O_2.

Group 14: Carbon group

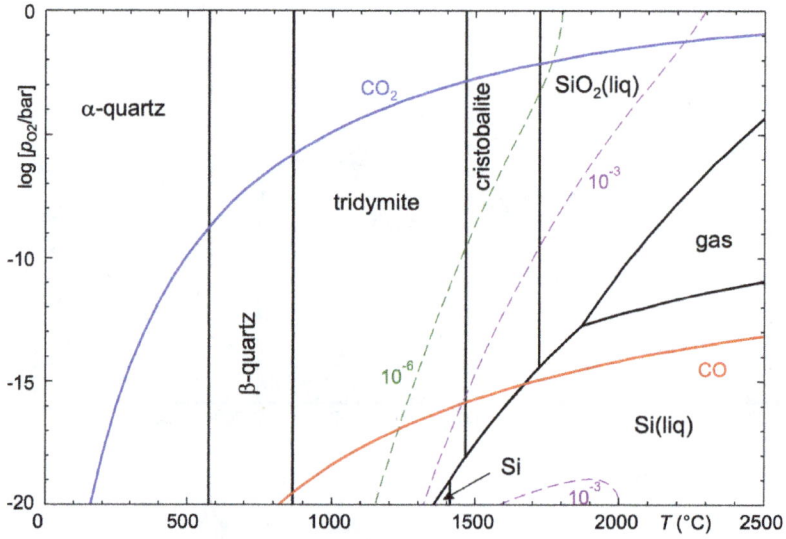

Figure C.38: The system $Si-O_2$.

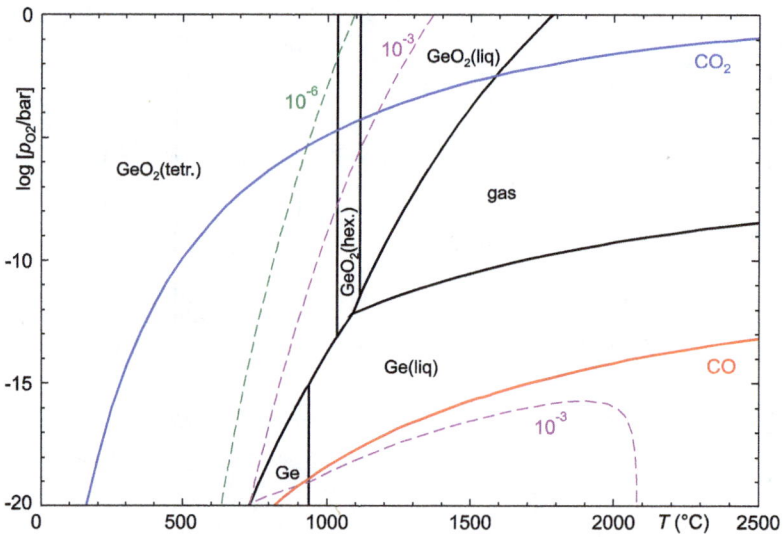

Figure C.39: The system $Ge-O_2$.

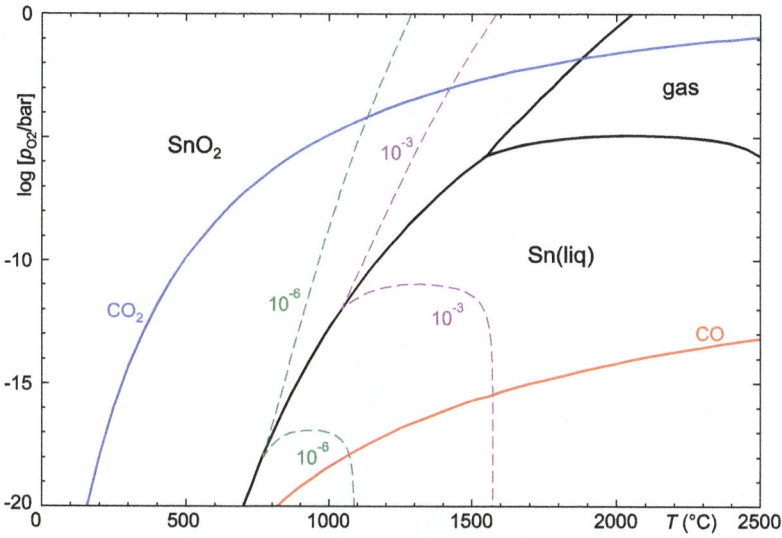

Figure C.40: The system Sn–O_2. Controversial data for T_f of SnO_2 were reported by different authors: 1630 °C (Barczak and Insley, 1962); 2000 °C (Cahen et al., 2003); Galazka et al. (2014) proved >2100 °C.

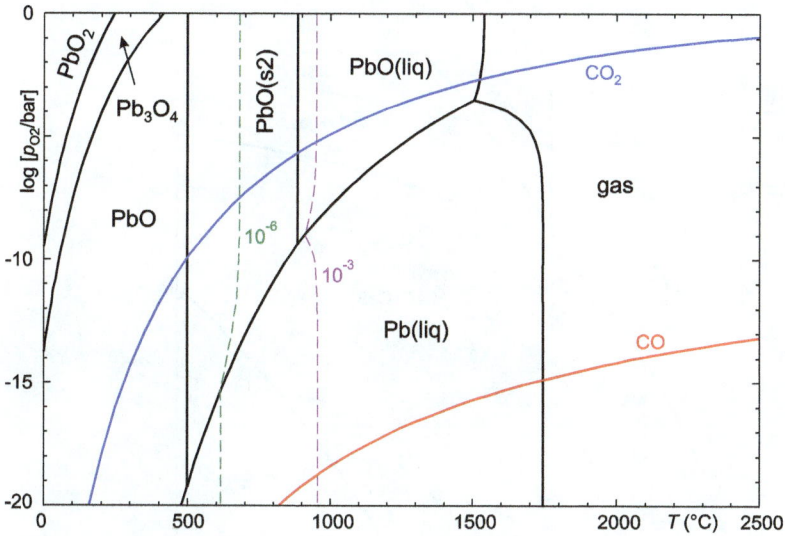

Figure C.41: The system Pb–O_2.

Group 15: Bi (pnictogen) and S (chalcogen)

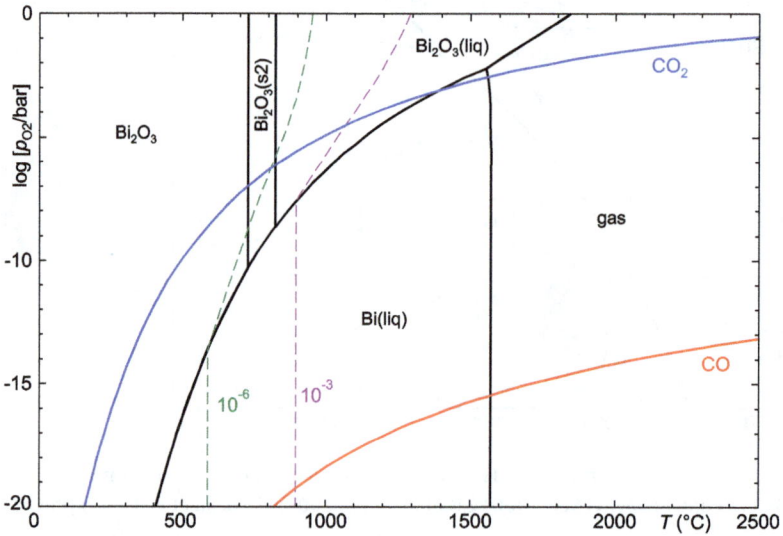

Figure C.42: The system Bi–O_2.

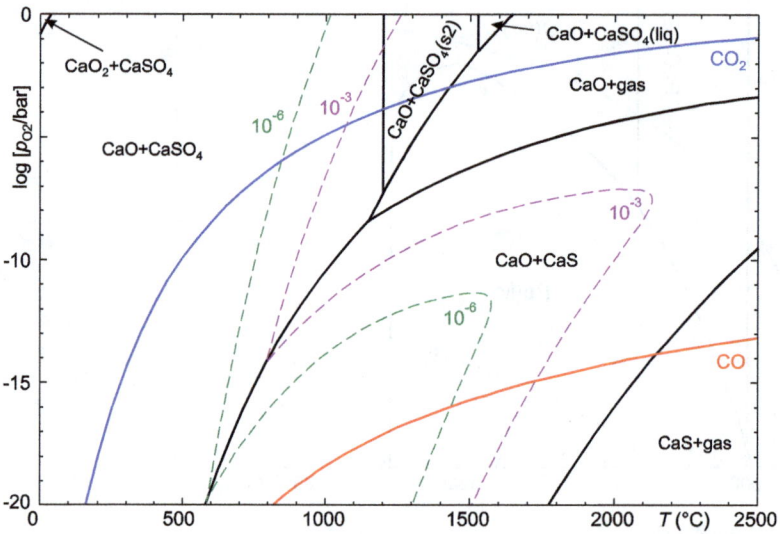

Figure C.43: The system Ca–S–O_2 ([Ca] : [S] = 9 : 1). The calcium excess stabilizes condensed phases containing sulfur, which would otherwise evaporate as an element or SO_2.

Bibliography

J. S. Abell, I. R. Harris, B. Cockayne, and J. G. Plant. A DTA study of zone-refined LiRF$_4$ (R = Y, Er). *J. Mater. Sci.*, 11:1807–1816, 1976. https://doi.org/10.1007/BF00708258.

S. C. Abrahams and J. L. Bernstein. Accuracy of an automatic diffractometer. Measurement of the sodium chloride structure factors. *Acta Crystallogr.*, 18:926–932, 1965. https://doi.org/10.1107/S0365110X65002244.

J. P. Abriata, J. C. Bolcich, and H. A. Peretti. The Hf–Zr (hafnium–zirconium) system. *Bull. Alloy Phase Diagr.*, 3:29–34, 1982. https://doi.org/10.1007/BF02873408.

ACerS-NIST. Phase Equilibria Diagrams, V. 4.0 (PC Database), 2014.

M. F. Acosta, R. I. Merino, S. Ganschow, and D. Klimm. Solidification of NaCl–LiF–CaF$_2$ ternary composites. *J. Mater. Sci.*, 52:5520–5530, 2017. https://doi.org/10.1007/s10853-017-0814-2.

A. Aharoni, E. H. Frei, Z. Scheidlinger, and M. Schieber. Curie temperature of some garnets by the differential thermal analysis technique. *J. Appl. Phys.*, 32(10):1851–1853, 1961. https://doi.org/10.1063/1.1728251.

AKTS. Thermokinetics Software (TK), 2021. https://www.akts.com.

C. B. Alcock, M. W. Chase, and V. P. Itkin. Thermodynamic properties of the group IA elements. *J. Phys. Chem. Ref. Data*, 23(3):385–497, 1994. https://doi.org/10.1063/1.555945.

P. Aldebert and J. P. Traverse. Etude par diffraction neutronique des structures de haute temperature de La$_2$O$_3$ et Nd$_2$O$_3$. *Mater. Res. Bull.*, 14:303–323, 1979. https://doi.org/10.1016/0025-5408(79)90095-3.

S. Arrhenius. Über die Reaktionsgeschwindigkeit bei der Inversion von Rohrzucker durch Säuren. *Z. Phys. Chem.*, 4U:226–248, 1889. https://doi.org/10.1515/zpch-1889-0416.

M. Avrami. Kinetics of phase change. I: General theory. *J. Chem. Phys.*, 7:1103–1112, 1939. https://doi.org/10.1063/1.1750380.

M. Avrami. Kinetics of phase change. II: Transformation-time relations for random distribution of nuclei. *J. Chem. Phys.*, 8:212–224, 1940. https://doi.org/10.1063/1.1750631.

H. L. J. Bäckström. The thermodynamic properties of calcite and aragonite. *J. Am. Chem. Soc.*, 47:2429–2431, 1925. https://doi.org/10.1021/ja01687a002.

J. M. Badie. Phases et transitions de phases à haute température dans les systèmes Sc$_2$O$_3$–Ln$_2$O$_3$ (Ln = lanthanide et yttrium). *Rev. Int. Hautes Temp. Réfract.*, 15:183–199, 1978.

V. J. Barczak and R. H. Insley. Phase equilibria in the system Al$_2$O$_3$–SnO$_2$. *J. Am. Ceram. Soc.*, 45:144, 1962. https://doi.org/10.1111/j.1151-2916.1962.tb11106.x.

I. Barin. *Thermodynamic Data of Pure Substances*. VCH, Weinheim, 1995. https://doi.org/10.1002/9783527619825.

A. F. M. Barton. Solubility data series. In A. S. Kertes, editor, *15: alcohols with water*, Pergamon, Oxford, 1984. ISBN 0-08-025276-1.

G. Bate. A survey of recent advances in magnetic recording materials. *IEEE Trans. Magn.*, 14:136–142, 1978. https://doi.org/10.1109/TMAG.1978.1059769.

T. Bauer, C. Odenthal, and A. Bonk. Molten salt storage for power generation. *Chem. Ing. Tech.*, 93:534–546, 2021. https://doi.org/10.1002/cite.202000137.

S. W. Benson and J. H. Buss. Additivity rules for the estimation of molecular properties. thermodynamic properties. *J. Chem. Phys.*, 29:546–572, 1958. https://doi.org/10.1063/1.1744539.

S. Bernal, G. Blanco, J. M. Gatica, J. A. Pérez-Omil, J. M. Pintado, and H. Vidal. Chemical reactivity of binary rare earth oxides. In G. Adachi, N. Imanaka and Z. C. Kang, editors, *Binary Rare Earth Oxides*, pages 9–55. Kluwer, Dordrecht, 2004. https://doi.org/10.1007/1-4020-2569-6_2.

R. Bertram and D. Klimm. Assay measurements of oxide materials by thermogravimetry and ICP–OES. *Thermochim. Acta*, 419:189–193, 2004. https://doi.org/10.1016/j.tca.2004.03.003.

https://doi.org/10.1515/9783110743784-007

O. Bianchi, R. V. B. Oliveira, R. Fiorio, J. D. N. Martins, A. J. Zattera, and L. B. Canto. Assessment of Avrami, Ozawa and Avrami-Ozawa equations for determination of EVA crosslinking kinetics from DSC measurements. *Polym. Test.*, 27:722–729, 2008. https://doi.org/10.1016/j.polymertesting.2008.05.003.

R. L. Blaine and H. E. Kissinger. Homer Kissinger and the Kissinger equation. *Thermochim. Acta*, 540: 1–6, 2012. https://doi.org/10.1016/j.tca.2012.04.008.

J. Bohm. Symmetriebeziehungen ohne Gruppeneigenschaft in der Kristallographie (Eine Verallgemeinerung der „OD-Strukturen"). *Wiss. Z. Humboldt Univ. Berlin, Math. Nat. R.*, XVI:821–831, 1967.

J. Bohm. Antisymmetrische OD-Strukturen. *Z. Kristallogr.*, 126:190–198, 1968. https://doi.org/10.1524/zkri.1968.126.1-3.190.

H. J. Borchardt and F. Daniels. The application of differential thermal analysis to the study of reaction kinetics. *J. Am. Chem. Soc.*, 79:41–46, 1957. https://doi.org/10.1021/ja01558a009.

W. Borchardt-Ott. *Crystallography*. Springer, Berlin, 2012. ISBN 978-3-642-16452-1.

M. E. Brown. *Introduction to Thermal Analysis – Techniques and Applications*. Kluwer, New York, 2001. ISBN 1-4020-0472-9.

G. Buchera, M. Kampe, and J. F. Roelcke. Peroxides and chromium compounds – the ether test for identity. *Z. Naturforsch. B*, 60(1):1–6, 2005. https://doi.org/10.1515/znb-2005-0101.

M. J. Buerger. Crystallographic aspects of phase transformations. In R. Smoluchowski, J. E. Mayer and W. S. Weyl, editors, *Phase Transformations in Solids* Wiley and Sons, New York, 1951.

S. Cahen, N. David, J. M. Fiorani, A. Maître, and M. Vilasi. Thermodynamic modelling of the O–Sn system. *Thermochim. Acta*, 403:275–285, 2003. https://doi.org/10.1016/S0040-6031(03)00059-5.

B. Cantor, I. T. H. Chang, P. Knight, and A. J. B. Vincent. Microstructural development in equiatomic multicomponent alloys. *Mater. Sci. Eng. A*, 375–377:213–218, 2004. https://doi.org/10.1016/j.msea.2003.10.257.

J. R. Carruthers, G. E. Peterson, M. Grasso, and P. M. Bridenbaugh. Nonstoichiometry and crystal growth of lithium niobate. *J. Appl. Phys.*, 42:1846–1851, 1971. https://doi.org/10.1063/1.1660455.

Y. A. Chang, S. Chen, F. Zhang, X. Yan, F. Xie, R. Schmid-Fetzer, and W. A. Oates. Phase diagram calculation: past, present and future. *Prog. Mater. Sci.*, 49:313–345, 2004. https://doi.org/10.1016/S0079-6425(03)00025-2.

M. W. Chase Jr., J. L. Curnutt, J. R. Downey Jr., R. A. McDonald, A. N. Syverud, and E. A. Valenzuela. JANAF thermochemical tables supplement. *J. Phys. Chem. Ref. Data*, 11:695–940, 1982. 1982. https://doi.org/10.1063/1.555666.

A. G. Christy. Multistage diffusionless pathways for reconstructive phase transitions: application to binary compounds and calcium carbonate. *Acta Crystallogr. B*, 49:987–996, 1993. https://doi.org/10.1107/S0108768193008201.

P. Claudy, J. C. Commerçon, and J. M. Lètoffé. Quasi-static study of the glass transition of glycerol by DSC. *Thermochim. Acta*, 128:251–260, 1988. https://doi.org/10.1016/0040-6031(88)85369-3.

A. W. Coats and J. P. Redfern. Kinetic parameters from thermogravimetric data. *Nature*, 201:68–69, 1964. https://doi.org/10.1038/201068a0.

A. W. Coats and J. P. Redfern. Kinetic parameters from thermogravimetric data. II. *Polym. Lett.*, 3:917–920, 1965. https://doi.org/10.1002/pol.1965.110031106.

O. M. Corbino. Thermische Oszillationen wechselstromdurchflossener Lampen mit dünnem Faden und daraus sich ergebende Gleichrichterwirkung infolge der Anwesenheit geradzahliger Oberschwingungen. *Phys. Z.*, 11:413–417, 1910.

O. M. Corbino. Periodische Widerstandsänderungen feiner Metallfäden, die durch Wechselströme zum Glühen gebracht werden, sowie Ableitung ihrer thermischen Eigenschaften bei hoher Temperatur. *Phys. Z.*, 12:292–295, 1911. https://digitalisate.sub.uni-hamburg.de/.

J. P. Coutures and M. H. Rand. Melting temperatures of refractory oxides — part II: Lanthanoid sesquioxides. *Pure Appl. Chem.*, 61:1461–1482, 1989. https://doi.org/10.1351/pac198961081461.

D. F. Craig and J. J. Brown Jr. Phase equilibria in the system CaF_2–AlF_3. *J. Am. Ceram. Soc.*, 60:396–398, 1977. https://doi.org/10.1111/j.1151-2916.1980.tb10714.x.

E. S. Davenport and E. C. Bain. Transformation of austenite at constant subcritical temperatures. *Metall. Trans. B*, 1:3503–3530, 1970. https://doi.org/10.1007/BF03037892.

W. P. Davey. Precision measurements of the lattice constants of twelve common metals. *Phys. Rev.*, 25:753–761, 1925. https://doi.org/10.1103/PhysRev.25.753.

DDBST. Dortmund Data Bank Software & Separation Technology GmbH, Oldenburg, Germany, 2021. http://www.ddbst.com.

A. M. de Jong and J. W. Niemantsverdriet. Thermal desorption analysis: Comparative test of ten commonly applied procedures. *Surf. Sci.*, 233:355–365, 1990. https://doi.org/10.1016/0039-6028(90)90649-S.

G. F. de Oliveira, R. C. de Andrade, M. A. G. Trindade, H. M. C. Andrade, and C. T. de Carvalho. Thermogravimetric and spectroscopic study (TG-DTA/FT-IR) of activated carbon from the renewable biomass source babassu. *Quím. Nova*, 40:284–292, 2017. https://doi.org/10.21577/0100-4042.20160191.

S. A. Degterov, A. D. Pelton, E. Jak, and P. C. Hayes. Experimental study of phase equilibria and thermodynamic optimization of the Fe–Zn–O system. *Metall. Mater. Trans. B*, 32:643–657, 2001. https://doi.org/10.1007/s11663-001-0119-2.

I. Dohnke, B. Trusch, D. Klimm, and J. Hulliger. A study on the influence of ytterbium and impurities on lattice parameters and the phase transition temperature of Czochralski-grown $LiNbO_3$. *J. Phys. Chem. Solids*, 65:1297–1305, 2004. https://doi.org/10.1016/j.jpcs.2004.02.010.

E. S. Domalski and E. D. Hearing. Estimation of the thermodynamic properties of C-H-N-O-S-Halogen compounds at 298.15 K. *J. Phys. Chem. Ref. Data*, 22:805–1159, 1993. https://doi.org/10.1063/1.555927.

K. Dornberger-Schiff. Grundzüge einer Theorie der OD-Strukturen aus Schichten. In *Abhandlungen der Deutschen Akademie der Wissenschaften, Klasse für Chemie, Geologie und Biologie*. Akademie-Verlag, Berlin, 1964.

I. A. Dos Santos, D. Klimm, S. L. Baldochi, and I. M. Ranieri. Thermodynamic modeling of the LiF–YF_3 phase diagram. *J. Cryst. Growth*, 360:172–175, 2012. https://doi.org/10.1016/j.jcrysgro.2011.11.009.

G. M. Drabkin, E. I. Zabidarov, Y. A. Kasman, and A. I. Okorokov. Investigation of a phase transition in nickel with polarized neutrons. *J. Exp. Theor. Phys.*, 29:261–266, 1969.

H. J. T. Ellingham. Reducibility of oxides and sulfides in metallurgical processes. *J. Soc. Chem. Ind.*, 63:125–160, 1944. https://doi.org/10.1002/jctb.5000630501.

G. Eriksson, P. Wu, M. Blander, and A. D. Pelton. Critical evaluation and optimization of the thermodynamic properties and phase diagrams of the MnO–SiO_2 and CaO–SiO_2 systems. *Can. Metall. Q.*, 33:13–21, 1994. https://doi.org/10.1179/cmq.1994.33.1.13.

O. Fabrichnaya and F. Aldinger. Assessment of thermodynamic parameters in the system ZrO_2–Y_2O_3–Al_2O_3. *Z. Metallkde*, 95:27–39, 2004. https://doi.org/10.3139/146.017909.

FactSage 8.0. GTT Technologies, Kaiserstr. 100, 52134 Herzogenrath, Germany, 2020. www.factsage.com.

P. P. Fedorov, B. P. Sobolev, L. V. Medvedeva, and B. M. Reiterov. Revised phase diagrams of LiF–RF_3 (R = La–Lu, Y) systems. In E. I. Givargizov and A. M. Mel'nikova, editors, *Growth of Crystals*, pages 141–154. #21. Kluwer Academic/Plenum Publishers, 2002. https://doi.org/10.1007/978-1-4615-0537-2_13.

E. B. Ferreira, M. L. Lima, and E. D. Zanotto. DSC method for determining the liquidus temperature of glass-forming systems. *J. Am. Ceram. Soc.*, 93:3757–3763, 2010. https://doi.org/10.1111/j.1551-2916.2010.03976.x.

J. H. Flynn and L. A. Wall. General treatment of the thermogravimetry of polymers. *J. Res. Natl. Bur. Stand.*, 70A:487–523, 1966. https://doi.org/10.6028/jres.070A.043.

E. S. Freeman and B. Carroll. The application of thermoanalytical techniques to reaction kinetics. The thermogravimetric evaluation of the kinetics of the decomposition of calcium oxalate monohydrate. *J. Phys. Chem.*, 62:394–397, 1958. https://doi.org/10.1021/j150562a003.

H. L. Friedman. Kinetics of thermal degradation of char-forming plastics from thermogravimetry. Application to a phenolic plastic. *J. Polym. Sci., C Polym. Symp.*, 6:183–195, 1964. https://doi.org/10.1002/polc.5070060121.

E. Füglein and D. Walter. Thermal analysis of lanthanum hydroxide. *J. Therm. Anal. Calorim.*, 110:199–202, 2012. https://doi.org/10.1007/s10973-012-2298-2.

Z. Galazka, R. Uecker, D. Klimm, K. Irmscher, M. Pietsch, R. Schewski, M. Albrecht, A. Kwasniewski, S. Ganschow, D. Schulz, C. Guguschev, R. Bertram, M. Bickermann, and R. Fornari. Growth, characterization, and properties of bulk SnO_2 single crystals. *Phys. Status Solidi A*, 211:66–73, 2014. https://doi.org/10.1002/pssa.201330020.

Z. Galazka, D. Klimm, K. Irmscher, R. Uecker, M. Pietsch, R. Bertram, M. Naumann, M. Albrecht, A. Kwasniewski, R. Schewski, and M. Bickermann. $MgGa_2O_4$ as a new wide bandgap transparent semiconducting oxide: growth and properties of bulk single crystals. *Phys. Status Solidi A*, 212:1455–1460, 2015. https://doi.org/10.1002/pssa.201431835.

Z. Galazka, K. Irmscher, S. Ganschow, M. Zupancic, W. Aggoune, C. Draxl, M. Albrecht, D. Klimm, A. Kwasniewski, T. Schulz, M. Pietsch, A. Dittmar, R. Grueneberg, U. Juda, R. Schewski, S. Bergmann, H. Cho, K. Char, T. Schroeder, and M. Bickermann. Melt growth and physical properties of bulk $LaInO_3$ single crystals. *Phys. Status Solidi A*, 218:2100016, 2021. https://doi.org/10.1002/pssa.202100016.

A. K. Galwey and P. Gray. Oxidation of a pyrophoric iron. Part 1. Kinetics of chemisorption of oxygen on the finely divided iron-carbon substrate. *J. Chem. Soc. Faraday Trans. I*, 68:1935–1946, 1972. https://doi.org/10.1039/F19726801935.

S. Ganschow, D. Klimm, P. Reiche, and R. Uecker. On the crystallization of terbium aluminium garnet. *Cryst. Res. Technol.*, 34:615–619, 1999. https://doi.org/10.1002/(SICI)1521-4079(199906)34:5/6<615::AID-CRAT615>3.0.CO;2-C.

S. Ganschow, A. Kwasniewski, and D. Klimm. Conditions for the growth of $Fe_{1-x}O$ crystals using the micro-pulling-down technique. *J. Cryst. Growth*, 450:203–206, 2016. https://doi.org/10.1016/j.jcrysgro.2016.06.033.

A. M. Gasanaliev, M. A. Dibirov, A. S. Trunin, and G. E. Shter. The $(KCl)_2–BaCl_2–CaMoO_4–BaMoO_4$ stable tetrahedron of the Ba, Ca, K ‖ Cl, MoO_4 reciprocal system. *Russ. J. Inorg. Chem.*, 27:115–118, 1982.

T. M. Gesing, R. Uecker, and J. -C. Buhl. Refinement of the crystal structure of praseodymium orthoscandate, $PrScO_3$. *Z. Kristallogr., New Cryst. Struct.*, 224:385–386, 2009. https://doi.org/10.1524/ncrs.2009.224.14.385.

J. W. Gibbs. On the equilibrium of heterogeneous substances. *Trans. Conn. Acad. Arts Sci.*, III:108–524, 1874–1878.

O. Glemser and H. Schröder. Über Manganoxyde. II. Zur Kenntnis des Mangan(VII)-oxyds. *Z. Anorg. Allg. Chem.*, 271:293–304, 1953. https://doi.org/10.1002/zaac.19532710507.

E. Gmelin. Classical temperature-modulated calorimetry: A review. *Thermochim. Acta*, 305(97):1–26, 1997. https://doi.org/10.1016/S0040-6031(97)00126-3.

N. A. Gokcen. The As (arsenic) system. *Bull. Alloy Phase Diagr.*, 10:11–22, 1989. https://doi.org/10.1007/BF02882166.

D. L. Graf. Crystallographic tables for the rhombohedral carbonates. *Am. Mineral.*, 46:1283–1361, 1961.

L. Grassia, Y. P. Koh, M. Rosa, and S. L. Simon. Complete set of enthalpy recovery data using Flash DSC: Experiment and modeling. *Macromolecules*, 51:1549–1558, 2018. https://doi.org/10.1021/acs.macromol.7b02277.

C. Guguschev, D. Klimm, M. Brützam, T. Gesing, M. Gogolin, H. Paik, T. Markurt,
D. Kok, A. Kwasniewski, U. Jendritzki, and D. Schlom. Czochralski growth and
characterization of perovskite-type (La, Nd) (Lu, Sc)O$_3$ single crystals with a
pseudocubic lattice parameter of about 4.09 Å. *J. Cryst. Growth*, 536:125526, 2020.
https://doi.org/10.1016/j.jcrysgro.2020.125526.

C. M. Guldberg and P. Waage. Ueber die chemische Affinität. *J. Prakt. Chem.*, 19:69–114, 1879.
https://doi.org/10.1002/prac.18790190111.

J. R. Gump, W. F. Wagner, and J. M. Schreyer. Preparation and analysis of barium ferrate(VI). *Anal. Chem.*, 26:1957, 1954. https://doi.org/10.1021/ac60096a027.

M. Guzik, J. Pejchal, A. Yoshikawa, A. Ito, T. Goto, M. Siczek, T. Lis, and G. Boulon. Structural
investigations of Lu$_2$O$_3$ as single crystal and polycrystalline transparent ceramic. *Cryst. Growth Des.*, 14:3327–3334, 2014. https://doi.org/10.1021/cg500225v.

A. Haase and G. Brauer. Hydratstufen und Kristallstrukturen von Bariumchlorid. *Z. Anorg. Allg. Chem.*, 441:181–195, 1978. https://doi.org/10.1002/zaac.19784410120.

H. M. Haendler, P. S. Sennett, and C. M. Wheeler Jr. The system LiF–LiCl, LiF–NaCl, LiF–KCl.
J. Electrochem. Soc., 106:264–268, 1959. https://doi.org/10.1149/1.2427319.

J. Hafner and D. Hobbs. Understanding the complex metallic element Mn. II. Geometric
frustration in β-Mn, phase stability, and phase transitions. *Phys. Rev. B*, 68:014408, 2003.
https://doi.org/10.1103/PhysRevB.68.014408.

T. Hahn, A. Looijenga-Vos, M. I. Aroyo, H. D. Flack, K. Momma, and P. Konstantinov. The space-group
tables. In M. I. Aroyo, editor, *International Tables for Crystallography. Volume A: Space-group symmetry*. International Union of Crystallography, 2016. ISBN 978-0-470-68575-4.
https://doi.org/10.1107/97809553602060000114.

J. P. Harvey, F. Lebreux-Desilets, J. Marchand, K. Oishi, A. F. Bouarab, C. Robelin, A. E.
Gheribi, and A. D. Pelton. On the application of the FactSage thermochemical software
and databases in materials science and pyrometallurgy. *Processes*, 8:1–31, 2020.
https://doi.org/10.3390/PR8091156.

P. Haseli, R. Jacob, M. Liu, P. Majewski, F. C. Christo, and F. Bruno. Experimental phase
diagram study of the binary KCl–Na$_2$CO$_3$ system. *Thermochim. Acta*, 695:178811, 2021.
https://doi.org/10.1016/j.tca.2020.178811.

T. Hattori, K. Igarashi, and J. Mochinaga. Enthalpies of fusion of intermediate compounds, KMgCl$_3$,
K$_2$MgCl$_4$, K$_2$BaCl$_4$, KCaCl$_3$, K$_2$SrCl$_4$, K$_2$LaCl$_5$, K$_3$PrCl$_6$, K$_3$NdCl$_6$, KGd$_3$Cl$_{10}$, and KDy$_3$Cl$_{10}$. *Bull. Chem. Soc. Jpn.*, 54:1883–1884, 1981. https://doi.org/10.1246/bcsj.54.1883.

E. Haussühl, L. Bayarjargal, and J. Ruiz-fuertes. Single-crystal elastic and thermodynamic properties
of γ-LiAlO$_2$. *J. Appl. Phys.*, 129:145108, 2021. https://doi.org/10.1063/5.0044998.

J. A. Hedvall. *Reaktionsfähigkeit fester Stoffe*. J. A. Barth, Leipzig, 1938.

F. H. Herbstein, M. Kapon, and A. Weissman. Old and new studies of the thermal
decomposition of potassium permanganate. *J. Therm. Anal.*, 41:303–322, 1994.
https://doi.org/10.1007/BF02549317.

J. Hidde, C. Guguschev, S. Ganschow, and D. Klimm. Thermal conductivity of rare-earth scandates
in comparison to other oxidic substrate crystals. *J. Alloys Compd.*, 738:415–421, 2018.
https://doi.org/10.1016/j.jallcom.2017.12.172.

M. Hillert. A solid-solution model for inhomogeneous systems. *Acta Metall.*, 9:525–535, 1961.
https://doi.org/10.1016/0001-6160(61)90155-9.

G. W. H. Höhne, W. Hemminger, and H. -J. Flammersheim. *Differential Scanning Calorimetry*.
Springer, Berlin, Heidelberg, New York, 1996. ISBN 3-540-59012-9.

S. Hull, S. T. Norberg, I. Ahmed, S. G. Eriksson, and C. E. Mohn. High temperature crystal
structures and superionic properties of SrCl$_2$, SrBr$_2$, BaCl$_2$ and BaBr$_2$. *J. Solid State Chem.*,
184:2925–2935, 2011. https://doi.org/10.1016/j.jssc.2011.09.004.

C. Hüter, G. Boussinot, E. A. Brener, and R. Spatschek. Solidification in syntectic and monotectic systems. *Phys. Rev. E*, 86: 021603, 2012. https://doi.org/10.1103/PhysRevE.86.021603.

ICDD. International Center for Diffraction Data, 2020. https://www.icdd.com/.

M. İlhan, A. Mergen, C. Sarıoğlu, and C. Yaman. Heat capacity measurements on $BaTa_2O_6$ and derivation of its thermodynamic functions. *J. Therm. Anal. Calorim.*, 128:707–711, 2017. https://doi.org/10.1007/s10973-016-5988-3.

T. Ishikawa, W. Y. Cai, and K. Kandori. Characterization of the thermal decomposition products of δ-FeOOH by Fourier-transform infrared spectroscopy and N_2 adsorption. *J. Chem. Soc. Faraday Trans.*, 88:1173–1177, 1992. https://doi.org/10.1039/FT9928801173.

IUCr. International Union of Crystallography, 2020. https://www.iucr.org.

G. Jaeger. The Ehrenfest classification of phase transitions: Introduction and evolution. *Arch. Hist. Exact Sci.*, 53:51–81, 1998. https://doi.org/10.1007/s004070050021.

W. Janke, D. A. Johnston, and R. Kenna. Properties of higher-order phase transitions. *Nucl. Phys. B*, 736:319–328, 2006. https://doi.org/10.1016/j.nuclphysb.2005.12.013.

M. M. Julian. *Foundations of Crystallography with Computer Applications*. CRC Press, Boca Raton, FL, USA, 2015. ISBN 978-1466552913.

J. S. Jur, V. D. Wheeler, D. J. Lichtenwalner, J. -P. Maria, and M. A. L. Johnson. Epitaxial growth of lanthanide oxides La_2O_3 and Sc_2O_3 on GaN. *Appl. Phys. Lett.*, 98:042902, 2011. https://doi.org/10.1063/1.3541883.

I. Kaban, M. Köhler, L. Ratke, R. Nowak, N. Sobczak, N. Mattern, J. Eckert, A. L. Greer, S. W. Sohn, and D. H. Kim. Phase separation in monotectic alloys as a route for liquid state fabrication of composite materials. *J. Mater. Sci.*, 47:8360–8366, 2012. https://doi.org/10.1007/s10853-012-6660-3.

F. X. Kayser and J. W. Patterson. Sir William Chandler Roberts-Austen – his role in the development of binary diagrams and modern physical metallurgy. *J. Phase Equilib.*, 19:11, 1998. https://doi.org/10.1007/BF02904718.

I. Keesmann and C. Schmitz. Analysen Tetraeder 1.3, 2021. https://www.monkeybreadsoftware.de/Analysen-Tetraeder/.

J. Kestin, S. T. Ro, and W. A. Wakeham. Viscosity of the noble gases in the temperature range 25–700 °C. *J. Chem. Phys.*, 56:4119–4124, 1972. https://doi.org/10.1063/1.1677824.

K. Kihara. An X-ray study of the temperature dependence of the quartz structure. *Eur. J. Mineral.*, 2:63–77, 1990. https://doi.org/10.1127/ejm/2/1/0063.

H. E. Kissinger. Reaction kinetics in differential thermal analysis. *Anal. Chem.*, 29:1702–1706, 1957. https://doi.org/10.1021/ac60131a045.

C. Kittel. *Introduction to Solid State Physics*, 8th edition. Wiley, 2004. ISBN 9780471415268.

W. Kleber, J. Bohm, D. Klimm, M. Mühlberg, and B. Winkler. *Einführung in die Kristallographie*. De Gruyter, Berlin, 2021. https://doi.org/10.1515/9783110460247.

W. Klement jun, R. H. Willens, and P. Duwez. Non-crystalline structure in solidified gold–silicon alloys. *Nature*, 187:869–870, 1960. https://doi.org/10.1038/187869b0.

D. Klimm. Unpublished results, 2010a.

D. Klimm. The melting behavior of lutetium aluminum perovskite $LuAlO_3$. *J. Cryst. Growth*, 312:730–733, 2010b. https://doi.org/10.1016/j.jcrysgro.2009.12.030.

D. Klimm. Electronic materials with a wide band gap: recent developments. *IUCrJ*, 1:281–290, 2014. https://doi.org/10.1107/S2052252514017229.

D. Klimm. Thermodynamic and kinetic aspects of crystal growth. In *Handbook of Solid State Chemistry*, pages 375–398. Wiley-VCH Verlag GmbH & Co. KGaA, Weinheim, Germany, 2017. https://doi.org/10.1002/9783527691036.hsscvol2022.

D. Klimm and S. Ganschow. The control of iron oxidation state during FeO and olivine crystal growth. *J. Cryst. Growth*, 275:e849–e854, 2005. https://doi.org/10.1016/j.jcrysgro.2004.11.080.

D. Klimm and P. Reiche. Comments on "Phase Equilibria in the Pseudo-Binary Systems LiF–CaAlF$_5$ and LiF–SrAlF$_5$" [J. Crystal Growth 235. *J. Cryst. Growth*, 249:596, 2002. 388–390, 2003. https://doi.org/10.1016/S0022-0248(02)02152-8.

D. Klimm, S. Ganschow, R. Bertram, J. Doerschel, V. Bermúdez, and A. Kłos. Phase separation during the melting of oxide borates LnCa$_4$O(BO$_3$)$_3$ (Ln = Y, Gd). *Mater. Res. Bull.*, 37:1737–1747, 2002. https://doi.org/10.1016/S0025-5408(02)00858-9.

D. Klimm, R. Uecker, and P. Reiche. Melting behavior and growth of colquiriite laser crystals. *Cryst. Res. Technol.*, 40:352–358, 2005. https://doi.org/10.1002/crat.200410350.

D. Klimm, S. Ganschow, A. Pajączkowska, and L. Lipińska. On the solubility of Nd^{3+} in Y$_3$Al$_5$O$_{12}$. *J. Alloys Compd.*, 436:204–208, 2007. https://doi.org/10.1016/j.jallcom.2006.07.001.

D. Klimm, S. Ganschow, Z. Galazka, R. Bertram, D. Schulz, and R. Uecker. Reactive atmospheres for oxide crystal growth. In N. P. Bansal, M. Kusnezoff and K. Shimamura, editors, *Advances in Solid Oxide Fuel Cells and Electronic Ceramics, Ceramic Engineering and Science Proceedings*, pages 157–167. John Wiley & Sons, Inc., Hoboken, NJ, USA, 2015. ISBN 978-1-119-21149-5. https://doi.org/10.1002/9781119211501.

J. M. Ko, S. Tozawa, A. Yoshikawa, K. Inaba, T. Shishido, T. Oba, Y. Oyama, T. Kuwabara, and T. Fukuda. Czochralski growth of UV-grade CaF$_2$ single crystals using ZnF$_2$ additive as scavenger. *J. Cryst. Growth*, 222:243–248, 2001. https://doi.org/10.1016/S0022-0248(00)00928-3.

O. Kondrat'eva, A. Tyurin, G. Nikiforova, A. Khoroshilov, M. Smirnova, V. Ketsko, and K. Gavrichev. Thermodynamic functions of magnesium gallate MgGa$_2$O$_4$ in the temperature range 0–1200 K. *Thermochim. Acta*, 641:49–54, 2016. https://doi.org/10.1016/j.tca.2016.08.015.

N. Kongkaew, W. Pruksakit, and S. Patumsawad. Thermogravimetric kinetic analysis of the pyrolysis of rice straw. *Energy Proc.*, 79:663–670, 2015. https://doi.org/10.1016/j.egypro.2015.11.552.

R. J. M. Konings, R. R. Van der Laan, A. C. G. Van Genderen, and J. C. Van Miltenburg. The heat capacity of Y$_3$Al$_5$O$_{12}$ from 0 to 900 K. *Thermochim. Acta*, 313:201–206, 1998. https://doi.org/10.1016/S0040-6031(98)00261-5.

I. Koštenská. Composition coordinate of the transition point in phase diagrams with polymorphous transformation of one component. *Chem. Zvesti*, 30:446–457, 1976.

A. Krell, G. M. Baur, and C. Dahne. Transparent sintered sub-μm Al$_2$O$_3$ with IR transmissivity equal to sapphire. In R. W. Tustison, editor, *Window and Dome Technologies VIII*, pages 199–207. #5078. International Society for Optics and Photonics, SPIE, 2003. https://doi.org/10.1117/12.485770.

J. Krogh-Moe, M. Vikan, and C. Krohn. K$_2$BaCl$_4$, another case of extreme ionic conductivity in solids. *Acta Chem. Scand.*, 21:309–311, 1967. https://doi.org/10.3891/acta.chem.scand.21-0309.

S. Krüger and J. Deubener. The TTT curves of the heterogeneous and homogeneous crystallization of lithium disilicate – a stochastic approach to crystal nucleation. *Front. Mater.*, 3:42, 2016. https://doi.org/10.3389/fmats.2016.00042.

P. Kumar. Theory of a higher-order phase transition: The superconducting transition in Ba$_{0.6}$K$_{0.4}$BiO$_3$. *Phys. Rev. B*, 68:064505, 2003. https://doi.org/10.1103/PhysRevB.68.064505.

E. E. Lakin, V. V. Kukol, and L. A. Syssev. Martensitic and diffusion-induced transitions in ZnS single crystals. *Neorg. Mater.*, 16:1175–1178, 1980.

L. Landau. The theory of phase transitions. *Nature*, 138:840–841, 1936. https://doi.org/10.1038/138840a0.

W. M. Latimer. Methods of estimating the entropies of solid compounds. *J. Am. Chem. Soc.*, 73:1480–1482, 1951. https://doi.org/10.1021/ja01148a021.

O. D. Lavrentovich and E. M. Terent'ev. Phase transition altering the symmetry of topological point defects (hedgehogs) in a nematic liquid crystal. *J. Exp. Theor. Phys.*, 64:1237–1244, 1986.

H. Le Chatelier. Über die Konstitution der Thone. *Z. Phys. Chem.*, 1U:396–402, 1887. https://doi.org/10.1515/zpch-1887-0143.

M. H. Le Chatelier. Sur la fusibilité des mélanges salins isomorphes. *C. R. Hebd. Séances Acad. Sci.*, 118:350–352, 1894.

J. Leitner, P. Voňka, D. Sedmidubský, and P. Svoboda. Application of Neumann–Kopp rule for the estimation of heat capacity of mixed oxides. *Thermochim. Acta*, 497:7–13, 2010. https://doi.org/10.1016/j.tca.2009.08.002.

C. H. Lin, S. Y. Chen, and P. Shen. Defects, lattice correspondence, and optical properties of spinel-like Cr_3O_4 condensates by pulsed laser ablation in water. *J. Phys. Chem. C*, 113:16356–16363, 2009. https://doi.org/10.1021/jp904288n.

A. L. Lindsay and L. A. Bromley. Thermal conductivity of gas mixtures. *Ind. Eng. Chem.*, 42:1508–1511, 1950. https://doi.org/10.1021/ie50488a017.

J. F. Löffler, J. Schroers, and W. L. Johnson. Time–temperature–transformation diagram and microstructures of bulk glass forming $Pd_{40}Cu_{30}Ni_{10}P_{20}$. *Appl. Phys. Lett.*, 77:681–683, 2000. https://doi.org/10.1063/1.127084.

A. F. Lopeandía, F. Pi, and J. Rodríguez-Viejo. Nanocalorimetric analysis of the ferromagnetic transition in ultrathin films of nickel. *Appl. Phys. Lett.*, 92(12):122503, 2008. https://doi.org/10.1063/1.2901166.

S. J. Louisnathan. Refinement of the crystal structure of a natural gehlenite, $Ca_2Al(Al, Si)_2O_7$. *Can. Mineral.*, 10:822–837, 1971.

M. Luisi. Characterizing the measurement uncertainty of a high-temperature heat flux differential scanning calorimeter. Master's thesis, TU Graz, 2014.

D. F. Lupton, J. Merker, and F. Schölz. The correct use of platinum in the XRF laboratory. In *19th Durham Conference on X-ray Analysis*, Durham, GB, 1995. www.heraeus.com/media/media/ hpm/doc_hpm/pt_labware/Correct-Use-of-Platinum.pdf.

A. Magnéli and B. M. Oughton. Studies on the vanadium pentoxide – molybdenum trioxide system I + II. *Acta Chem. Scand.*, 5:581–589, 1951. https://doi.org/10.3891/acta.chem.scand.05-0581.

C. G. Maier and K. K. Kelley. An equation for the representation of high-temperature heat content data. *J. Am. Chem. Soc.*, 54:3243–3246, 1932. https://doi.org/10.1021/ja01347a029.

A. M. Maitra. Determination of solid state basicity of rare earth oxides by thermal analysis of their carbonates. *J. Therm. Anal.*, 36:657–675, 1990. https://doi.org/10.1007/BF01914518.

M. Malinovský and J. Vrbenská. Phasendiagramm des Systems Li_3AlF_6–CaF_2. *Chem. Zvesti*, 21:806–817, 1967.

S. J. McCormack and A. Navrotsky. Thermodynamics of high entropy oxides. *Acta Mater.*, 202:1–21, 2021. https://doi.org/10.1016/j.actamat.2020.10.043.

J. P. Meehan and E. J. Wilson. Single crystal growth and characterization of $SrAlF_5$ and $Sr_{1-x}Eu_x^{2+}AlF_5$. *J. Cryst. Growth*, 15:141–147, 1972. https://doi.org/10.1016/0022-0248(72)90136-4.

H. Mehrer. Fast Ion Conductors. In *Diffusion in Solids: Fundamentals, Methods, Materials, Diffusion-Controlled Processes*, pages 475–490. Springer, Berlin, Heidelberg, 2007. https://doi.org/10.1007/978-3-540-71488-0_27.

R. A. Mendybaev, F. M. Richter, and A. M. Davis. Reevaluation of the åkermanite-gehlenite binary system. In *Lunar and Planetary Science XXXVII*, League City, Texas, page 2268, 2006.

Mettler Toledo. DSC2+, 2021. https://www.mt.com/.

T. Miyazawa, Y. Kano, Y. Nakayama, K. Ozawa, T. Iga, M. Yamanaka, A. Hashimoto, T. Kikuchi, and K. Mase. Improved pumping speeds of oxygen-free palladium/titanium nonevaporable getter coatings and suppression of outgassing by baking under oxygen. *J. Vac. Sci. Technol. A*, 37:021601, 2019. https://doi.org/10.1116/1.5074160.

A. F. Möbius. *Der barycentrische Calcul – ein neues Hülfsmittel zur analytischen Behandlung der Geometrie*. Johann Ambrosius Barth, Leipzig, 1827.

F. L. Mohler, V. H. Dibeler, and R. M. Reese. Mass spectra of fluorocarbons. *J. Res. Natl. Bur. Stand.*, 49:343–347, 1952.

L. V. Moiseeva. Crystals, glasses and melts of halide systems for active media for mid IR lasers. PhD thesis, Russian Academy of Sciences, Institute for General Physics, 2019 (in Russian).

A. Molchanov, J. Friedrich, G. Wehrhan, and G. Müller. Study of the oxygen incorporation during growth of large CaF_2-crystals. *J. Cryst. Growth*, 273:629–637, 2005. https://doi.org/10.1016/j.jcrysgro.2004.09.040.

K. Momma and F. Izumi. VESTA 3 for three-dimensional visualization of crystal, volumetric and morphology data. *J. Appl. Crystallogr.*, 44:1272–1276, 2011. https://doi.org/10.1107/S0021889811038970.

W. Moritz, S. Krause, U. Roth, D. Klimm, and A. Lippitz. Re-activation of an all solid state oxygen sensor. *Anal. Chim. Acta*, 437:183–190, 2001. https://doi.org/10.1016/S0003-2670(01)00993-X.

E. Moukhina and E. Kaisersberger. Temperature dependence of the time constants for deconvolution of heat flow curves. *Thermochim. Acta*, 492:101–109, 2009. https://doi.org/10.1016/j.tca.2008.12.022.

S. C. Mraw and D. F. Naas. The measurement of accurate heat capacities by differential scanning calorimetry comparison of d. s. c. results on pyrite (100 to 800 K) with literature values from precision adiabatic calorimetry. *J. Chem. Thermodyn.*, 11:567–584, 1979. https://doi.org/10.1016/0021-9614(79)90097-1.

S. Musić, S. Krehula, and S. Popović. Thermal decomposition of β-FeOOH. *Mater. Lett.*, 58:444–448, 2004. https://doi.org/10.1016/S0167-577X(03)00522-6.

mymathtables. convolution calculator, 2020. https://www.mymathtables.com/calculator/number/convolution-calculator.html.

I. Nagata, Y. Kawamura, Y. Ogasawara, and S. Tokuriki. Excess enthalpies of (chloroform + cyclohexane), (propan-2-ol + toluene + cyclohexane), and (propan-2-ol + chloroform + cyclohexane) at 298.15 K. *J. Chem. Thermodyn.*, 12:223–228, 1980. https://doi.org/10.1016/0021-9614(80)90040-3.

H. P. Nair, Y. Liu, J. P. Ruf, N. J. Schreiber, S. L. Shang, D. J. Baek, B. H. Goodge, L. F. Kourkoutis, Z. K. Liu, K. M. Shen, and D. G. Schlom. Synthesis science of $SrRuO_3$ and $CaRuO_3$ epitaxial films with high residual resistivity ratios. *APL Mater.*, 6:046101, 2018a. https://doi.org/10.1063/1.5023477.

H. P. Nair, J. P. Ruf, N. J. Schreiber, L. Miao, M. L. Grandon, D. J. Baek, B. H. Goodge, J. P. Ruff, L. F. Kourkoutis, K. M. Shen, and D. G. Schlom. Demystifying the growth of superconducting Sr_2RuO_4 thin films. *APL Mater.*, 6:101108, 2018b. https://doi.org/10.1063/1.5053084.

N. Nakagawa, H. Ohtsubo, Y. Waku, and H. Yugami. Thermal emission properties of $Al_2O_3/Er_3Al_5O_{12}$ eutectic ceramics. *J. Eur. Ceram. Soc.*, 25:1285–1291, 2005. https://doi.org/10.1016/j.jeurceramsoc.2005.01.031.

G. H. G. Nakamura, D. Klimm, and S. L. Baldochi. Thermal analysis and phase diagram of the $LiF–BiF_3$ system. *Thermochim. Acta*, 551:131–135, 2013. https://doi.org/10.1016/j.tca.2012.10.005.

A. Navrotsky. Progress and new directions in calorimetry: A 2014 perspective. *J. Am. Ceram. Soc.*, 97:3349–3359, 2014. https://doi.org/10.1111/jace.13278.

A. Navrotsky, L. Benoist, and H. Lefebvre. Direct calorimetric measurement of enthalpies of phase transitions at 2000–2400 °C in yttria and zirconia. *J. Am. Ceram. Soc.*, 88:2942–2944, 2005. https://doi.org/10.1111/j.1551-2916.2005.00506.x.

A. A. Nayeb-Hashemi and J. B. Clark. The Mg–Ni (magnesium–nickel) system. *Bull. Alloy Phase Diagr.*, 6:238–244, 1985. https://doi.org/10.1007/BF02880406.

NETZSCH. NETZSCH-Gerätebau GmbH, Wittelsbacherstr. 42, 95100 Selb, Germany, 2020. www.netzsch-thermal-analysis.com/en/.

NETZSCH. NETZSCH Kinetics Neo, V. 2.5.0, 2021. https://kinetics.netzsch.com/en/.

H. Neumann, G. Kühn, and W. Möller. Heat capacity and lattice anharmonicity in Cu–III–VI_2 chalcopyrite compounds. *Phys. Status Solidi B*, 144:565–573, 1987. https://doi.org/10.1002/pssb.2221440215.

NIST. NIST National Institute of Standards and Technology, US Dept. of Commerce, 2020. https://webbook.nist.gov/chemistry/form-ser/.

K. Nitsch, A. Cihlář, D. Klimm, M. Nikl, and M. Rodová. Na–Gd phosphate glasses – preparation, thermal and scintillating properties. *J. Therm. Anal. Calorim.*, 80:735–738, 2005. https://doi.org/10.1007/s10973-005-0722-6.

J. J. Nye. *Physical Properties of Crystals*. Clarendon, Oxford, 1957. ISBN 978-0198511656.

V. M. Orera, J. I. Peñña, P. B. Oliete, R. I. Merino, and A. Larrea. Growth of eutectic ceramic structures by directional solidification methods. *J. Cryst. Growth*, 360:99–104, 2012. https://doi.org/10.1016/j.jcrysgro.2011.11.056.

T. Ozawa. A new method of analyzing thermogravimetric data. *Bull. Chem. Soc. Jpn.*, 38:1881–1886, 1965. https://doi.org/10.1246/bcsj.38.1881.

T. Ozawa, T. Sunose, and T. Kaneko. Historical review on research of kinetics in thermal analysis and thermal endurance of electrical insulating materials. I. Thermal endurance test and isoconversion methods. *J. Therm. Anal. Calorim.*, 44:205–216, 1995. https://doi.org/10.1007/bf02547149.

L. Paama, I. Pitkänen, H. Halttunen, and P. Perämäki. Infrared evolved gas analysis during thermal investigation of lanthanum, europium and samarium carbonates. *Thermochim. Acta*, 403:197–206, 2003. https://doi.org/10.1016/S0040-6031(03)00038-8.

S. A. Palomares-Sanchez, B. E. Watts, D. Klimm, A. Baraldi, A. Parisini, S. Vantaggio, and R. Fornari. Sol-gel growth and characterization of In_2O_3 thin films. *Thin Solid Films*, 645:383–390, 2018. https://doi.org/10.1016/j.apcatb.2004.11.022.

Pandat. CompuTherm LLC, 8401 Greenway Blvd Suite 248, Middleton, WI 53562, USA, 2020. https://computherm.com.

P. Paufler. *Phasendiagramme*. Akademie-Verlag, Berlin, 1981. ISBN 978-3528068653.

P. Paufler. *Physikalische Kristallographie*. Akademie-Verlag, Berlin, 1986. ISBN 978-3527264544.

A. D. Pelton. Thermodynamics and phase diagrams of materials. In R. W. Cahn, P. Haasen and E. J. Kramer, editors, *Phase Transformations in Materials*, volume 5, pages 1–73. VCH, Weinheim, 1991. ISBN 978-3527268184.

A. D. Pelton. *Phase Diagrams and Thermodynamic Modeling of Solutions*. Elsevier, Amsterdam, 2019. ISBN 978-0128014943.

A. Petit and P. L. Dulong. Recherches sur quelques points importants de la théorie de la chaleur. *Ann. Chim. Phys.*, 10:395–413, 1819.

A. G. Petrosyan, V. F. Popova, V. L. Ugolkov, D. P. Romanov, and K. L. Ovanesyan. A study of phase stability in the Lu_2O_3–Al_2O_3 system. *J. Cryst. Growth*, 377:178–183, 2013. https://doi.org/10.1016/j.jcrysgro.2013.04.054.

C. E. Porter and F. D. Blum. Thermal characterization of PMMA thin films using modulated differential scanning calorimetry. *Macromolecules*, 33:7016–7020, 2000. https://doi.org/10.1021/ma000302l.

A. Prakash, P. Xu, A. Faghaninia, S. Shukla, J. W. Ager, C. S. Lo, and B. Jalan. Wide bandgap $BaSnO_3$ films with room temperature conductivity exceeding 10^4 S cm^{-1}. *Nat. Commun.*, 8:15167, 2017. https://doi.org/10.1038/ncomms15167.

E. G. Prout and M. E. Brown. Thermal decomposition of irradiated calcium azide. *Nature*, 205:1314–1315, 1965. https://doi.org/10.1038/2051314a0.

E. G. Prout and F. C. Tompkins. The thermal decomposition of potassium permanganate. *Trans. Faraday Soc.*, 40:488–498, 1944. https://doi.org/10.1039/tf9444000488.

C. R. Quick, J. E. K. Schawe, P. J. Uggowitzer, and S. Pogatscher. Measurement of specific heat capacity via fast scanning calorimetry – accuracy and loss corrections. *Thermochim. Acta*, 677:12–20, 2019. https://doi.org/10.1016/j.tca.2019.03.021.

A. P. Ramirez, A. Hayashi, R. J. Cava, R. Siddharthan, and B. S. Shastry. Zero-point entropy in 'spin ice'. *Nature*, 399(110):333–335, 1999. https://doi.org/10.1038/20619.

I. M. Ranieri, S. L. Baldochi, A. M. E. Santo, L. Gomes, L. C. Courrol, L. V. G. Tarelho, W. de Rossi, J. R. Berretta, F. E. Costa, G. E. C. Nogueira, N. U. Wetter, D. M. Zezell, N. D. Vieira Jr., and S. P. Morato. Growth of LiYF$_4$ crystals doped with holmium, erbium and thulium. *J. Cryst. Growth*, 166:423–428, 1996. https://doi.org/10.1016/0022-0248(95)00506-4.

I. M. Ranieri, S. L. Baldochi, and D. Klimm. The phase diagram GdF$_3$–LuF$_3$. *J. Solid State Chem.*, 181:1070–1074, 2008. https://doi.org/10.1016/j.jssc.2008.02.017.

S. Rastogi, M. Newman, and A. Keller. Unusual pressure-induced phase behavior in crystalline poly-4-methyl-pentene-1. *J. Polym. Sci., Part B, Polym. Phys.*, 31:125–139, 1993. https://doi.org/10.1002/polb.1993.090310202.

R. G. Reddy. Molten salts: Thermal energy storage and heat transfer media (editorial). *J. Phase Equilib. Diff.*, 323:269–270, 2011. https://doi.org/10.1007/s11669-011-9904-z.

O. Redlich and A. T. Kister. Algebraic representation of thermodynamic properties and the classification of solutions. *Ind. Eng. Chem.*, 40:345–348, 1948. https://doi.org/10.1021/ie50458a036.

O. Regen and G. Brandes. *Formelsammlung Physikalische Kristallographie*. Deutscher Verlag für Grundstoffindustrie, Leipzig, 1979. ISBN 978-3342006541.

M. Reuß. Beeinflussung der feinstkristallinen Ausscheidung von Calciumcarbonat (CaCO$_3$) durch Verfahrensparameter und die Dotierung mit ausgewählten Chloriden der Seltenen Erden. PhD thesis, Universität zu Köln, 2003. https://kups.ub.uni-koeln.de/691/.

C. M. Rost, E. Sachet, T. Borman, A. Moballegh, E. C. Dickey, D. Hou, J. L. Jones, S. Curtarolo, and J. -P. Maria. Entropy-stabilized oxides. *Nat. Commun.*, 6:8485, 2015. https://doi.org/10.1038/ncomms9485.

S. Rudtsch. Uncertainty of heat capacity measurements with differential scanning calorimeters. *Thermochim. Acta*, 382:17–25, 2002. https://doi.org/10.1016/S0040-6031(01)00730-4.

L. Rycerz. Practical remarks concerning phase diagrams determination on the basis of differential scanning calorimetry measurements. *J. Therm. Anal. Calorim.*, 113:231–238, 2013. https://doi.org/10.1007/s10973-013-3097-0.

N. Saadatkhah, A. Carillo Garcia, S. Ackermann, P. Leclerc, M. Latifi, S. Samih, G. S. Patience, and J. Chaouki. Experimental methods in chemical engineering: Thermogravimetric analysis – TGA. *Can. J. Chem. Eng.*, 98:34–43, 2020. https://doi.org/10.1002/cjce.23673.

K. Sakata. Note on the phase transition in NbO$_2$. *J. Phys. Soc. Jpn.*, 26:582, 1969. https://doi.org/10.1143/JPSJ.26.582.

E. K. H. Salje. *Phase Transitions in Ferroelastic and Co-Elastic Crystals*. University Press, Cambridge, 1990. https://doi.org/10.1017/CBO9780511586460.

J. Sangster and A. D. Pelton. Phase diagrams and thermodynamic properties of the 70 binary alkali halide systems having common ions. *J. Phys. Chem. Ref. Data*, 16:509–561, 1987. https://doi.org/10.1063/1.555803.

E. Sani, M. Rabe, G. Reck, P. Becker, M. Roßberg, R. Bertram, and D. Klimm. Growth and characterization of LiCaGaF$_6$. *Cryst. Res. Technol.*, 40:26–31, 2005. https://doi.org/10.1002/crat.200410304.

Y. Sato and T. Taira. Study on the specific heat of Y$_3$Al$_5$O$_{12}$ between 129 K and 573 K. *Opt. Mater. Express*, 11:551–558, 2021. https://doi.org/10.1364/OME.416480.

J. F. Schairer. Some aspects of the melting and crystallization of rock-forming minerals. *Am. Mineral.*, 29:75–91, 1944.

J. E. K. Schawe and J. F. Löffler. Existence of multiple critical cooling rates which generate different types of monolithic metallic glass. *Nat. Commun.*, 10:1337, 2019. https://doi.org/10.1038/s41467-018-07930-3.

J. E. K. Schawe, T. Hütter, C. Heitz, I. Alig, and D. Lellinger. Stochastic temperature modulation: A new technique in temperature-modulated DSC. *Thermochim. Acta*, 446:147–155, 2006. https://doi.org/10.1016/j.tca.2006.01.031.

J. E. K. Schawe, S. Pogatscher, and J. F. Löffler. Thermodynamics of polymorphism in a bulk metallic glass: Heat capacity measurements by fast differential scanning calorimetry. *Thermochim. Acta*, 685:178518, 2020. https://doi.org/10.1016/j.tca.2020.178518.

M. Schick, A. Watson, M. to Baben, and K. Hack. A modified Neumann–Kopp treatment of the heat capacity of stoichiometric phases for use in computational thermodynamics. *J. Phase Equilib. Diff.*, 40:104–114, 2019. https://doi.org/10.1007/s11669-019-00708-0.

A. Schindler, M. Doedt, Ş. Gezgin, J. Menzel, and S. Schmölzer. Identification of polymers by means of DSC, TG, STA and computer-assisted database search. *J. Therm. Anal. Calorim.*, 129:833–842, 2017. https://doi.org/10.1007/s10973-017-6208-5.

D. G. Schlom, L. -Q. Chen, C. J. Fennie, V. Gopalan, D. A. Muller, X. Pan, R. Ramesh, and R. Uecker. Elastic strain engineering of ferroic oxides. *Mater. Res. Soc. Bull.*, 39:118–130, 2014. https://doi.org/10.1557/mrs.2014.1.

C. L. Schmidt, R. Dinnebier, U. Wedig, and M. Jansen. Crystal structure and chemical bonding of the high-temperature phase of AgN_3. *Inorg. Chem.*, 46:907–916, 2007. https://doi.org/10.1021/ic061963n.

J. M. Schneider. How high is the entropy in high entropy ceramics? *J. Appl. Phys.*, 130:150903, 2021. https://doi.org/10.1063/5.0062523.

I. Schröder. Über die Abhängigkeit der Löslichkeit eines festen Körpers von seiner Schmelztemperatur. *Z. Phys. Chem.*, 11:449–465, 1893. https://doi.org/10.1515/zpch-1893-1134.

A. H. Schultz and V. S. Stubican. Separation of phases by spinodal decomposition in the systems $Al_2O_3–Cr_2O_3$ and $Al_2O_3–Cr_2O_3–Fe_2O_3$. *J. Am. Ceram. Soc.*, 53:613–616, 1970. https://doi.org/10.1111/j.1151-2916.1970.tb15984.x.

D. Schultze. *Differentialthermoanalyse*. Verlag Chemie, Weinheim/Bergstr, 1969.

D. Schultze, U. Steinike, J. Kussin, and U. Kretzschmar. Thermal oxidation of ZnS modifications sphalerite and wurtzite. *Cryst. Res. Technol.*, 30:553–558, 1995. https://doi.org/10.1002/crat.2170300422.

D. Schulz, R. Bertram, D. Klimm, T. Schulz, and E. Thiede. Segregation of Mg in $Zn_{1-x}Mg_xO$ single crystals grown from the melt. *J. Cryst. Growth*, 334:118–121, 2011. https://doi.org/10.1016/j.jcrysgro.2011.08.036.

J. H. Schulze. Scotophorus pro phosphoro inventvs seu experimentum curiosum de effectu radiorum solarium. In J. C. Franck, editor, *Bibliotheca Novissima Observationvm Ac Recensionvm. Halle/Saale*, 1719. http://digitale.bibliothek.uni-halle.de/vd18/content/pageview/4921290.

E. R. Segnit and A. E. Holland. The system $MgO–ZnO–SiO_2$. *J. Am. Ceram. Soc.*, 48:409–413, 1965. https://doi.org/10.1111/j.1151-2916.1965.tb14778.x.

R. Senesi, M. Nardone, F. P. Ricci, M. A. Ricci, and A. K. Soper. Microscopic structure of the hydrogen-xenon mixture. *Phys. Rev. E*, 56:2993–2999, 1997. https://doi.org/10.1103/PhysRevE.56.2993.

O. N. Senkov, J. D. Miller, D. B. Miracle, and C. Woodward. Accelerated exploration of multi-principal element alloys with solid solution phases. *Nat. Commun.*, 6:1–10, 2015. https://doi.org/10.1038/ncomms7529.

R. Serrano-López, J. Fradera, and S. Cuesta-López. Molten salts database for energy applications. *Chem. Eng. Process. Process Intensif.*, 73:87–102, 2013. https://doi.org/10.1016/j.cep.2013.07.008.

M. Shamsuddin. In *Roasting of Sulfide Minerals*, pages 39–69. John Wiley & Sons, Ltd, 2016. Chapter 2. ISBN 9781119078326. https://doi.org/10.1002/9781119078326.ch2.

R. D. Shannon. Revised effective ionic radii and systematic studies of interatomic distances in halides and chalcogenides. *Acta Crystallogr. A*, 32:751–767, 1976. https://doi.org/10.1107/S0567739476001551.

D. Shechtman, I. Blech, D. Gratias, and J. W. Cahn. Metallic phase with long-range orientational order and no translational symmetry. *Phys. Rev. Lett.*, 53:1951–1953, 1984. https://doi.org/10.1103/PhysRevLett.53.1951.

E. V. Shevchenko, E. V. Charnaya, E. N. Khazanov, A. V. Taranov, and A. S. Bugaev. Heat capacity of rare-earth aluminum garnets. *J. Alloys Compd.*, 717:183–189, 2017. https://doi.org/10.1016/j.jallcom.2017.05.106.

L. T. Shi and K. N. Tu. Thermogravimetric study of the recovery of oxygen-deficient superconducting $YBa_2Cu_3O_{7-\delta}$ oxides in ambient oxygen. *Appl. Phys. Lett.*, 55:1351–1353, 1989. https://doi.org/10.1063/1.102474.

A. N. Shirsat, M. Ali, K. N. G. Kaimal, S. R. Bharadwaj, and D. Das. Thermochemistry of $La_2O_2CO_3$ decomposition. *Thermochim. Acta*, 399:167–170, 2003. https://doi.org/10.1016/S0040-6031(02)00459-8.

C. H. Shomate. A method for evaluating and correlating thermodynamic data. *J. Phys. Chem.*, 58:368–372, 1954. https://doi.org/10.1021/j150514a018.

A. A. A. P. Silva, M. S. Lamoglia, G. Silva, J.-M. Fiorani, N. David, M. Vilasi, G. C. Coelho, C. A. Nunes, and L. T. Eleno. Heat capacity measurements of the Fe_2Nb and Fe_7Nb_6 intermetallic compounds. *J. Alloys Compd.*, 878:160411, 2021. https://doi.org/10.1016/j.jallcom.2021.160411.

R. Simura and H. Yamane. Crystal structure of lutetium aluminate (LUAM), $Lu_4Al_2O_9$. *Acta Crystallogr. E*, 76:752–755, 2020. https://doi.org/10.1107/S2056989020005757.

B. J. Skinner and P. B. Barton Jr. The substitution of oxygen for sulfur in wurtzite and sphalerite. *Am. Mineral.*, 45:612–625, 1960.

S. W. Smith. *The Scientist and Engineer's Guide to Digital Signal Processing*. California Technical Publishing, San Diego, 1997. ISBN 978-0966017632.

B. P. Sobolev, P. P. Fedorov, D. B. Shteynberg, B. V. Sinitsyn, and G. S. Shakhkalamian. On the problem of polymorphism and fusion of lanthanide trifluorides. I. The influence of oxygen on phase transition temperatures. *J. Solid State Chem.*, 17:191–199, 1976. https://doi.org/10.1016/0022-4596(76)90220-6.

O. A. Sofekun and L. K. Doraiswamy. High-temperature oxidation of zinc sulfide: Kinetic modeling under conditions of strict kinetic control. *Ind. Eng. Chem. Res.*, 35(9):3163–3170, 1996. https://doi.org/10.1021/ie960013e.

M. Sulaiman, A. A. Rahman, and N. S. Mohamed. Sol-gel synthesis and characterization of Li_2CO_3–Al_2O_3 composite solid electrolytes. *Ionics*, 22:327–332, 2016. https://doi.org/10.1007/s11581-015-1548-2.

I. P. Swainson, M. T. Dove, W. W. Schmahl, and A. Putnis. Neutron powder diffraction study of the åkermanite–gehlenite solid solution series. *Phys. Chem. Miner.*, 19:185–195, 1992. https://doi.org/10.1007/BF00202107.

A. Szysiak, D. Klimm, S. Ganschow, M. Mirkowska, R. Diduszko, L. Lipińska, A. Kwasniewski, and A. Pajączkowska. The investigation of $YAlO_3$–$NdAlO_3$ system, synthesis and characterization. *J. Alloys Compd.*, 509:8615–8619, 2011. https://doi.org/10.1016/j.jallcom.2011.06.049.

G. Tammann. *Kristallisieren und Schmelzen*. Johann Ambrosius Barth, Leipzig, 1903. https://jscholarship.library.jhu.edu/bitstream/handle/1774.2/35747/31151027815343.pdf.

G. Tammann. Über die Anwendung der thermischen Analyse in abnormen Fällen. *Z. Anorg. Chem.*, 45:24–30, 1905. https://doi.org/10.1002/zaac.19050450104.

O. Teppo, J. Niemelä, and P. Taskinen. The copper–lead phase diagram. *Thermochim. Acta*, 185:155–169, 1991. https://doi.org/10.1016/0040-6031(91)80126-4.

The Engineering Toolbox. Properties of technical materials, 2021. www.engineeringtoolbox.com.

Thermo-Calc. Thermo-Calc Software AB, Råsundavägen 18, 169 67 Solna, Sweden, 2020. https://thermocalc.com.

R. E. Thoma, C. F. Weaver, H. A. Friedman, H. Insley, L. A. Harris, and H. A. Yakel Jr. Phase equilibria in the system LiF–YF_3. *J. Phys. Chem.*, 65:1096–1099, 1961. https://doi.org/10.1021/j100825a003.

R. E. Thoma, G. D. Brunton, R. A. Penneman, and T. K. Keenan. Equilibrium relations and crystal structure of lithium fluorolanthanate phases. *Inorg. Chem.*, 9:1096–1101, 1970. https://doi.org/10.1021/ic50087a019.

D. Thomas. Thermodynamische und kinetische Untersuchungen im System Lithium-Silicium. PhD thesis, TU Bergakademie Freiberg, 2015 (in German).

TraGMin 5.1. U. Steiner, HTW, Friedrich-List-Platz 1, 01069 Dresden, Germany, 2008. https://www.htw-dresden.de/en/luc/forschung/chemieingenieurwesen/festkoerperchemie/translate-to-english-tragmin.

M. -H. Tsai and J. -W. Yeh. High-entropy alloys: A critical review. *Math. Res. Lett.*, 2:107–123, 2014. https://doi.org/10.1080/21663831.2014.912690.

R. Uecker, B. Velickov, D. Klimm, R. Bertram, M. Bernhagen, M. Rabe, M. Albrecht, R. Fornari, and D. G. Schlom. Properties of rare-earth scandate single crystals (Re = Nd-Dy). *J. Cryst. Growth*, 310:2649–2658, 2008. https://doi.org/10.1016/j.jcrysgro.2008.01.019.

R. Uecker, D. Klimm, R. Bertram, M. Bernhagen, I. Schulze-Jonack, M. Brützam, A. Kwasniewski, T. M. Gesing, and D. G. Schlom. Growth and investigation of $Nd_{1-x}Sm_xScO_3$ and $Sm_{1-x}Gd_xScO_3$ solid-solution single crystals. *Acta Phys. Pol. A*, 124:295–300, 2013. https://doi.org/10.12693/APhysPolA.124.295.

R. Uecker, R. Bertram, M. Brützam, Z. Galazka, T. M. Gesing, C. Guguschev, D. Klimm, M. Klupsch, A. Kwasniewski, and D. G. Schlom. Large-lattice-parameter perovskite single-crystal substrates. *J. Cryst. Growth*, 457:137–142, 2017. https://doi.org/10.1016/j.jcrysgro.2016.03.014.

S. V. Ushakov and A. Navrotsky. 2021. Private communication.

S. V. Ushakov, S. Hayun, W. Gong, and A. Navrotsky. Thermal analysis of high entropy rare earth oxides. *Materials*, 13, 2020. https://doi.org/10.3390/ma13143141.

J. J. van Laar. Die Schmelz- oder Erstarrungskurven bei binären Systemen, wenn die feste Phase ein Gemisch (amorphe feste Lösung oder Mischkristalle) der beiden Komponenten ist. *Z. Phys. Chem.*, 63:216–253, 1908. https://doi.org/10.1515/zpch-1908-6314.

M. A. van Spronsen, J. M. N. Frenken, and W. M. Groot. Observing the oxidation of platinum. *Nat. Commun.*, 8:429, 2017. https://doi.org/10.1038/s41467-017-00643-z.

D. Velazquez and R. Romero. Calorimetric study of spinodal decomposition in β-Cu–Al–Mn. *J. Therm. Anal. Calorim.*, 2020. https://doi.org/10.1007/s10973-019-09234-0.

B. Veličkov, V. Kahlenberg, R. Bertram, and R. Uecker. Redetermination of terbium scandate, revealing a defect-type perovskite derivative. *Acta Crystallogr., Sect. E*, 64:i79, 2008. https://doi.org/10.1107/S1600536808033394.

A. F. Venero and E. F. Westrum. Heat capacities and thermodynamic properties of $Li_{0.5}Fe_{2.5}O_4$ and $Li_{0.5}Al_{2.5}O_4$ from 5 to 545 K. *J. Chem. Thermodyn.*, 7:693–702, 1975. https://doi.org/10.1016/0021-9614(75)90010-5.

M. Venkatesh, P. Ravi, and S. P. Tewari. Isoconversional kinetic analysis of decomposition of nitroimidazoles: Friedman method vs Flynn–Wall–Ozawa method. *J. Phys. Chem. A*, 117:10162–10169, 2013. https://doi.org/10.1021/jp407526r.

J. P. R. D. Villiers. Crystal structures of aragonite, strontianite, and witherite. *Am. Mineral.*, 56:758–767, 1971.

J. Vince. Barycentric Coordinates. In *Mathematics for Computer Graphics*, pages 193–221. Springer, London, 2006. https://doi.org/10.1007/1-84628-283-7_11.

N. N. Vinogradova, B. I. Galagan, L. N. Dmitruk, L. V. Moiseeva, V. V. Osiko, E. E. Sviridova, M. N. Brekhovskikh, and V. A. Fedorov. Growth of rare-earth-doped K_2LaCl_5, K_2BaCl_4, and K_2SrCl_4 single crystals. *Inorg. Mater.*, 41:654–657, 2005. https://doi.org/10.1007/s10789-005-0185-y.

J. Vrbenská and M. Malinovský. Phasendiagramm des Dreistoffsystems Li_3AlF_6–LiF–CaF_2. *Chem. Zvesti*, 21:818–825, 1967.

S. Vyazovkin, A. K. Burnham, J. M. Criadoc, L. A. Pérez-Maquedac, C. Popescud, and N. Sbirrazzuoli. ICTAC kinetics committee recommendations for performing kinetic computations on thermal analysis data. *Thermochim. Acta*, 520:1–19, 2011. https://doi.org/10.1016/j.tca.2011.03.034.

K. Walenta, B. Lehmann, and M. Zwiener. Colquiriit, ein neues Fluoridmineral aus der Zinnlagerstätte von Colquiri in Bolivien. *Tschermak's Mineral. Petrogr. Mitt.*, 27:275–281, 1980.

S. Wang. Atomic structure modeling of multi-principal-element alloys by the principle of maximum entropy. *Entropy*, 15:5536–5548, 2013. https://doi.org/10.3390/e15125536.

Wolfram Research. Barycentric coordinates, 2020. mathworld.wolfram.com/BarycentricCoordinates.html.

WolframAlpha. Calculus & Analysis, 2021. www.wolframalpha.com/examples/mathematics/calculus-and-analysis/.

W. L. Worrell. Estimation of the entropy of formation at 298 K for some refractory metal carbides. *J. Phys. Chem.*, 68:954–955, 1964. https://doi.org/10.1021/j100786a504.

J. B. Wrzesnewsky. Über die Schmelzbarkeit und den Fließdruck isomorpher Salzgemische. *Z. Anorg. Chem.*, 74:95–121, 1912. https://doi.org/10.1002/zaac.19120740112.

P. Wu and A. D. Pelton. Coupled thermodynamic-phase diagram assessment of the rare earth oxide–aluminium oxide binary systems. *J. Alloys Compd.*, 179:259–287, 1992. https://doi.org/10.1016/0925-8388(92)90227-Z.

T. Wu, E. Moosavi-Khoonsari, and I. -H. Jung. Estimation of thermodynamic properties of oxide compounds from polyhedron method. *Calphad*, 57:107–117, 2017. https://doi.org/10.1016/J.CALPHAD.2017.03.002.

R. W. G. Wyckoff. *Crystal Structures*. Interscience, New York, 1963.

Y. Yang, H. Jingtao, and W. Laixing. DMC-PID cascade control algorithm for the constant rate temperature rising process in differential scanning calorimeter. *J. Phys. Conf. Ser.*, 1213:032008, 2019. https://doi.org/10.1088/1742-6596/1213/3/032008.

Q. Ye and B. H. T. Chai. Crystal growth of $YCa_4O(BO_3)_3$ and its orientation. *J. Cryst. Growth*, 197:228–235, 1999. https://doi.org/10.1016/S0022-0248(98)00947-6.

J. W. Yeh, S. K. Chen, S. J. Lin, J. Y. Gan, T. S. Chin, T. T. Shun, C. H. Tsau, and S. Y. Chang. Nanostructured high-entropy alloys with multiple principal elements: Novel alloy design concepts and outcomes. *Adv. Eng. Mater.*, 6:299–303, 2004. https://doi.org/10.1002/adem.200300567.

J. M. Yeomans. *Statistical Mechanics of Phase Transitions*. University Press, Oxford, 1992. ISBN 978-0198517306.

Y. Yin and D. A. Keszler. Crystal chemistry of colquiriite-type fluorides. *Chem. Mater.*, 4:645–648, 1992. https://doi.org/10.1021/cm00021a028.

T. Yonezawa, K. Matsuo, J. Nakayama, and Y. Kawamoto. Behaviors of metal-oxide impurities in CaF_2 and BaF_2 single-crystals grown with PbF_2 scavenger by Stockbarger's method. *J. Cryst. Growth*, 258:385–393, 2003. https://doi.org/10.1016/S0022-0248(03)01567-7.

M. Zeilinger, V. Baran, L. van Wüllen, U. Häussermann, and T. F. Fässler. Stabilizing the phase $Li_{15}Si_4$ through lithium–aluminum substitution in $Li_{15-x}Al_xSi_4$ $(0.4 < x < 0.8)$–single crystal X-ray structure determination of $Li_{15}Si_4$ and $Li_{14.37}Al_{0.63}Si_4$. *Chem. Mater.*, 25(20):4113–4121, 2013. https://doi.org/10.1021/cm402721n.

S. Žemčužny and F. Rambach. Schmelzen der Alkalichloride. *Z. Anorg. Allg. Chem.*, 65:403–428, 1909. https://doi.org/10.1002/zaac.19100650125.

M. Zinkevich. Thermodynamics of Rare Earth Sesquioxides. *Prog. Mater. Sci.*, 52:597–647, 2007. https://doi.org/10.1016/j.pmatsci.2006.09.002.

Subject index

https://doi.org/10.1515/9783110743784-008

Formula index

https://doi.org/10.1515/9783110743784-009

www.ingramcontent.com/pod-product-compliance
Lightning Source LLC
Chambersburg PA
CBHW082110220326
41598CB00066BA/6022